ENVIRONMENTAL DIPLOMACY

Environmental Diplomacy

*An Examination
and a Prospective of
Canadian–U.S.
Transboundary
Environmental
Relations*

John E. Carroll
Foreword by Maxwell Cohen

*Published under the auspices of the
C. D. Howe Institute*

**Ann Arbor
The University of Michigan Press**

Copyright © by The University of Michigan 1983
All rights reserved
Published in the United States of America by
The University of Michigan Press and simultaneously
in Rexdale, Canada, by John Wiley & Sons Canada, Limited
Manufactured in the United States of America

1986   1985   1984   1983   5   4   3   2   1

Library of Congress Cataloging in Publication Data

Carroll, John E. (John Edward), 1944–
    Environmental diplomacy.

    "Published under the auspices of the C. D. Howe
Institute."
    Includes bibliographical references and index.
    1. Environmental policy—International cooperation.
2. Environmental policy—Canada.   3. Environmental
policy—United States.   I.   Title
HC79.E5C37   1983   363.7'0526   82-17363
ISBN 0-472-10029-7

For my mother and father

Foreword

The literature of the last ten years, building upon an already developing scholarship, demonstrates that interest in Canadian-U.S. relations in general, and environmental resource questions in particular has become a dominant theme of research and publication, as well as popular writings, in both countries. A brief retrospect of scholarly approaches to the bilateral relationship will show that important changes have taken place. These are timely in the light of the fundamental importance of environmental questions and the difficulties the two countries are experiencing in addressing them in a spirit of cooperation.

Until a generation ago the emphasis in Canadian-U.S. studies tended toward the political-historical with the prime focus on the long story of the peaceful frontier about which so many "unguarded words have been spoken"—to remember Lester Pearson's touch of irony. Indeed, although the Carnegie Endowment series of the 1938–41 period included Percy Corbett's small classic on the settlement of Canadian-U.S. disputes, and although there were many volumes reaching for a general rationale that would explain the overall character of this unique dyadic relationship, research in depth in particular areas was spotty, infrequent, and occasionally explosive, as in the critical literature that arose, for example, in the late 1950s and the 1960s out of the Columbia River dispute and the 1963–64 treaty that followed.

Furthermore, there was always a noticeable deficiency of reciprocal concentration or scope in most Canadian-U.S. studies. Generally Canadians knew much more about their great neighbor than the reverse, a natural condition where the shadow of a world power knew no continental boundaries as it shaded the northern acres that are Canada. Almost everything Canadians did or thought as North Americans reflected so much of the United States' power, culture, wealth, technology, and policy that the unceasing Canadian search for "soul," identity, independence, and the rest became a chronic challenge to Canadian psychic stability. By contrast, the average informed American seemed willing to accept the vague reality of a Canadian presence on the continent without much awareness of the details, or much anxiety about a benign and minor neighbor—so alike in political-cultural appearance that the real differences were often unknown and certainly untroubling.

The research position over the years had been molded in response to the differences in mutual awareness. For while the Canadian sensitivity was both popular and academic to all things American, the actual numbers of formal Canadian centers or programs studying the United States as such were negligible if any indeed existed until recent years. Canadian spontaneity and a sense of relevance combined to provide a steady output of "objective" or "nationalist" literature concerning the United States—its continental dominance and world role. On the other hand in the United States there had been developing since the late twenties a more formal interest in matters Canadian, from its politics to its geography and resources, sometimes indigenously American and sometimes through the binational eyes of a transplanted Canadian now rooted in an American academy and viewing the north from this strategic and often better-funded academic base—as with Burt in Minnesota, Corbett at Yale and Princeton, and Brebner at Columbia.

That generation, however, pre–World War II in its perspective and training, should be distinguished from the more recent character of interactions leading to fresh forms of scholarly investigation. For World War II produced a new level of Canadian vitality, competence, and industrial and military growth, and in other ways provided a new plane of significance from which the country would be viewed by an unworried neighbor, now the most powerful nation in the world—pace the U.S.S.R. since 1970— and the leader of the industrialized West.

Perhaps more important than this wartime reshaping of the Canadian fact for the United States were the vital changes that occurred between Canada and the United Kingdom in terms of their older political intimacy and, more important, of the economic role that Britain (and the Commonwealth) were to play henceforth in Canadian trade and investment. By the 1950s it was evident that henceforth it would be the United States whose markets and capital would become the dominant economic influences penetrating deep into the structure and operations of the Canadian economy. To all of this must be added the now established Americanization of Canadian industrial and commercial patterns and styles through the primary position of American technology and investment. The mediators of that process, of course, were the U.S. subsidiaries in Canada with their parents in Chicago, New York, and elsewhere treating much of Canadian primary and secondary industry as branch operations of these often multinational U.S. parents.

Side by side with the radical change in the wartime and postwar political, military, and economic relations of the two countries was the transformation of the level of scholarly curiosity from the casual to the concentrated. An increasing awareness of the coming of U.S. resource dependency turned both corporate and scholarly eyes northward—a move predicted by the findings of the U.S. Paley Commission Report of 1952. While some university centers such as Duke had quite intensive Canadian studies well developed by the early 1960s, and outstanding U.S. students such as Mason Wade were now widely known in Canada, it was not really until the end of that decade and the beginning of the 1970s that Canadian studies as a formal part of U.S. university structures began to take on their present and ever-enlarging character. Indeed, it is quite interesting to contemplate the significance of this rapid rise in the number of U.S. institutions with Canadian studies programs, which today number about thirty or more and now have their own national association, with a membership that is steadily on the increase.

The growth of formal U.S. scholarship dealing with Canadian affairs may have originated in military and political-historical curiosity and moved from there to economic questions of high common interest. But today concern ranges over the broad spectrum of sociopolitical life and styles in Canada to detailed studies of the resource situation that now dominates so much of Canadian-U.S. relations and recent political and academic discourse. From this resource context to the environmental consequences of resource questions was a logical and inevitable move for policy and for studies about policy.

Dr. Carroll's book is the first major study of the totality of Canadian-U.S. environmental relations on three oceans and along the great midcontinent landmass. It is therefore a mirror of the times that a vital interest in shared resource areas, and equally vital concerns with adjacent and interacting environmental systems, inevitably have prompted a holistic look at the Canadian-U.S. environmental interface and beyond.

The chapters of this volume represent perhaps the most exhaustive effort to date to examine hydrocarbons and the common oceanic neighborhood together with the pollution-resource-navigation complex of issues now so conspicuous in all Canadian-U.S. seacoast dealings. Despite the literature arising from the existence of the Boundary Waters Treaty of 1909 and the International Joint Commission since 1911, these chapters probably provide the most succinct, but not superficial, examination of the shared midcontinent environmental systems that both unite and divide the two neighbors.

Canada and the United States occupy and inevitably have to jointly manage a vast continental region that extends from Maine–New Brunswick to the Pacific and then northwestward to the Yukon-Alaskan border—an additional thousand-mile-long frontier. Hence some five thousand miles of boundary are the major fact of sovereign life for both countries. That boundary has about 150 lakes and rivers crossing it; the longest water boundary in the world from Cornwall on the St. Lawrence River to the Lake of the Woods at the Manitoba-Ontario-Minnesota boundary; five great watersheds demarking the continental landmass for both countries and, within each of these watersheds, numbers of rivers and lakes crossing the boundary; and, perhaps most dramatic, joint possession of the Great Lakes, the world's largest freshwater system—with Lake Michigan, a U.S.

lake, both in and out of these resource calculations and their joint legal frameworks.

This unprecedented joint management challenge called for unprecedented institutions. From the very beginning the creation of the International Joint Commission, as well as the principles and procedures set out in the Boundary Waters Treaty of 1909, was viewed as something of "a noble experiment." What no one could have foreseen was that the determination by the United States and Canada to use this treaty to develop dispute avoidance and dispute settlement machinery, wherever boundary and transboundary water quantity questions were involved, should have led, in due course, to a vast responsibility for water quality–pollution issues based upon a modest thirty-six-word paragraph in Article IV of the treaty. Nor could it have been forecast by the draftsmen or the first commissioners that some day the interaction between transboundary air pollution and water quality would give the commission both a handle on fresh jurisdiction as well as a new and more dynamic perspective on the scope of its own geophysical responsibilities.

A valuable but hazardous feature, therefore, of this volume is the attempt to deal with all of the great issues that have touched the coastline and the landmass-water frontiers, either shared directly when they are boundaries, or shared in a river basin system crossing the common frontier. Much of the study inevitably is concerned with a single vital resource on that landmass, namely, fresh water. However in an indirect fashion the concern with water systems on the continent, and covered by the 1909 treaty, necessarily has drawn into the orbit of awareness the other resource debates that directly or indirectly touch on these fresh waters. Such issues as use of coal, development of water power, and urban and rural agricultural growth and development practices, have all been inexorably drawn within the perspective with which both countries have viewed these joint systems since water, itself a vital resource, often becomes a measure of how these other resources were being utilized, or urban areas developed.

In short, the boundary waters system and the International Joint Commission (IJC) became not merely an agency for a kind of common management of boundary and transboundary watercourses but also a sensitive barometer for the uses to which these waters were being put in the chemical-industrial-agricultural development along the common frontier and beyond. How significant this barometer has become is demonstrated today by the dispute over acid rain to which this volume devotes a necessary chapter. Indeed, the sensitivity to acid rain of the boundary waters as well as of interior rivers and lakes gives the commission a jurisdictional interest even though, unhappily, the commission's role has not yet been fully defined by the governments nor its experience adequately utilized in the present dispute and negotiations over the acid rain question.

Parallel to acid rain in profile is the emergence of transboundary movement of toxics by air, land, and water where the classical role of the IJC is more firmly established. Certainly under the Great Lakes Water Quality Agreements of 1972 and 1978 the toxics position of the commission was more clearly delineated because of its direct involvement with the water quality effects of toxics. By contrast the acidity of large boundary waters does not appear to have become, for a variety of technical-physical reasons, as significant as the issue now appears to be for many small lakes beyond the boundary on both sides of the frontier. Indeed, the acid rain issue, inevitably outracing in complexity and controversy any attempt at definitive treatment, is about to become, perhaps, the "acid test" for the management by both countries of their future environmental relations.

A new administration in the United States now stands for quite different philosophies of government and environmental management than those held by its predecessor, headed by President Carter. In fact, it may be argued that the entire range of Canadian-U.S. differences now emerging under President Reagan may have as a centerpiece the environmental resource complex of questions. The new Canadian energy policy; the unfinished and frustrating arguments over East Coast fisheries; the slow movement in dealing with West Coast salmon problems (although the recently agreed upon tuna arrangement is a plus); the revival of the Skagit–High Ross Dam dispute; the stalemate resulting from the commission's report on flood control and environmentally related problems in the Lake Champlain–Richelieu River reference; and, finally, the ambiguities now surrounding the future of the Great Lakes clean-up program in view of the less sanguine ambitions as well as the possible reduction of funds that some leaders in the United States seem to be opting for—all of these are possible parameters for a new era in Canadian-U.S. relations. And of course, this new period and policies go well beyond the environmental and the resource based controversies.

Indeed, one irony of this volume is that it is being written at a time when an observer coursing his eye over the 1970s might have said, as Dr. Carroll recognizes, that both environmental philosophy and the principal actors in the two countries were approaching levels of similarity in substance and organization that had not been the case a decade or two before. The established strength of the U.S. voluntary sector—such organizations as the Audubon Society and the Sierra Club, for example—had far exceeded that of their Canadian counterparts in authority and in influence on politics and politicians, perhaps in part because they predated their Canadian analogues by almost a generation. It is not a little sad that just when some equality of perception, as well as of human and institutional resources, were being reached in both countries as to their common and separate environmental problems, the Canadian commitment should seem to be moving steadily forward while that of the United States, at the time of the publication of this volume, suggests some reconsideration of its former devotion to preventing the Silent Spring from being only a metaphor for an ominous future. What will happen in the 1980s to restore the Canadian-U.S. equilibrium of weight and commitment to their joint environmental interests will best be known in the next twelve to twenty-four months, when many of the issues discussed in these pages will have reached some point of crucial debate and decision making.

It was, of course, not possible for Dr. Carroll to examine every aspect of the issues his twelve chapters have raised for the benefit of students and actors alike in both countries. An example of a topic calling for further reflection is the fact that despite the long and quite orderly boundary waters relationship under the 1909 treaty, it failed to lead to systematic joint planning for these waters. Indeed, in a recent study prepared for a conference of the world's waterways commissions sponsored by the United Nations and held in Dakar, Senegal, in May, 1981, it was evident that the Canadian-U.S. experience under the 1909 treaty, while unique in many respects, had led in fact to very little common planning in the classical sense of that term—*pace* perhaps the St. Lawrence Seaway and Power Development, and the Columbia River Treaty.

In fact, it may be argued that while the varied uses or conditions of a river or a lake being studied and later monitored by the commission, or whose level was controlled under a commission order, may have led to some joint understanding and agreement about the best uses for the waters concerned—domestic, sanitation, power, irrigation, navigation, fisheries, habitat, recreation, etc.—a specific joint long-range uses plan was rarely if ever the actual basis of a Canadian-U.S. approach to the solution of a given dispute over any of the waters in question. It was inevitable, however, that a significant degree of less than conscious planning would enter into a reference or a study by the commission. This was certainly true with respect to the recently concluded reference on the Poplar River discussed in chapter 9. Here an old 1948 reference to the commission did invite the IJC to consider the best way to manage the scarce water supplies of the dry midwest region which included the Montana-Saskatchewan Poplar River Basin. Several reports by the commission's boards as well as its own reports are frequently concerned here with the optimum uses of the waters in the region where the competition now developing between water for power (generated by Saskatchewan from nearby coal deposits), and water for agriculture and livestock, were clearly in competition along with historic U.S. Indian rights cutting across any short- or medium-term planning process.

Indeed, almost all commission studies and reports, whether concerned with its quasi-regulatory powers to deal with levels and flows and therefore with flooding, erosion, etc., or whether resulting from matters referred to it (references), demonstrate the inevitability of the IJC having to consider the economic (and alternative) uses inherent in the water-land interface of the given region and to make observations or recommendations about priorities and comparative benefits among competing uses. Varying degrees of planning therefore have entered into the long relationship of the two countries represented by the work of the commission and in the very nature of the commission's orders (with conditions), studies, and reports, even though it may not have been a conscious or deliberate effort at planning as such. Certainly the great IJC studies leading to the Columbia River Treaty (in its final version) of 1963–64, the St. Lawrence Seaway Orders of Approval of 1952 and 1956, and the Great Lakes Water Quality Agreements of 1972 and 1978 were in all cases a kind of planning process even if the final actors were governments and agencies on both sides of the common frontier.

What could not have been foreseen however, was the degree to which planning was required as an urgent and sensible preventative to the chemical-industrial impact of economic growth in the Great Lakes Basin. Much of the Great Lakes Water Quality Agreements of 1972 and 1978 are therefore a late but commendable effort to prevent the further deterioration of the lower lakes and connecting channels, which have been observably polluted, and badly so, since the third decade of this century. At the same time this joint Canadian-U.S. exercise in rescuing these waters also has been aimed at preserving the very high quality of the upper lakes where with some exceptions industrial-chemical man has not yet penetrated enough to diminish their essentially pristine quality.

Inevitably this volume will also be read as an exercise in comparing the techniques and records of both countries in managing their land-based environmental frontier with their efforts in the three oceanic neighborhoods they share in the Arctic, the Pacific, and the Atlantic. From the environmental point of view, the legal-administrative systems operating for so long under the Boundary Waters Treaty are conspicuously different from the lack of joint systems, on any level of comprehensiveness and authority, on the three oceans. No Canadian-U.S. seaward boundary commission has been established as yet to anticipate the consequences of massive joint and conflicting interests, actual and potential, in the Atlantic, the Pacific, and the Arctic. It is true that among the high achievements of Canadian-U.S. diplomacy were the West Coast halibut and salmon exercises which, until recently, were entirely bilateral. Indeed, for a very long while the management of halibut and salmon fishing stocks and shares seemed to be working well under commissions established with the power to make quota recommendations determining the take of both countries. Until the recent disputes with respect to tuna (now more or less settled), and the as yet unfinished and newly acrimonious conflict over salmon, it was generally believed that West Coast oceanic questions in the case of fisheries at least were reasonably under control.

Similarly, the East Coast fisheries problem, one of the oldest sources of Canadian-U.S. arbitration-treaty relationships, was until recent years thought to be reasonably controlled within the managerial regulation of both countries. Here the not unsuccessful years of quotas determined by the International Commission for the North Atlantic Fisheries (ICNAF) seemed to provide a certain stability in the sixties and early seventies before it was recently replaced because of the new jurisdiction resulting from the two-hundred-mile Exclusive Economic Zone claims.

Crystallized by the Third Law of the Sea Conference (LOS III), the two-hundred-mile Exclusive Economic Zone, and the parallel debate over the continental shelf and the continental margin, have raised new questions affecting both the living resources of the sea as well as the exploitation of the seabed and subsoil of the shelf-margin areas. For the first time it became important to ascertain the Canadian-U.S. seaward boundary in the Georges Bank area, both for reasons of fisheries control within the new two-hundred-mile limit and for the purpose of exploiting hydrocarbons on the continental shelf. The present jurisdictional uncertainties in the Atlantic coastal region coincide with the imminent and expected completion of the Third Law of the Sea Conference. The scheduling of the draft treaties had been interrupted momentarily by the U.S. withdrawal in early 1981 from the proceedings pending a reconsideration by the Reagan administration of the U.S. position on this massive new regime for all the seas of the world. The return of the United States to the conference table has made it likely that this immensely complex and codified regime of the seas will soon achieve the successful completion long hoped for by the entire family of nations.

In the Arctic the combination of an old dispute with respect to jurisdiction—the Arctic island waters and the Beaufort Sea line—and the new oil and gas exploratory-developmental activities by both countries (and particularly by Canada in the Beaufort Sea) have focused attention on common environmental and boundary questions in that area. In this instance the two countries can rely on no previous common juridical or administrative experience in jointly dealing over time with Arctic matters. As Dr. Carroll describes it, this combination of the jurisdictional and environmental elements—with certain unique activities by Canadians, e.g., in the Beaufort Sea, in contrast with the special interest of the United States in transportation by tankers through the Northwest Passage of production from the Prudhoe Bay area—will in due course have to be resolved by some kind of agreement. Joint monitoring mechanisms may be necessary to insure the effectiveness of any permanent understanding.

If any institutional exercise unites the chapters of this volume as one of the first major studies of Canadian-U.S. environmental relations, it is the International Joint Commission that occupies that high place of interest even though its role has been absent from any of the oceanic conflicts. In Professor Carroll's discussion of the commission he perhaps places too much reliance upon the views of commission members, present and past, as well as on diplomats and civil servants, for his generally favorable, but not uncritical, description of the commission's achievements.

The conflict of philosophies he detects between former commissioners, particularly on the appropriate scope for the commission's role, may be somewhat exaggerated. What seems to be a view by some commissioners advocating a broader range of activities and initiatives for the commission, in contrast to those commissioners and students of the IJC who warn against an overextension of its functions, may in fact only be a difference in the syntax and tone used, not in actual jurisdiction claimed. The views of most past commissioners seem to have much more in common regarding the value of the IJC and the range of its proper role than the differences Dr. Carroll tends to emphasize. Nevertheless, some differences do exist among commissioners, civil servants, and scholars, and essentially these relate to an appraisal of how to maintain secure and stable relations between an independent and autonomous commission and both federal governments.

Despite the absence of sustained commission leadership within recent months—which may have been encouraged by some worried about sporadic IJC activism—both governments, throughout the first half of 1981, have demonstrated a regrettable lack of concern for the commission's minimum needs. They left five of its six memberships (three U.S. and two Canadian) vacant from January to September, 1981, so that the commission, as of October 1, 1981, had not been able to function effectively for over eight months in the absence of chairmen, a quorum, and decision-making capacity. Rarely has the commission been subjected to the stresses of the relative indifference of both governments to safeguard this, one of the most durable and valuable of Canadian-U.S. institutions.

It would be unwise to suggest that Professor Carroll's serious and commendable effort to master the mass of data in the totality of Canadian-U.S. environmental relations has succeeded on every page. That would be unfair to any reasonable expectations on the one hand and to the complexity and variety of the questions facing both countries on the other. At the very least, however, this volume examines with great detail in some cases, and with less in others, the immense network of oceanic and land-based environmental conflicts, long- and short-term, that now confront Canada and the United States. These have behind them a frequently successful but sometimes contentious and discouraging past, and today they present a mix of technically complex, if hopeful, futures qualified by many political unknowns. Not unexpectedly, this study contains many references to the International Joint Commission. Indeed, any volume on Canadian-U.S. environmental relations that excluded frequently turning to the record of the commission would have been—with some overstatement—Hamlet minus the Prince. More relevant is the thought that the IJC model and achievements, particularly its technique of common fact-finding and fact evolution, may have some urgent lessons for other areas and possible agencies to manage Canadian-U.S. joint problems and disputes today and in the future.

The rise of some neoconservative policies in the U.S., financial constraints in both countries, and the emergence of a general reluctance, now appearing in Canada and already present in the United States, to rely upon extensive regulatory devices to save or salve the economy may all be conspiring to render common environmental problems more difficult to manage to the joint satisfaction of both countries in the future. The movement from cost-benefit to risk-benefit analysis; the rise of coal and its demands for water (over and above the chronic crises of the dry belts) while creating at the same time more dangerous levels of acid rain as old coal-burning plants reopen and new ones begin; the uncertainty surrounding U.S. funding for environmental protection projects including the massive and forever unfinished challenges of the Great Lakes; the resource-pollution complex of present and forecastable disputes on the three oceans, all affected now by the new and prospective Law of the Sea regime; these and many other factors set out in this valuable book will dominate Canadian-U.S. relations for years to come. Dr. Carroll has made an able and timely contribution to the better appreciation of the depth, the variety, and the detail of these issues and has provided some guidance to their partial resolution as both countries move forward in the daily business of getting along together as oceanic and continental neighbors who are joint custodians of immense land, water, and air resources and uses for themselves and their hemisphere.

Maxwell Cohen
Ottawa
January 15, 1982

Preface

Over the last twenty years people have become increasingly aware that decisions taken elsewhere affect the quality of the air, land, and water around them. Since Canada and the United States share a common ecosphere along an extended land and water border, each perceives that decisions crucial to their environment are being taken in the other country. Eighty percent of all Canadians live within a hundred miles of the border, while some of the major industrial cities and most treasured recreation areas in the United States lie due south of it. Growing environmental sensitivity coincides, however, with a desire to increase or to strengthen industrial development on both sides of the border. Canadians who are still building up their industrial structure hope to establish industry in new areas while some of the U.S. border states feel the need to reinforce their industrial base threatened by the shift of economic activity to the southwest. Thus border concerns are multiplying and drawing in new actors while more recent awareness of the long-range environmental effects of airborne pollution is creating deep anxiety far from the border. The term *acid rain,* practically unknown a few years ago, has almost become a household expression. Solution of the problems it creates will require far-reaching political and economic trade-offs in both countries. Indeed, transboundary environmental relations may well dominate Canadian-U.S. affairs in the 1980s.

In 1980, the C. D. Howe Institute, which has traditionally devoted a large part of its resources to furthering understanding of Canadian-U.S. relations, commissioned the first comprehensive overview of transboundary environmental issues. John E. Carroll, the author, addresses both environmental and diplomatic considerations in the study that follows. His systematic examination of the initiation, buildup, and resolution (or deadlock) of these issues shows how the bilateral relationship works, or does not work, in the environmental field. One of the main conclusions of this study, as of many others the institute has released, is that despite the apparent closeness of the two countries, harmonious relations between Canada and the United States require continued inventiveness as well as attention. Part 3 of this book is an example of such inventiveness.

The institute was fortunate in that Maxwell Cohen, one of the foremost experts in the field and former Canadian Cochairman of the International Joint Commission, accepted an invitation to write a foreword to this book.

In addition to acknowledging Professor Cohen's continued support, the institute and the author wish to acknowledge the expert assistance they received from the members of the Canadian-U.S. Transboundary Environmental Relations Project Advisory Committee. The committee, the members of which are listed below, met twice during the life of the project, first to discuss the research plan and methodology and later to review and comment on the author's findings prior to drafting. Members of the committee also gave generously of their time to familiarize the author with information at the industry level and to review individual chapters of the manuscript. The institute and the author warmly acknowledge the committee's invaluable assistance. They also wish to point out that the committee's task was limited to providing expert information and commentary; the conclusions of the study are those of the author.

Douglas C. Calder
Manager, Right-of-Way and
    Environmental Affairs
TransCanada PipeLines

Harvey H. Clare
Environmental Protection
    Coordinator
Imperial Oil Limited

Creighton Cross
Vice-President, Business
    Planning and Development
Alcan Aluminium Limited

John C. Cuthbertson
Director, Environmental
    Affairs
American Can Company

George Francis
Faculty of Environmental
    Studies
University of Waterloo

Frank Frantisak
Director of Environmental Services
Noranda Mines Limited

Charles Ferguson
Director, Environmental
    Affairs
INCO Limited

Ronald H. Hall
Advisor-Environmental Affairs
Gulf Canada Limited

Newell B. Mack
Energy & Environmental
    Policy Center
Jefferson Laboratory
Harvard University

O. P. Mukheja
Manager, Occupational Health
Union Carbide Canada
    Limited

Scott Munro
Environmental Officer
Fiberglas Canada
    Incorporated

Don Munton
Research Director
Canadian Institute for
    International Affairs

Milton Steinmueller
Department of Resource
    Development
Michigan State University

David Stewart
Canadian Steel Environmental
    Association

John B. Sweeney
Vice President, Technical
    Development
    and Environmental Affairs
Consolidated-Bathurst Inc.

Caroline Pestieau
Project Coordinator
C. D. Howe Institute, Montreal

Acknowledgments

The author is indebted to the University of New Hampshire for providing a sabbatical leave to write this book, to his research assistant, Roy Stever, for invaluable assistance throughout the conduct of his study, and to his wife, Diana C. Carroll, for complete sharing in all aspects of the work. It is her effort as much as it is his.

Contents

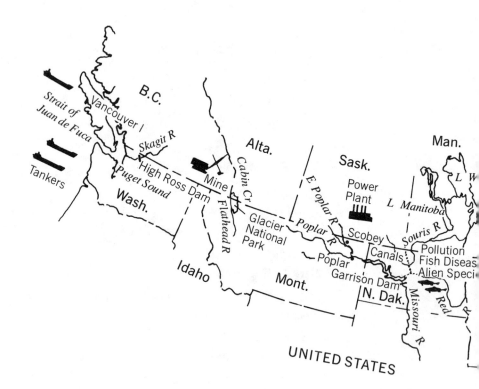

Tankers

Strait of
Juan de Fuca

Vancouver I

B.C.

Skagit R

Puget Sound

High Ross Dam

Wash.

Mine

Cabin Cr

Alta.

Flathead R

Glacier
National
Park

Idaho

Mont.

E Poplar R

Poplar R

Poplar

Sask.

Power
Plant

Scobey

Canals

Garrison Dam

N. Dak.

Man.

L W

L Manitoba

Souris R

Pollution
Fish Diseas
Alien Speci

Missouri R

Red

UNITED STATES

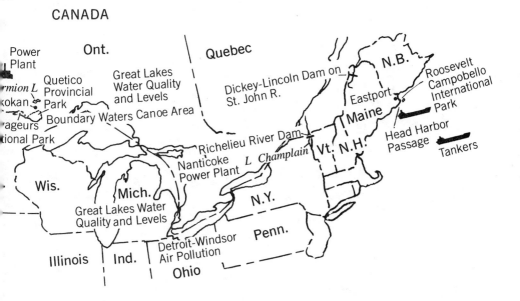

CANADA

Power Plant
Ont.
Quebec
N.B.
Roosevelt Campobello International Park

Quetico Provincial Park
Great Lakes Water Quality and Levels
Dickey-Lincoln Dam on St. John R.

rmion L
kokan
ageurs
ional Park
Boundary Waters Canoe Area
Eastport
Maine

Richelieu River Dam
Head Harbor Passage

Nanticoke Power Plant
L Champlain
Vt. N.H.
Tankers

Wis.
Mich.

Great Lakes Water Quality and Levels
N.Y.

Detroit-Windsor Air Pollution
Penn.

Illinois
Ind.
Ohio

Overview—Main Points of Bilateral Environmental Frictions. (Adapted from *National Parks & Conservation Magazine*, March, 1978. Copyright © 1978 by National Parks & Conservation Association. Used by permission of the publishers.)

Introduction

Perhaps no other relationship between two countries anywhere in the world seems smoother flowing or is as often taken for granted as that between the United States and Canada. President John Kennedy's dictum, "Geography has made us neighbors. History has made us friends. Economics has made us partners. Necessity has made us allies"[1] was evocative of the fact that part of this free-flowing relationship stems from conditions of necessity and reality, both geographical and economic. But a real part also stems from the values held in common by the two peoples, their alikeness, their friendship, and their regard for one another as individuals over a long period.

Yet there have always been certain philosophical and institutional differences between the two peoples and the relationship has not always been a warm, familial one. The history of Canadian-U.S. relations has been viewed as being divided into nine historical periods: a relationship of enmity and rather poor relations prior to 1910; a decade of pulling together for the Great War; a decade of reveling in the prosperity of the 1920s; a decade of sharing the misery of the 1930s; a decade of meeting the global wartime and postwar challenge (1940s); a decade of passivity and joint development, both planning and carrying out megascale projects together (1950s); a decade of coming apart on large and small issues (1960s); a decade of serious problems and the end of the "special relationship" in diplomacy (1970s); and a continuation of those serious[2] problems today (1980s).[3]

It is only in the present century that the relationship has been founded on the kind of trust and admiration one has come to associate with Canadian-U.S. relations. During more than half its history (1750s–1910), the United States' relations with Canada were characterized by mistrust and suspicion, whether in planning for war (or in a few instances conducting it) or in being concerned over "manifest destiny" in the form of expansionism or annexation. Hence, it is both inaccurate and perhaps dangerous to go as far as both peoples do in taking the bilateral relationship for granted. It may well need much more attention and even nurturing than is commonly realized. Preventive medicine may be called for in greater doses lest curative medicine someday be found necessary.

In the past, bilateral frictions usually arose from political, military, or economic issues. The possibility of disagreement in these fields has not been eliminated, but while some of the traditional issues lie dormant, a whole new sphere of interaction has recently come to the fore. Of the many issues inhibiting a harmonious relationship, the environment is rapidly becoming prominent. Bilateral environmental disagreements are becoming more numerous, more serious, and less tractable and are beginning to represent greater stakes than ever before. They now constitute a substantial portion of all of the bilateral difficulties between the two nations and are strongly represented on any list of the top five or ten most serious diplomatic differences outstanding at any one time.

Part of the problem relates, of course, to geographical circumstance—a shared ecosphere along an extended border—for the land, the water, and the atmosphere of North America are shared by both countries. Both must cope with this reality and its inherent challenges and opportunities whether they wish to or not. The reality includes shared water pollution, shared air masses and the pollution thereof, and shared land, soil, vegetation, and wildlife to the degree they cross the border or are affected by pollution crossing the border. Solving the problems identified in this book will thus benefit both countries. As a matter of necessity, therefore, the environmental question is no longer the peripheral matter it once was but is now taking the center of the stage.

This book is unique in providing a first comprehensive overview of Canadian-U.S. transborder environmental relations. It is being written equally for both peoples, so that while recalling the facts of shared geography (of river basins, of air masses, of polluting activities) it also has to stress that the two societies are at different stages in their historical development and maturing processes. They do not always share the same economic and social goals and philosophies, the lip service and rhetoric of some in both countries notwithstanding. Writing for both peoples simultaneously provides a substantive challenge, given the great difference in the level of knowledge they have of each other and the different ways in which they conduct their domestic affairs.

This first comprehensive introduction to Canadian-U.S. bilateral environmental relations is addressed to a wide and growing audience of government officials, corporate decision makers, academics, citizens' groups, students, residents of the border regions, and interested readers on both sides of the border. It is intended to serve the needs of those practicing, or interested in, the environmental aspects of international diplomacy, as well as the international diplomatic aspects of environmental policy, in four major sectors of modern society: government, business, citizen interest groups, and academe.

The book breaks new ground in combining an interdisciplinary approach—addressing environmental and diplomatic considerations—with an overview of the major bilateral environmental problems. It thus pulls together a substantial quantity of material on past and present transboundary environmental issues along the lengthy Canadian-U.S. border and will serve as a reference for both practitioners and concerned citizens. At the same time, its systematic examination of the initiation, buildup, and resolution (or deadlock) of these issues provides basic insights into the way in which the bilateral relationship works, or does not work, in this field. As transboundary environmental problems multiply in the 1980s, an understanding of the interactive processes will be crucial if unnecessary and destructive tensions are to be avoided.

The word *environmental* can of course cover a very broad field and may, indeed, include many topics extraneous to this study. The following chapters are essentially concerned with the quality (or lack thereof) of air, water, and, in some cases, land. Natural resources including wildlife, forests, agricultural crops, commercial fisheries, and soils, which are more appropriately treated under the equally broad term *conservation*, are largely ignored here. The study attempts to present a reasonably comprehensive general introduction to all aspects of, first, water quality and quantity at the border (while not delving into the issues of continental water management, or of international water export or import), and, second, air quality at the border and across the continent as well. The latter is fast developing as an international issue.

Part 1 sets the stage by introducing the different groups of actors in each country. It describes the atmosphere at the interface including some of the pertinent differences in background and approaches of the two societies. It also examines the crucial Boundary Waters Treaty of 1909 and its creation, the International Joint Commission.

Part 2 presents a series of case studies. Issues were chosen as case studies with reference to five criteria:

- their role and importance within the bilateral environmental relationship;
- geographical balance—coastlines, Great Lakes, prairies, western mountain regions, the Arctic;
- coverage of the diverse types of problems—water quantity, water quality, air quality over limited and over continental expanses, coastal pollution;
- the nature of the decision-making process involved; and
- their diplomatic importance and relevance to the overall bilateral relationship.

All of the major Canadian-U.S. transboundary environmental problems, both historical and contemporary, have been included. Forthcoming fric-

TABLE 1. Transboundary Environmental Issues by Functional Characteristic

	Water Quality	Water Quantity	Air Quality	Marine Environment	IJC Involvement	Acid Rain Implications	Toxic Substances	Direct Corporate Concern	Directly Energy Related
Beaufort Sea				X				X	X
West Coast tankers				X				X	X
Skagit		X			X				X
Trail Smelter								X	
Cabin Creek	X					X		X	X
Poplar	X	X	X		X	X	X	X	X
Garrison	X				X				
Souris River	X	X			X				
Red River	X	X			X			X	
Roseau River	X	X			X				
Lake of the Woods	X	X			X				
Atikokan		X	X		X	X		X	X
Great Lakes	X	X			X		X	X	X
Detroit-Windsor			X		X	X	X	X	X
Nanticoke			X			X		X	X
Cornwall Island			X		X		X	X	X
Champlain-Richelieu		X			X				
Eastport			X	X	X	X		X	X
St. John River	X	X			X			X	
St. Croix River	X				X			X	
Acid rain	X		X		X	X		X	X
Toxic substances	X		X	X	X		X	X	
Georges Bank				X				X	X

Note: The Canadian complaint concerning U.S. acid rain sources relates to the Ohio Valley and various other nonborder localities and thus is not incorporated in this chart per se.

tion points have also been pinpointed where possible. The issues are cross-referenced in table 1. The unifying theme of case study selection is to show how issues have been initiated, how they developed, and how decisions have been reached in transboundary bilateral matters.

In view of the large number of issues described and analyzed in this book, the reader might well wonder what constitutes a bilateral issue. Put simply, any project or undertaking representing a benefit to one country and a cost to another and concerning which the latter raises objection is a bilateral or potential bilateral issue. The cases represented in Part 2 are all bilateral (though they may also have significant domestic components). They were all identified by one or both diplomatic communities as problems in the international relationship, in the present or in the recent past; or, in a few cases, have been identified by some governmental or nongovernmental actors, and by the media, as likely to become diplomatic problems in the future. Hence, the core focus of this book is on actual or potential diplomatic matters. Since only governmental officials at the federal levels in both countries are constitutionally authorized to negotiate or make decisions in this area, these federal actors are naturally the central focus of attention. This is not to say, however, that other governmental and private sector actors are not important (indeed they are often key players) and thus their role is also treated in some detail. Their official bilateral impact, however, is channeled through the federal actors in both countries.

Part 3 presents the study's conclusions and discusses the hypothesis of an increased formalization of Canadian-U.S. environmental relations. Such a hypothesis has been advanced by a number of those working in, or concerned with, this field so that it is now possible to describe some of the necessary characteristics of a more formalized arrangement. The thrust of this final chapter is that the time has come to consider such a step and that the preceding eleven chapters provide us with much of the information needed to do so constructively.

In preparing this book, the author could not rely on extensive documentation in the key area of bilateral environmental interaction. A large number of comprehensive interviews were therefore necessary and, because of the novelty of the field, almost all of the principal actors were interviewed. Standard questions posed in these interviews and a summary of the responses can be found in chapter 2. The majority of central actors, as described in chapter 1, are government officials. Yet what they do affects others in the private sector, which then reacts and can in fact itself initiate actions that have a bilateral environmental impact.

This study thus inevitably focuses on the governmental process but demonstrates how other groups, such as agriculturalists, manufacturers, and residential communities, often play an active role in all stages of the development and maturation of environmental issues.

NOTES

1. President John F. Kennedy, Address to the Parliament of Canada, Ottawa, Ontario, May 17, 1961.

2. The word *serious* as used in this context is, of course, relative to other periods in Canadian-U.S. relations and not relative to relations between other pairs of countries.

3. Senator George C. VanRoggen, Senate of Canada, interviews, Ottawa, Ontario, October 9, 1980, and Vancouver, British Columbia, December 22, 1980.

Part 1
The Stage

CHAPTER 1

The Actors

As environmental questions take the center of the stage in the bilateral relationship, it is timely to look at the actors on each side of the border. Who are they? What is their background? What are their traditional decision-making mechanisms? Such an examination allows us to draw several important conclusions about the actors on the stage. First, there are relatively few people officially responsible for the transboundary environmental relationship. Second, the nature of their work is extremely challenging when viewed from a traditional diplomatic viewpoint. Third, there is a greater investment in human resources on the Canadian side. Fourth, the awareness of institutional differences is important to an understanding of the expectations and behavior of the actors on each side of the border.

The Canadian-U.S. bilateral relationship is a complex, extensive, mature, and well-developed relationship between two advanced industrialized societies—one a superpower, the other a middle power. It is an intense relationship carried on at many different levels simultaneously. Each country is critical to the other for defense and national security. Each is economically dependent upon the other. Yet in spite of all of this, a very small number of actors, either in or out of government, participate on a full-time or even on a significant-time basis in the conduct or management of the bilateral relationship. Some might argue that this is because the two nations are so similar in nature and purpose, have such great understanding of and respect for each other, and, in effect, get on so well that there is little need to devote manpower to the maintenance of the relationship. Lack of attention might be positive, not negative, in the sense of not exacerbating things. Others argue, however, that such an apparent underinvestment in protecting and optimizing the relationship, especially on the part of that nation so often accused of taking the relationship for granted, the United States, has led to needless friction between the two countries. It has also led to potentially dangerous erosion of the health of the relationship and to the loss of opportunities for joint endeavors and of chances of optimizing benefits for both peoples.

Diplomats in the environmental field have to deal with an unaccustomed and recurrent problem. This is the difficulty of working with an evolving scientific data base. Government researchers and regulators who develop and interpret the information for the use of diplomats wield substantial power. At the same time, environmental research in the private sector is a highly developed form of lobbying and it is encumbent upon diplomats to keep up with or even out in front of continuing research, lest they be caught unawares as the data base changes.

A study of the actors in Canadian-U.S. environmental relations shows that Canadians are more interested in, involved in, and consequently more specialized in bilateral relations than their U.S. peers. Canadian diplomats and other government officials are also more numerous in this area and of a higher level of expertise, while U.S. counterparts, less experienced in bilateral relations, rely more on specialized agencies for expert support. Provincial actors in Canada play a far stronger role than do state actors in the United States. Finally, the network of actors concerned with bilateral environmental questions is widening, but perhaps not fast enough to keep up with the interest the media is catalyzing, particularly in Canada.

There are also differences between the institutions in each country. Not the least of these is the greater importance and role of lobbying and litigation (and hence of the judiciary) in the United States and the greater media (and specifically public broadcasting) role in Canada. These and other institutional differences in the environmental area dictate a different configuration of actors in each nation and a different set of expectations about their respective behavior.

This chapter situates the actors and thus provides a necessary context for the understanding of the initiation and development of bilateral environmental issues. Primary attention is given to governmental actors who are officially responsible for bilateral relations and who provide the framework for private sector interests. In the following pages, we therefore first examine four sets of governmental actors: federal officials on each side of the border, Canadian provincial officials, and U.S. state officials.

The Canadian-U.S. relationship is, however, conducted more on a private nongovernmental level than is typical of other relationships between two different nations. In the latter part of the chapter we turn to nongovernmental actors: the corporate sector, citizen environmentalists, and the media. Throughout the discussion we attempt to highlight the common and the contrasting characteristics of actors in each of the respective countries.

Canadian Federal Government Actors

Canadian federal actors may be divided into two sets of people: diplomats and other officials, mainly environmental. The diplomatic actors are within the Department of External Affairs, the agency which conducts Canada's foreign relations. Consistent with the general tendency of the Canadian government (and reflecting Canadian society as a whole), External Affairs accords great weight to its bilateral relationship with the United States, and thus has organized a bureau devoted solely to this U.S. relationship, the Bureau of U.S.A. Affairs.[1] The bureau is further subdivided into a General Relations Division, handling the political, economic, and military relationship, and a Transboundary Relations Division, handling all transboundary environmental problems. The latter division incorporates fisheries resources, wildlife (particularly migratory species such as waterfowl, caribou, and marine mammals), and parks and wilderness responsibilities in its mandate, as well as all water quality and apportionment and air quality questions, marine and Arctic concerns, and relations with the International Joint Commission (IJC). It is staffed by five transboundary environmental specialists, and the workload is divided geographically (coastlines, prairies, Great Lakes, etc.) and substantively (air, water, marine, etc.). Further assistance is provided by a separate supporting arm, the Legal Division.

The field arm of this Ottawa group is found at the Canadian embassy in Washington, which employs one full-time diplomat with the rank of Environmental Counsellor. His area of responsibility corresponds to the duties of the five environmental specialists in Ottawa, and he provides not only a more direct link with U.S. diplomats but more importantly a direct link with U.S. environmental officials and legislators. In recent years this officer has had credentials in the environmental and diplomatic areas. The field arm is further extended by the consular posts in a great number of U.S. cities. These act as the eyes and ears of External Affairs in the U.S. regions in their data-gathering and information transfer functions and are also encouraged and expected to provide analysis. Indeed, few Americans are aware of the size or effectiveness of operation of many of these "mini-embassies" in their midst and may think of them as tourism promotion and trade assistance offices, if they think of them at all. Consular staff often attend public hearings, legislative sessions, and so on when matters of interest to Canada are on the agenda, gathering information and providing analysis for the Washington embassy and for Ottawa.

Outside the diplomatic community, the most actively involved people in the Ottawa government in transboundary environmental issues are found in the Department of the Environment, the federal agency which has environmental policy, research, and regulatory responsibility. The U.S. Group within the Intergovernmental Affairs Directorate of Environment Canada often consists of former diplomats or officials with some foreign relations experience. This U.S. Group is environmentally knowledgeable in a technical sense and is able, therefore, to advise the nontechnical generalists in External Affairs, while still having appropriate knowledge of, and sensitivity toward, the intricacies of international diplomacy and Canadian-U.S. relations. Their role also includes liaison with the numerous scientists throughout Environment Canada and, through that linkage, with scientists at the U.S. Environmental Protection Agency (EPA) and the U.S. Department of the Interior. There is, therefore, a liaison between

Environment Canada and the U.S. EPA. This liaison circumvents diplomatic channels, but is known to and accepted by diplomats.

Officials of Environment Canada's Environmental Protection Service (EPS) (air, water, and toxics pollution), Inland Waters Directorate (IWD) (water resources), and Atmospheric Environment Service (AES) (weather and research aspects of acid rain) all play a role with External Affairs as major Ottawa actors. For example, the IWD is involved in virtually every freshwater-related transborder environmental quality issue, including the Great Lakes issues, either directly or indirectly through participation on IJC boards, and it is this agency that played a significant role in laying the foundation and participating in behind-the-scenes negotiations on both Great Lakes Water Quality Agreements (1972 and 1978) (see chap. 6). IWD regional offices are especially involved in transboundary questions on a constant basis. The EPS provides the technical (and a not inconsiderable proportion of the total Ottawa) leadership on the acid rain question, while the AES provides a major research lead on the acid rain question. The Canadian Wildlife Service of Environment Canada is also occasionally involved in bilateral environmental matters.

Mention must be made of other federal departments. Fisheries and Oceans has critical responsibility in the Eastport and West Coast oil pollution issues and some Arctic responsibility. Indian and Northern Affairs has major Arctic marine and Yukon terrestrial responsibilities. Energy, Mines and Resources has energy-related decision-making authority with often significant transborder environmental ramifications. Less directly involved are the Department of Transport (including the Canadian Coast Guard and vessel traffic management responsibility—see chap. 4), and the National Energy Board.

The Canadian Parliament likewise reflects broad knowledge of the United States, and is often pressured by constituents in border areas to take action to protect Canadian interests from real or perceived U.S. threats. The prime minister and cabinet and their offices (especially the Office of Federal-Provincial Relations) are likewise under pressure to be knowledgeable about and active on these matters. The ministers of external affairs and of the environment, however, bear most of this responsibility.

In general, Canadian federal officials dealing directly with U.S. relations are widespread in the government, often organized into institutional entities such as sections or divisions, well staffed with the best available personnel, and have considerable knowledge of the United States. Senior decision makers (particularly deputy and assistant deputy ministers[2]) and senior policy advisers often have considerable bilateral experience. They are thus well suited to make measured bilateral judgments, notwithstanding the fact that their own job responsibility might be purely domestic in nature. The investment made by the Canadian government in this area is therefore considerable.

Canadian Provincial Government Actors

Provinces on average are much larger than U.S. states in both land area and economic wealth, and thus actors on the provincial level play a much more significant role than most Americans generally realize. Provincial governments in Canada are very powerful actors within the Canadian confederation and are in many ways semiautonomous, occupying a place in the power structure far superior to that of state governments in the United States.[3] The provincial government role is well established in Canada's constitution, the British North America Act of 1867 (BNA Act). The BNA Act and subsequent judicial decisions have given the provinces ownership of their resources, including water and air. They are thus able to maintain jurisdiction which in the United States is often within the authority of the federal government. There has been an effort in Canada to "patriate the constitution" by bringing home the BNA Act (from London to Ottawa) and, in effect, amending it in ways which would give the federal government greater leverage over the provinces. After much debate and confrontation, the effort succeeded in late 1981. Its probable consequences for federal-provincial relations remain to be seen and are the subject of extensive controversy.

Environmental jurisdiction in Canada is constitutionally a gray area under the BNA Act (since patriation known as the Constitution Act) since it is an area of shared responsibility. Much which represents a threat to health or is transboundary in nature is under federal jurisdiction. Navigable water is federal, as in the United States, but Ottawa has not been able or willing to exert as wide a mandate as Washington in this area. Fisheries are also federal. In the air resource area, ambient air and local sources of emissions are a provincial responsibility, as long as there is no health danger.[4] In the United States, virtually all the real authority in these areas resides in Washington.[5]

The broader environmental or pollution regulatory and control agencies are the principal provincial actors in transborder environmental questions. Ottawa has the lead responsibility in all international matters, including environmental, while the provinces have the lead responsibility in the domestic environmental area, including regulation and management of air and water resources. However, the political power of the provinces has increased in recent years as has the practice of federal-provincial negotiation on an almost equal partner basis. Given the frequent de facto need for provincial consent for federal initiatives, the actor role of the provinces, and particularly of their environmental departments, has increased immeasurably.

Officially, the provincial environment departments do not contain U.S. relations or international relations staff. However, the deputy ministerial level contains a wealth of both experience and expertise in the area of

U.S. relations and an interest in, and ability (if not authority) to conduct, international negotiation. This stems from experience in almost constant federal-provincial negotiation, from often deep knowledge of adjacent U.S. states and regions, and from frequent IJC-related experience, either in dealings with the commission itself or from participation on or with its boards of inquiry. These public officials often have a sharp awareness and understanding of the Boundary Waters Treaty, superior not only to that of their state government counterparts but often superior to that of U.S. federal officials within the adjacent states and regions. This knowledge and expertise is often found at the divisional level as well, and sometimes farther down through the ranks. Senior policy advisers in the provincial environmental bureaucracies also often have specialized U.S. expertise similar to that of their federal counterparts.

Given that the matters described here involve a province's external relations, the premier and cabinet may also often take an interest in the evolution of events in the Environment Department. But such direct interest has been subsumed in recent years by the newly created departments of intergovernmental affairs which are found in all of the larger provinces and are beginning to appear even in the smaller ones. These departments are "mini–foreign offices" which, although primarily established to coordinate interprovincial and federal-provincial relations, have been increasingly striking out internationally. British Columbia, Saskatchewan, and Ontario are particularly active in this area environmentally, and Quebec and Alberta are well equipped to so move should the need arise. Hence, the provinces now have an institutional counterpart to External Affairs as well as a wealth of U.S.-related expertise within these intergovernmental relations bureaucracies. If they work cooperatively with their own Environment Department colleagues, they may well come to equal Ottawa in capability as transboundary environmental actors. (Some believe Ontario is already equal to Ottawa in such capability.)

Other agencies of provincial government also become involved in the transboundary environmental relationship as the need arises. The most important of these are the provincial water resource management agencies which work in the quantity and apportionment rather than quality area, and find themselves at the environmental interface in connection with transborder flooding, allocation of water for municipalities and industry diversions, or simply maintenance of agreed levels and supplies. By definition, they are closely involved with the IJC on numerous apportionment and flooding questions and often contain significant IJC expertise. In addition, such provincially owned crown utilities as Ontario Hydro, Saskatchewan Power Corporation, and British Columbia Hydro have been concerned with large-scale coal, hydro, or nuclear projects having transboundary effects. Although technically government-owned, crown utilities have behaved as nongovernmental corporate actors and thus will be further discussed in that section.

U.S. Federal Government Actors

Distinguishing the U.S. federal actors from their Canadian counterparts is not so much a difference in the inventory of agencies as it is a difference of approach. The U.S. government has all the expertise it needs to manage or solve these problems, but that expertise is not mobilized or applied to Canada.

The foreign office of the United States, the Department of State, is a giant foreign relations bureaucracy, managing as it does the extensive international involvements of one of the planet's two great superpowers. However, the area of Canadian affairs is an exception to the rule, for the Office of Canadian Affairs (EUR/CAN) housed within the important Bureau of European Affairs is a small effort, with only one diplomatic officer assigned full-time to transborder environmental responsibility. This stems in part from the relatively low level of importance the United States assigns to the relationship and in part from a different division of responsibilities in the U.S. government. Coastal environmental issues are handled by an economics officer who also handles coastal fishery and energy matters, while international wildlife, parks, and global environment questions are handled by a different bureau at the State Department, the Bureau of Ocean, Environment, and Scientific Affairs (OES). The environmental officer in the Office of Canadian Affairs concentrates mainly on conventional transborder pollution problems, the Boundary Waters Treaty, and IJC liaison. This is a tall order, given the length of the border and the number and complexity of the issues, and the extra assistance provided within the department by Legal Affairs still does not bring its capability up to that of its Canadian counterpart.

Assisting the State Department effort directly is the small Office of International Activities at the Environmental Protection Agency. International Activities is a new actor on the scene and one which appears to have already reached the peak of its power and been substantially reduced in role for budgetary reasons. This office provides one officer full-time to Canadian affairs. In addition, the regional offices of EPA have devoted personnel to the study and solution of transboundary environmental questions. EPA has been a major contributor to IJC boards and, through its Great Lakes National Program Office in Chicago, to U.S. responsibilities under the Great Lakes Water Quality Agreements. Its national research laboratories in various regions have played a major role in providing technical assistance to the State Department and U.S. negotiators on a variety of questions. They are currently playing a large role in acid rain research with direct bilateral ramifications.

A great number of other U.S. specialists assist the Office of Canadian Affairs in a variety of ways. Civilian and military personnel of the Army Corps of Engineers contribute directly, or indirectly through the IJC, where they often serve on boards. The Interior Department's Bureau of

Reclamation, Geological Survey, and Fish and Wildlife Service and, less often, the Department of Agriculture's Forest Service and the Department of Transportation's Coast Guard likewise play a role.

The U.S. government has more than sufficient technical expertise in any bilateral environmental issue that may arise in Canadian-U.S. relations. However, it differs from the Canadian government in that, while its personnel are more numerous and diversified, it does not have as much insight into the bilateral relationship. Few U.S. officials or personnel have any Canadian knowledge or expertise. Also, assigned work in this area is often added on to other already burdensome responsibilities while there is little incentive through reward or recognition to do such work. And, the level of expertise the United States assigns to Canadian work tends closer to the average in ability of all government workers in these areas, there being no reason to select out the best from other higher priority programs. Priorities, in other words, are more often elsewhere, for reasons that have to do with the totality of the bilateral relationship or of U.S. global involvements.

U.S. officials who spend most of their time on Canada are middle level and

> since matters concerning Canadian issues are relatively of lesser importance, the communication flows concerning Canadian issues are generally horizontal, or among these middle-level officials.[6]

When an important issue arrives it is handled on a high level by officials who lack familiarity with things Canadian, often resulting in policy distortions. Most Washington officials are not interested in Canada in general and only develop interest through a specific issue. On the other hand, issues emanating from the United States are handled at higher levels in Ottawa. Swanson concludes that

> given the relative permanence of most of the senior levels of the Canadian civil service, there is greater continuity and professional expertise brought to bear on a given issue than is the case with the more transient political appointees in Washington.[7]

The field arm of the Office of Canadian Affairs, the U.S. embassy in Ottawa, and a small number of consular posts in major Canadian cities contribute a very small effort on the transboundary environmental scene. The embassy assigns one diplomatic officer less than half time to such endeavors, and he has virtually no assistance available. The consular posts are small and sufficiently overburdened with traditional consular work that, even when there is interest, there is little capability to perform transbound-

ary environmental work, the latter being largely limited to simple information gathering and transferal with little or no analytical attention. The bulk
of the U.S. effort, therefore, is left to Washington where it is spread widely
through the bureaucracy, with the State Department coordinating the
whole.

U.S. State Government Actors

U.S. states are not accustomed to dealing across international borders.
Thus they have little history and little expertise in this area. There exists a
strong tendency to defer to the State Department on international questions
or, if this fails, to bring pressure to bear on the congressional delegation
(particularly senators) to intervene with the State Department or to get laws
passed or regulations adopted in the state's interests. For example, Montana
has called for the negotiation of an international air quality agreement, an
effort which appears to have been successful.[8]

All states have environmental quality and water resources or natural
resources bureaucracies, although the size and capability of those bureaucracies and the means they have at their disposal varies greatly. The values
and political ideologies of the various states differ immensely. Rarely does
a state environmental official have significant Canadian, bilateral, or general international experience or expertise, nor is such an official likely to be
given time or opportunity to develop them. His external dealings are likely
to be limited primarily to the federal EPA or the Department of the Interior
and to his peers in neighboring states. Most of the expertise to be found in
this area derives from some level of contact with the IJC and hence with the
Boundary Waters Treaty.

There is no institutional counterpart in state government to provincial
intergovernmental affairs departments. Generally, intergovernmental (including external) matters are handled by the governor's office through a
special assistant assigned to this task among many others. In one state,
Maine, an Office of Canadian Affairs has been established, but it is largely
promotional and culturally oriented and has not dealt at all with Maine-
Canada environmental issues. A few state legislatures (including Maine)
have also given some attention to Canada, but this attention is sporadic and
generally rises and falls on a specific issue. A limited number of officials of
other state agencies from agriculture and forestry to fish and game also
occasionally become involved in issues of import to their responsibilities.[9]

State governments have thus been minor actors on the scene and
have deferred to Washington on many substantive matters. With attention
to states' rights increasing in the United States in the 1980s this situation
may change slightly, but it is unlikely that the State Department will be
prone to yield on any of its constitutional prerogatives in this area.

The Role of Citizen Organizations

Citizen environmentalist organizations, as collectivities of common environmental interests and as political lobbies, have been active in the United States since the nineteenth century. It is only since the late 1960s, however, that such organizations as the National Audubon Society, the Sierra Club, Friends of the Earth, the National Parks and Conservation Association, the Izaak Walton League, the Wilderness Society, and other national as well as regional organizations have grown sufficiently in stature, membership, and expertise to be able to have a significant impact on public decision making. These general environmental groups, combined with hundreds of special interest organizations that often rise and fall on a single issue, do constitute a force on the overall environmental scene. To date they have played a small role in Canadian-U.S. environmental relations except perhaps on a very issue-specific basis, and then principally on the domestic aspects of the question (albeit with international spillovers).

The effective work of the National Audubon Society in delaying and somewhat modifying the Garrison Diversion Unit in North Dakota, almost totally for domestic reasons, has certainly had its effects on Canada, as has the work of the National Parks and Conservation Association in publicizing the alleged threat of the Atikokan power plant to Minnesota's Boundary Waters Canoe Area. The work of Great Lakes Tomorrow on lakes issues regionally, and of the National Clean Air Coalition on transboundary air pollution and acid rain nationally, are further examples.[10]

The nearest citizen movements have come to such a bilateral effort is the formation in the mid-1970s of the joint Canada-U.S. Environment Committee (CUSEC), chaired in Washington by the Wilderness Society and in Ottawa by the Canadian Nature Federation (CNF). Even this ambitious undertaking, an across-the-border affiliation of a large number of national and regional organizations in both countries, has had to struggle to do no more than hold an annual joint meeting and to pass jointly agreed upon resolutions on a variety of transboundary issues, including major international wildlife and Arctic concerns as well as pollution questions. CUSEC has been very short on means, and its officers, who are the officers of member organizations, are greatly pressured by compelling domestic priorities and are not able to devote much time to international issues. Furthermore, this effort too has neglected to focus on the nature of the environmental relationship, and how it is carried on. Of course, such organizations raise most of their funds from sources interested in and concerned about individual issues, and attention thereto on the part of the groups is understandable—by their nature this must be their priority.

The history of Canadian citizen environmental organizations is a much shorter one, with Canadian groups being smaller, less affluent, and less experienced in the area of lobbying. There are national and regional groups in Canada, such as the Canadian Nature Federation, Pollution Probe/Energy Probe, the Canadian Environmental Law Association, Scientific Pollution and Environmental Control Society, the Sierra Club, and Friends of the Earth Canada, among others. But the structure of the Canadian legal and legislative systems is not such as to permit easy access to litigation or to be vulnerable to lobbying. With less prospect of success in these areas, Canadian groups have therefore not grown as have their U.S. counterparts nor have they achieved the level of support of groups south of the border.

In Canada, with the responsible minister sitting in the House of Commons and vulnerable to a daily question period, a small, poorly organized special interest group need find only one sympathetic member of the opposition willing to raise questions to embarrass the government into action. In the United States, with the EPA administrator isolated from the legislative branch, an extremely powerful lobby is required to stir Congress to force changes at the executive level.

In recent years, groups in both countries have begun cooperation and collaboration with joint planning of political strategy to achieve shared objectives. Great Lakes Tomorrow in the United States and the Conservation Council of Ontario actually set up a joint structure. The Garrison Diversion and Atikokan power plant issues were among the first site-specific issues to experience such transborder collaboration and represented the emergence of organized citizen groups as actors on the trans-boundary environmental scene.[11] The establishment in 1981 of the Canadian Coalition Against Acid Rain, with offices in Washington and Toronto and with significant Canadian federal and Ontario provincial funding support, marks the beginning of a whole new era in transborder environmental collaboration and will undoubtedly provide a test case and a challenge for U.S. diplomacy. It is a U.S. lobby funded in part by a foreign government and requires classification as an agent of a foreign government under U.S. law. Maxwell Cohen has written:

> Until recently, at least, a significant difference in public awareness as between Canada and the United States was observable to any "environmentalist" in the private or public sector. . . . Indeed, before 1970 it would have been difficult to find a sustained and national Canadian ethos, or tradition, about environmental protection that transcended the regional or sectoral experience concerned. The contrast with the United States, therefore, was striking. The presence of strong organizations such as the Audubon Society or the Sierra Club with well developed links to the legislative process . . . as well as members of pressure groups with funds and personalities able to exercise pressure, these were the envy of Canadians only beginning to muster resources and organize attitudes.[12]

Cohen contends that the social framework for environmental awareness in Canada has matured swiftly and substantially, and while it has not yet reached the level of effectiveness of U.S. groups, "the Canadian-U.S. differences are now less in the area of awareness than they are in the means at their disposal to translate awareness into political results."[13]

Some of the basic reasons for the differences are obvious: the greater tendency of Canadians to respect (or at least refrain from questioning) authority and to accept the status quo; the lower population pressure and lower levels of pollution characteristic of the vast Canadian environment and resource base, and the consequent desire to develop this resource base, even if at some environmental cost; and, some environmentalists would argue most importantly, the fact that not having access to the courts or litigation, Canadian environmental organizations have not often been able to demonstrate the clout necessary to attract a large membership and significant funding. Their activities are essentially educational in nature, and that is simply not exciting to most people. Hence, the citizen environmentalist impact south of the border has likely been much greater. Cross-border collaboration, however, will certainly work toward reducing the difference.

This new era of the citizens' environmental aspect of the relationship is yet to be chronicled, but it does appear that cross-border environmentalist cooperation will increase. It may ultimately focus on reform in the conduct of the relationship and reform in process and in institutions that should be of concern to all involved in bilateral environmental relations.

The Role of Corporate Actors

Perhaps the least public attention has thus far been given to the growing role of U.S. and Canadian corporations in the transnational environmental relationship. Yet these participants could come to play a role of far greater importance in the future in guiding and influencing the course of the bilateral relationship. This is due in part to the declining role of the U.S. government in regulation at the outset of the 1980s. In part, it is a reaction to the increased role of citizen environmental groups in the conduct of the relationship and the concurrent recognition that some of the involvement of this new citizen actor is inimical to the interests of industry. But perhaps most of all it is a corporate recognition that the environmental relationship is in need of order, of stability, and of predictability, values of great importance to business and industry, and values which the two federal governments have had increasing difficulty in protecting.

Although Canadian and U.S. businessmen have much in common in their concerns, attitudes, and perceptions vis-à-vis environmental regulation and the role of government thereto, there are important differences. The role of government is considerably greater in Canada than in the

United States, for Canada is much more tolerant of government intervention. Although the role of government in the United States, particularly at
the federal level, has broadened in recent years, Americans still value
private enterprise and the protection of private entrepreneurial options to a
higher degree than do the Canadians. (The United States may well be
returning to an even stronger market orientation in the 1980s.) Thus south
of the border one finds a stronger, more outspoken private sector often not
only ready to lobby strongly and publicly for its interests, but also willing
to litigate, if need be, directly and through strong industry associations. To
lobby to attack regulation it perceives as unfair and inimical to its interests,
to litigate publicly and before the media are characteristics of U.S. industry
which do not seem to be shared to the same degree north of the border.
Canadian corporate leaders appear less willing to attempt public lobbying
or to try litigation with its attendant media exposure. The hardball attitude
of some U.S. industry on some issues may have been the cause of stiffer
and less compromising enforcement on the part of the U.S. Environmental
Protection Agency and stands in sharp contrast to the Canadian corporate
(and government) preference for low-profile negotiation.

The corporate actors in Canadian-U.S. environmental relations are
summarized in table 2. They may be categorized into eleven groups, each
group with its own sets of concerns.

There are obviously many kinds of private entrepreneurial or corporate actors involved in Canadian-U.S. bilateral environmental relations,
and for many the stakes are considerable. The involvement itself may be
through a common trade association representing a collectivity of common
interests within one sector (for example, the Canadian Manufacturers'
Association, the American Petroleum Institute), or may involve the direct
participation of the affected company. If the latter, the personnel assuming
responsibility for a company's interests on these questions are drawn from
top management, legal staffs, or environmental staffs. Canadian businessmen are generally more aware of the bilateral nature of these issues as well
as the diplomatic processes applied to them than are their U.S. counterparts. Many more businessmen will undoubtedly become involved in future years, either directly or through trade groups. Quite possibly more
firms distant from the border in the United States as well as in Canada will
be drawn into the mélange of bilateral environmental actors and will make
their concerns known.

Some Canadian businessmen believe that their country neglects the
Canadian corporate viewpoint, but that closer attention is paid to the corporate viewpoint in the United States. Many Canadian and U.S. businessmen
welcome charging governments with the responsibility for balancing society's economic objectives. They support the application of rigorous cost-
benefit analysis to environmental decision making, and the use of socioeconomic impact statements.

TABLE 2. Corporate Actors

Category	Examples	Concerns
Hydrocarbon exploration and development	Oil and natural gas firms, pipeline and allied industries	OCS oil and gas reserves in disputed territory or constituting a threat to the fishery or ecology of Georges Bank, Bay of Fundy, Beaufort Sea, Pacific Coast
Oil transport	Oil firms, shippers	Perceived threat to marine and shoreline ecology, commercial fisheries (i.e., Eastport, Maine; Strait of Juan de Fuca; offshore British Columbia)
Fossil fuel–fired electricity production	Private and crown utilities, coal mining interests, railroads	Near-border developments contributing to air or water pollution, acid precipitation, etc.
Hydro power development	Utilities	Near-border dams, cross-border flooding, water quality deterioration, fisheries, salmon runs, etc.
Hard rock mining	Copper, nickel, gold-mining interests, etc.	Dredging, sediment runoff, erosion, toxic metals, other pollution
Toxic substances manufacturers, users, by-product producers	Chemical industry, other numerous manufacturers and processors	Pollution impacts on border or transboundary waters or air masses

Emitters of SO_2, NO_x, ozone	Metal smelters, power plants, pulp and paper mills, refineries, etc.	Acid precipitation from near-border or more distant sources
Near-border manufacturing and processes	Pulp and paper mills, etc.	Local water pollution, air pollution of transboundary and boundary waters and air masses
Agribusiness interests	Irrigators, food processors, farmers	Dependence on irrigation from transboundary or border water, on effluent removal capability of border waters, farmers needing flood protection, farmers using toxic sprays, etc.
Great Lakes shipping	Iron and steel, grain, coal, shipping interests	Maintenance of high lake levels for navigation and IJC decisions thereto; diversions into and out of the lakes system; amelioration of competing demands by other lake users and shoreline interests
Great Lakes manufacturing and processing	Iron and steel, chemical, pulp and paper, coal interests, etc.	Great Lakes water quality regulation; recommendations and decisions of IJC, U.S. EPA, Environment Canada, Ontario Environment, states; competition with Detroit, Cleveland, Hamilton, etc., for effluent absorptive capacity
Great Lakes electricity production	Electric power plants (nuclear, coal, etc.)	Dependent on high lake levels and on cooling abilities of the Great Lakes and other boundary waters, problems of accelerating eutrophication

Canadians have additional concerns symptomatic of the unequal relationship. For example, there is feeling in the Canadian corporate community that U.S.-type environmental regulations are excessive for Canada but that Canadian governments have imported such regulations without making their own independent analysis. Thus there has been a transboundary movement of ideas, some of which may be inappropriate for Canada. One corporate executive has expressed the view that there are cases where provincial regulators drew directly on U.S. precedent without consultation with Canadian interests.[14] Like many other transborder influences, such behavior may well be a function of the size difference and ease of communications between the two countries.

The Role of the Media as Actors

Contrary to popular belief or traditional practice, the media does more than simply report the news. It does act and, in a sense, becomes a newsmaker itself.

Thanks to the print and broadcast media, Canadians are more knowledgeable about the United States than the average U.S. resident is about Canada. High-profile media attention (particularly where it occurs as saturation or is carried on with a high level of commitment, even passion) has resulted in a rapidly informed, angered, and even inflamed public north of the border. This public becomes ever more frustrated given its perception of ignorance and indifference south of the border, thus laying the foundation for basic misunderstanding between the two peoples. Further divisiveness is the result. However, high media attention has also meant increased pressure on Canadian politicians to "clean up their own house" and on Canadian diplomats to take a hard negotiating line with the United States, possibly causing other problems in the relationship. It also creates a climate for additional research and investigation into highly complex and often insufficiently understood environmental issues of potentially great importance to the future of society.

U.S. media, print and broadcast, has a long history of ignoring Canada and Canadian-U.S. relations. This practice even extends to the two nationally read newspapers, the *New York Times* and the *Washington Post*, and especially to the three major broadcast networks. A number of studies have been done indicating the sparseness (and superficiality) of Canadian content coverage to U.S. readers and viewers, and the transboundary environmental area is no exception.[15] There has been substantial environmental coverage in the U.S. media for at least ten years, but invariably of a domestic nature, with transborder aspects viewed superficially or not at all.

The real actors have been the Canadian media. This is especially true of two TV networks (CBC and CTV), one radio network (CBC), and two newspapers (the *Toronto Globe and Mail* and the *Toronto Star*). However, the willingness of the Canadian national wire service, Canadian Press (CP), to carry a large quantity of material in this area, the interest of virtually all major big-city English-language dailies to print this material even when it does not directly affect their own readership area, and the interest of these and smaller newspapers to report and editorialize on national and regional transboundary environmental issues have insured very heavy coverage accessible to all Canadians. In addition to extensive news reporting, the networks have also aired longer in-depth feature specials on many issues, produced expensive "docudramas" involving professional actors, and devoted time on discussion and commentary programs to debate on these questions. Furthermore, Canadian media tend to assign a larger number of correspondants to analyze and report from the United States than vice versa, and also devote much effort to science (including environmental) coverage.

It is probably to the credit of the media that many of the transborder issues which Americans consider regional most Canadians accept as national. There is a surprisingly high level of awareness of Garrison, for example, on the West Coast and in the Maritimes, far removed from the impacted area. And many more Maritimers know the long history of the Skagit controversy four thousand miles to the west in British Columbia than most non-Canadians realize. These essentially regional issues (albeit with national implications) do have a way of becoming cause célèbre throughout the country with such media attention, and the attention itself increases the frequency of debate in the House of Commons, which in turn causes more media coverage—a full circle.

Nothing in recent years, however, matches the role of the media on two serious questions currently subject to controversy: acid precipitation and toxic substances in the Great Lakes. The treatment of these two subjects has been extensive in amount of time and space devoted to them as well as the breadth and depth of the treatment. And, while a good deal of attention is domestic in nature, much is bilateral and concentrates on the bilateral processes, as any survey of the Toronto press will verify.

The subject of media as actors should not be concluded without mention being made of the question of journalistic bias. The suggestion has been made in a CBC documentary that the North Dakota press has been so heavily biased in favor of the Garrison project that it has avoided or at times deliberately misrepresented the positions of Manitoba and Canada in its coverage of the issue.[16] CBC television itself has been accused of misrepresentation in its docudrama on Garrison, "Someday Soon."[17] There is some evidence that supports both accusations. Finally, the Toronto press has been viewed by some businessmen and U.S. officials as overreactive and perhaps too emotional in its treatment of acid rain. Whether or not the role of media actors is viewed positively or negatively, there can be no question that they are actors, not onlookers. They do influence decision making, and their role in the bilateral environmental relationship may well increase.

Conclusion

The actors participating in bilateral environmental relations are drawn broadly from many areas of Canadian and U.S. society. At the core, there are a small number of diplomats and other supporting federal officials, surrounded by provincial and state officials who vary with the issue. Farther out on the periphery, but often very influential, are the concerned and affected nongovernmental actors, both citizen public interest groups and corporate interests, and those who interpret the whole, the media and academic researchers. Through all levels the Canadian presence is more obvious, more knowledgeable, more intense, as befitting the role of the smaller and more vulnerable of the two countries.

There are implications in this chapter for most other chapters in this book, for all of the actors introduced here reappear in subsequent chapters. The theme of lack of U.S. attention to Canadian matters runs as a thread throughout the following pages, as does the assumption that with greater knowledge and appreciation of transborder interests and responsibilities people would be less likely to do things that cause transboundary environmental problems.

The reader has thus far become acquainted with some of the prevailing differences between the Canadian and U.S. actors. Chapter 2 introduces further differences of philosophy and process as they exist at the Canadian-U.S. interface.

NOTES

1. The bureau was upgraded in the late 1970s from a U.S. division which, as Swanson notes, "had greater difficulty in defining its mandate and tended to be residual in its activities given the high rate of participation of other specialized organizational units in U.S.-related matters. However, the bureau now has greater centrality as a coordinating and monitoring unit in which it maintains an overview of U.S. issues and of other organizational units active in these issues" (Roger F. Swanson, *Intergovernmental Perspectives on the Canada-U.S. Relationship* [New York: New York University Press, 1978], p. 33). This upgrading from divisional to bureau status is further evidence of the importance Canada attaches to the relationship.

2. In the Canadian system, it is deputy ministers who are the chief appointed officials, answerable directly to a politically elected minister. They are approximately equivalent to assistant secretaries in the U.S. system, although the latter are most often political appointees in contrast to the deputy ministers, who are frequently career civil servants who rose through the ranks. Canadian assistant deputy

ministers are senior bureaucrats approximately equivalent to U.S. deputy assistant secretaries. According to Swanson, "deputy ministers and assistant deputy ministers constitute the hinge between the cabinet and the departmental working level. . . . [I]t is they who constitute the rather sharp Ottawa distinction between political and professional" (Swanson, *Intergovernmental Perspectives,* p. 31).

3. The provinces have control over the implementation of treaties dealing with powers within their jurisdiction and therefore have greater autonomy in conducting transborder relations. However, in spite of this, Swanson concludes that "any attempt on the part of the U.S. government to operate outside mutually acceptable channels in dealing directly with the provinces, or for the Canadian federal government to deal directly with the U.S. states, is both undesirable and ultimately counterproductive" (Swanson, *Intergovernmental Perspectives,* p. 231).

4. Michel LaPointe, Senior Policy Advisor, Office of Federal-Provincial Relations, personal interview, Ottawa, Ontario, September 3, 1980.

5. For further treatment of Canadian-U.S. constitutional differences and their impact on federal-provincial and federal-state relations, see J. Carroll, "Differences in the Environmental Regulatory Climate of Canada and the United States," *Canadian Water Resources Journal* 4 (1979):16–25.

6. Swanson, *Intergovernmental Perspectives,* p. 11.

7. Ibid., p. 12.

8. Swanson reports that direct state-provincial interactions "are more comprehensive and cover a greater range of activities than might be expected," and that formal state-provincial agreements, understandings and arrangements totaled 766 in 1978 (*Intergovernmental Perspectives,* p. 222). The vast majority of these agreements, however, are routine, mundane reciprocal arrangements, such as recognition of drivers licenses, auto insurance, etc.

9. One further outlet for state governors to become involved in Canadian-U.S. issues, including but not commonly those environmental in nature, has been through the regional commissions (economic development) and river basin commissions (resource planning), both of which are federal-interstate compacts which have not been oblivious to border problems but have not given them major attention either. Both institutional mechanisms are likely to be phased out in the near future, however.

10. See *The Garrison Diversion Unit* by John E. Carroll and Roderick Logan (Montreal: C. D. Howe Institute, 1979), *National Audubon Magazine, National Parks and Conservation Magazine,* and the newsletters of Great Lakes Tomorrow and the National Clean Air Coalition for further detail.

11. For further discussion of the role of citizen environmental groups in Canadian-U.S. environmental relations, see John E. Carroll, "Environmental Factors and Actors," in *Natural Resources in U.S.-Canadian Relations,* 3 vols., ed. C. E. Beigie and A. O. Hero (Boulder, Colo.: Westview Press, 1980–), vol. 3, *Perspectives, Prospects, and Policy Options* (forthcoming).

12. Maxwell Cohen, "Transboundary Environmental Attitudes and Policy—Some Canadian Perspectives" (Paper presented to the Harvard Center for International Affairs, Harvard University, Cambridge, Massachusetts, October 21, 1980), p. 17.

13. Ibid.

14. R. H. Hall, Gulf Canada, Ltd., Toronto, personal correspondence, Ontario, August 6, 1980.

15. Donald R. Hall, *Press Conceptions of Americans' Knowledge of Canada: The 1979 Federal Election* (Tucson: University of Arizona, 1980). *See also* Jim A. Hart, "The Flow of News Between the U.S. and Canada," *Journalism Quarterly*, Winter 1963, pp. 70–74.

16. Canadian Broadcasting Corporation, "Garrison: The Fight Goes On," television documentary, 1979.

17. Canadian Broadcasting Corporation, "Someday Soon," television docudrama, 1977.

CHAPTER 2   Canadian-U.S. Differences

In chapter 1 the various actors in Canadian-U.S. environmental relations were introduced and their differences brought to light. This chapter highlights the differences in background, approach, and techniques that characterize each country's actors and the way each country views bilateral environmental problems at the interface—at the point where the two nations meet geographically, institutionally, and philosophically.

To most of the peoples of the world, there is very little difference between Canada and the United States. Alleged differences or disputes of any type tend to be discounted as of no real consequence almost everywhere except within Canada, a nation all too aware of its small population and vulnerability relative to its neighbor to the south. But there are innate differences in all areas, including environmental questions. The geographical realities of sharing a continent inevitably give rise to tensions and disputes. Such disputes are then exacerbated by differences in social organization. This chapter reviews the pertinent differences constitutionally, institutionally, and environmentally. All such differences existing at the interface contribute to the transboundary environmental problems described in this volume.[1]

Differences in Background and Approach

The United States and Canada are not at equivalent stages in their history or economic development. The United States may be looked at as an older, more mature economy, as a society which has long experienced the material fruits of urban-industrial development, has long had and used the capacity to damage the environment.

In spite of many seeming similarities between the two societies in the environmental and other areas, differences are in fact greater than imagined. U.S. governmental institutions are a product of eighteenth-century liberal thinking, in many ways opposed to the British system, and evolved through war and bloodshed. Canada's institutions are a product of nineteenth-century Victorian thinking and its independence evolved under very different conditions. The United States has long been moving on a course of increasing federal powers at the expense of the states (increasing centralization), while Canada has continued to be decentralized, with much power in the provinces. Governmental structures and the role of political parties differ sharply from one country to the other, resulting in a strong Canadian central authority weakened with respect to the provinces, while the United States has an inherently weaker central executive authority that is more likely to be checked from within than by the states (though states at times can frustrate federal policy).

This difference in constitutional powers leads to a problem in the area of federal-subfederal relations. The reality is that power under the constitution is in federal hands in one nation and largely in provincial hands in the other. And yet, the two federal governments must deal directly with each other and not in any direct manner with subnational levels of government. More than one government official on each side has described the frustration that results from this situation. Hence some officials in the United States would prefer to deal more directly with provincial governments,[2] in spite of the constitutional problems this creates.[3] Regional differences are substantial in both countries but more extreme in Canada though growing in the United States.

One area of apparent similarity is that of laws, institutions, and public policy in the environmental field. Both countries have federal laws, and public agencies to administer them, in most environmental quality areas, as do the respective states and provinces. Often the laws and agencies even bear the same name and express their policy missions and goals in the same way. A little study of the situation, however, soon reveals that these similarities are no more than skin deep; in fact the climate under which laws and policies are made and administered is different. One striking example is that Canadians generally view litigation as a failure of the system.

Canada's underlying differences from the United States have been attributed to three factors: the time lag in independence, colonialism through the imperial tie with the United Kingdom, and the "French fact." Further, there is a triple tension in Canadian life which affects the bilateral relationship: English-French feelings, an enormously broad landmass along a long frontier leading to physical and psychic alienation, and north-south polarities leading to strong regional Canadian-U.S. ties. Basic similarities are attributed to the sharing of one continent, the sharing of geophysical realities, Canada's learning from the lessons of U.S. federalism, its copying of some social and educational institutions, and the influence of U.S. media.

The demographic difference between the two countries introduces a further problem area which cannot be ignored. The population ratio in the boundary water basins is three Americans to one Canadian, while the overall national ratio is ten to one. Does this mean that consumptive use of water or rights to pollute should be split fifty-fifty (based on shared geography and equity) or three to one (based on population ratio)? For obvious reasons

Canada often argues for geographical criteria (especially on the Great Lakes) while the United States argues for population criteria[4] and such arguments permeate national thinking. This imbalance creates a condition of Canadian environmental vulnerability vis-à-vis the United States.

The Environmental Interface

Today, many United States residents are aware of and sensitive to the ecological, economic, and health problems which have resulted from their ability to damage their environment. They are, perhaps, relatively more interested than their Canadian counterparts in sacrificing some economic development by cleaning the air and water or at least preventing further significant deterioration and in designating wilderness parks, endangered species and critical habitats, and wild and scenic rivers for protection from development.[5] The election of 1980 and political events of 1981 may indicate a slowing down in this direction, but polling results indicate that the U.S. public is as concerned about environmental questions as ever, so that the actions of the Reagan administration and some state governments may not represent a fundamental turnabout in the environmental area.[6]

Canada appears to be at an earlier stage of economic and industrial development than the United States. For much of its history, Canada has been denied the opportunity to develop economically (and thereby pollute the environment) at the pace of its neighbor to the south. Now, for the first time, many regions of Canada, from the Lake Erie shoreline of Ontario to the prairies of southern Saskatchewan and on through most of the territory of Alberta have expectations of large-scale industrial development and the jobs, the urban amenities, the material goods, and the pollution that goes with it.

The possibility that these expectations may be realized comes at a time when the world is becoming more ecologically sensitive, and when the inhabitants of the border states to the south are more likely to feel ecologically threatened than ever before. The probable conflict is compounded by the fact that U.S. development interests are concerned about domestic pollution laws which assign incremental "rights" to pollute until a certain level of pollution is reached. Such laws place U.S. developers in competition with Canadians to use up the same increments of available clean air and clean water. Even lightly populated rural states like Montana, Minnesota, Washington, and Maine exhibit great environmental sensitivities, as do their more heavily populated, and more polluted counterparts such as New York and Michigan. Hence, many border states now threaten Canadian expectations of developing its own border region resources for industrial purposes. However, significant changes are taking place. Ironically, Canada's generally defensive stance of some years ago (mid to late 1970s), induced by numerous U.S. diplomatic complaints over border power plants, is now shared by the United States. Acid rain and Great Lakes toxics questions are leading to a reversal of roles. ·

Pollution Controls

With the advent in the United States of the Water Pollution Control Act Amendments of 1972 and roughly concurrent federal clean air legislation, the United States established clear point source standards (i.e., at the point of emission) for clean water and clean air. These standards have been implemented largely regardless of the location or condition of the receiving water or air, in contrast to the Canadian system of objectives that relate to the conditions of the receiving water or air. A second difference in the two countries' control methods is that Canadian objectives are negotiable between government (provincial or federal as the case may be) and industry, while U.S. standards are much less flexible. The concept of considering the quality of the recipient water or air, the use currently being made of those bodies, and the idea of negotiation on a case-by-case basis, while sometimes given informal consideration are by no means integral to the U.S. law, as they are in Canada. Furthermore, what often passes for guidelines and recommendations without the force of law north of the border often bears the full force of law south of the border. Hence, a given province may actually have stricter environmental criteria than a neighboring state, but the question one must ask is whether that stricter criterion has the force of law behind it and, if so, is it litigable in court. From a narrow viewpoint it may seem that definitive legislation and regulations based on source control results in better protection of the environment. Results, however, based on pollutant loading reductions suggest that control by objectives may work as well, at least in Canada.[7]

Ontario is moving gradually toward U.S.-type effluent standards, but the present situation is substantially different from the U.S. approach. The general approach in Canada is for the environmental quality agencies to adopt standards and guidelines, usually following public hearings, but these are not given the force of law (with a few exceptions in Ontario). Emissions sources are then licensed by the agency with particular conditions imposed in the licenses that are to conform so far as practicable to the standards and guidelines. The terms and conditions of the licenses are legally enforceable. Considerable bargaining takes place between the agency and the applicant for a pollution control license. Much litigation is thus avoided. Don Munton illustrates the situation as follows:

> In the case of water pollution, Ontario's traditional position has been that what matters most is the overall quality of the "receiving waters" and the use being made of those waters. These questions are consid-

ered and negotiated largely on a case-by-case basis. Thus . . . there would be less pressure for pollution controls put on the marginal-profit pulp mill that was the only industry on a river in a remote area where there were no downstream users than on a highly profitable mill that was one of a number of industries on a lake in a heavily populated area. . . . Critics of Ontario's performance . . . argue that the industry by industry or company by company negotiation process does not provide the provincial government officials with much "clout" to enforce standards and secure agreement on speedy action.[8]

Canadians often find they do not need nor can they afford the more extreme steps the United States has taken in the environmental control area.[9] However, one must question if this finding will be relevant for the 1980s, particularly in view of possible changing values in the United States toward environmental regulation if the change in the U.S. Congress and the initial support for the Reagan administration is indicative.[10] It is possible that as the United States moves toward the use of incentives and disincentives and away from regulatory power to achieve environmental control, the two countries might well move closer together in their thinking and approach.[11] Since Canada may be moving toward greater environmental sensitivity, any ideological gap between the two countries, if such exists, might be closed. Thus the United States might become less environmentally sensitive while Canada becomes more so.

Until 1970 both public and governmental opinion in Canada in the environmental area was a decade behind that of the United States. U.S. environmental groups were creatures of the 1960s and were highly visible, whereas Canadian groups were creatures of the 1970s and were much less visible. As a result, pressures on Canadian civil servants were less severe than those on their U.S. counterparts; hence the former were slower to act with a narrow disciplinary approach, and traditional in their approach. The enormous task of the IJC on the Great Lakes was a catalyst for change. With the evolution of the Great Lakes perspective, the growth of Canadian environmental groups, and the evolution of some regional concerns, environmental concerns in Canada began to change. By the late 1970s, the threat from toxics and public health fears emerged and saved the environmental movement from loss of widespread support which it might otherwise have suffered as a result of the combination of pressures from the economy and energy concerns. Environment and health preoccupations then converged, and the environmental cause broadened from recreation and aesthetics to health. Even if the U.S. environmental movement satisfactorily responds to the acid rain challenge, henceforth environmental protection will no longer be viewed as regional but continental in scope. Ultimately a large core of shared continental perceptions must develop, although we will always face some issues that are unique to one country or the other but not to both. In the light of this evolving situation a survey was taken of a number of key actors to obtain their views.

Viewpoints at the Interface

Four standard questions were posed to a relatively large number of inter-viewees and are presented here as an illustration of the outlook of these many interviewees, all of whom are working at the interface of Canadian-U.S. environmental relations.[12]

In answer to the question "What do you see as THE issues or problems in Canada-United States environmental relations?" acid precipita-tion was overwhelmingly cited as a real issue of great substance, and one which held much potential for serious Canadian-U.S. diplomatic problems. This answer occurred often enough to place acid precipitation on a level apart from all other issues. Other issues which were commonly cited include Great Lakes water quality, transborder air pollution problems aside from acid precipitation, Arctic Ocean issues (marine pollution from hydrocarbon development and transport, and sovereignty questions), water apportion-ment, especially in the dry prairie regions, and the Garrison Diversion issue and the diplomatic implications of its completion.

Though less commonly referred to, concerns over matters of process most often focused on the differences in the two water quality control philosophies of standards (U.S.) versus objectives (Canada); numerous concerns (and misunderstandings) over the IJC and its role; too rapid turnover of diplomatic personnel; disagreement over the proper roles of subnational government; disagreement over the role of nongovernmental institutions and entities; the lengthy time period necessary to accomplish anything; and the requirements of diplomatic protocol (including secrecy and lack of access to the process).

In answer to the question *"What do you see as coming down the pike with respect to issues?"* once again many cited acid precipitation as a serious potential as well as actual issue. Others cited hazardous wastes and toxic substances, both as spillage into waters at or crossing the border, and as cargo transported across the border to dumping sites. However, most tended to agree with the diplomats who basically view this as more an international extension of a domestic issue than a bilateral issue in and of itself. However, domestic or international, it is clearly on the minds of many people. Others identified rather broadly increasing energy problems, dra-matic water shortages, and even food issues that would ultimately appear on the bilateral environmental agenda. Some also labeled Arctic development issues, winter navigation on the Great Lakes, and further energy develop-ment at numerous near-border locations (especially in the prairies) as among those to be tackled in future.

In answer to the question *"Are we experiencing an evolution of principles in this area?"* the response that prevailed was a clear "No." Sometimes this negative response was emphatic, and occasionally it was expanded to the point of demonstrating retrenchment (for example, "we are farther away from reliance on principles and mechanisms and more heavily reliant on ad hoc arrangements now than we were in the past"). Occasionally, it was accompanied by signs of regret. Some feeling was evident that the two nations are farther away from progress toward such evolution than had been the case a decade or two ago. Of the minority who answered the question affirmatively, most cited the continued functioning of the Boundary Waters Treaty of 1909 and the continuing work of the International Joint Commission and, most especially, the Great Lakes Water Quality Agreements of 1972 and 1978 as signs of evolution. Lesser numbers cited the "early warning and consultation" understanding in effect at the State and External Affairs departments, the bilateral monitoring agreement and resultant committee at the Poplar River site, the informal steering committee at the Cabin Creek–Flathead River site, and the U.S.-Canada Memorandum of Intent setting in motion negotiations on a bilateral air quality and acid rain agreement and creating technical working groups to lay the foundation. Critics applaud these measures, but downplay their significance, and note that each, while perhaps a positive effort in itself, represents an ad hoc approach which does not tie into a uniform and consistent whole. The prevailing perception is that "adhockery" is the order of the day.

In answer to the question *"What are important principles or mechanisms which you feel need to be established?"* trends were much less clear. Many had no developed thoughts on this subject. However, the sizable minority which did respond in this area made a number of suggestions. These included: formal implementation of systems of early warning and consultation; significant strengthening in, and expansion of, the mandate of the IJC, particularly in the air quality area; increased involvement in the bilateral process, an expression particularly raised by state government officials in the United States and by citizen environmentalist leaders in both countries; a general increase of decision making at the local level and by those directly involved in the issues; a reduction in diplomatic protocol and easier access to meetings and information (and therefore a mechanism to provide for a local role in the decision making); a mechanism to provide for reciprocity in the area of transborder damages, and an underlying principle that would guarantee that transborder issues be dealt with in the courts as would purely domestic issues; and a general call for more guidelines and rules and a more organized approach to Canadian-U.S. environmental relations that would guarantee affected parties and those who may be so affected recourse to established rules and therefore stability in the relationship. A small number of interviewees felt no change is called for, and that present laws are being used effectively. This latter group also cited the values of flexibility in the present system, a value that would be lost if a more formal regime were put in place. This argument may ultimately boil down to a question of flexibility versus stability, and while a system which optimizes both may be most desirable, this might be an impossible goal.

Conclusion

By the end of the 1970s it could be said that Canada, being a nation of low population density and relatively underindustrialized and underdeveloped compared to its southern neighbor, exhibited behavioral traits more akin to those of a frontier society hungering for industrial development. This is illustrated in the Poplar, Cabin Creek, and Atikokan case study chapters. Thus it seemed that the United States, which had always had the greater ability to impact the transborder or continental environment, would in future be more sinned against than sinner, especially in light of increasingly higher levels of environmental control south of the border. If directions of the later 1970s had been followed, the tables might well have thus been turned. Not calculated in this scenario, however, was the far greater ability of the United States to pollute the continental commons[13]—aquatic and especially atmospheric—including Canada. Coincidentally, increasing U.S. self-awareness of serious economic problems may well preclude substantial movement toward further environmental regulation and cleanup for some time. This tendency is especially evident in the acid rain, toxics, and Great Lakes case study chapters. In retrospect, therefore, the beginning of a reversal in the historic pattern may have only been an aberration of the 1970s rather than a permanent change, and Canada may still have more to fear from U.S.-induced harm to its environment than from U.S.-inspired frustration of its industrial development plans. Of course, Canadian impact on the United States may well increase, but it is never likely to be greater than U.S. impact on Canada. A seesaw is in motion but the heaviest weight is always on the U.S. side.

   The group of chapters in Part 2 illustrates virtually all the inherent differences between the Canadian and U.S. systems. Each is predicated upon one or more of these differences which have caused problems, problems which may be overcome or at least alleviated by the approaches described in Part 3.

   However, a knowledge of the actors and the differing national approaches is not sufficient to understand the various case studies of Part 2. One must also understand the basic Boundary Waters Treaty of 1909 and its vehicle, the International Joint Commission (IJC), and their often fundamental role in the bilateral environmental relationship.

CHAPTER 3

The International Joint Commission

The best known institution active in the Canadian-U.S. bilateral environmental relationship is the International Joint Commission (IJC). It is now over seventy years old, offspring of the historic Boundary Waters Treaty of 1909. Smaller than most realize, in theory it has a wide mandate, in practice a constrained but broadening one. It continues to study, decide, monitor, and advise, on a varied number of transboundary environmental problems from Atlantic to Pacific. The work of the IJC will be referred to in a number of chapters in this book, for an analysis of many transboundary water and air problems could not be made without reference to its work. This chapter, however, focuses on the commission itself, describing its origins and evolution, examining its present-day role, and reviewing some of the strengths and weaknesses that make it what it is.

The International Joint Commission (IJC) is the only permanent joint public institution operating in Canadian-U.S. environmental relations. It has been a key actor playing a central role in the relationship. Furthermore, its successes and failures may foreshadow successes and failures of other institutional mechanisms that may be created. It is a model to the world and is unique not only by definition, being a joint, unitary yet bilateral body, but also because it stands as an exception to traditional U.S. repugnance toward placing limitations on its sovereignty.

Much has been written on the history of the Boundary Waters Treaty of 1909, the provisions of the treaty, and the body to which it gave rise, the IJC. The reader is referred to a number of solid studies of these matters in much greater depth than can be offered in these pages.[1] A copy of the full text of the treaty itself is included in the appendix. It is, however, the intent of this chapter to present the broad framework—the origins, the treaty, the commission, and divergent perspectives on it—to enable the reader to better understand the overall role of the treaty and commission in bilateral relations, as well as their specific application to the case studies presented in Part 2.

Early Origins

Toward the end of the nineteenth century, a number of bilateral issues coalesced to set the foundation for action culminating in the 1909 treaty.

Such matters as the drive to establish a St. Lawrence Seaway for navigation; the unilateral (U.S.) construction of the Chicago Diversion Canal; the St. Mary River–Milk River irrigation dispute on the western prairies; hydroelectric project proposals for the St. Marys River at Sault Ste. Marie, Michigan and Ontario, which required bilateral action; concern over protection of Niagara Falls and interest in hydroelectricity diversions from the falls; and a proposed damming of the outlet of Lake Erie all demonstrated the need for concerted joint binational action. It was the matter of irrigation on the western prairies and a Montana-Alberta conflict over water allocation, coupled with joint interest in establishing an international commission to investigate the conditions and uses of waters in the Great Lakes–St. Lawrence system that culminated in the first institutional action. This was the establishment in 1905 of the weak, but symbolically significant, International Waterways Commission (IWC), immediate predecessor to the IJC.

Though largely limited to the Great Lakes–St. Lawrence system, the main value of the IWC was that it provided a platform for new ideas on bilateral boundary water relations and that it had access to the top levels of government.[2] The IWC, though only advisory, is credited with developing certain important concepts accepted by both countries—equitable distribution of water between the two countries and the paramountcy of navigation to all other water uses after domestic supply needs were met. It also advanced the idea of extending jurisdiction beyond the Great Lakes–St. Lawrence system to encompass all boundary waters, an hypothesis not embraced by the United States until the later establishment of the IJC. With the general bilateral acceptance of the first two concepts and other basic tenets, including prohibiting the diversion of streams crossing the boundary, as well as opposing pollution in one country that would cause injury in the other, an important foundation was laid and serious negotiation of a treaty could begin. The treaty was drafted and the International Joint Commission established by two men, Sir George C. Gibbons (representing Canada) and Chandler P. Anderson (representing the United States), working closely under Prime Minister Sir Wilfrid Laurier and Secretary of State Elihu Root, the latter of whom took strong interest and maintained a direct role.[3]

The Canadian negotiators wanted a powerful treaty and a strong commission, essentially a supranational institution to be given charge of all boundary and cross-boundary matters. Dreiziger reports, "no doubt it was this aspect of the [treaty] draft which appealed to [Prime Minister] Laurier and his advisers who tended to regard the Americans as bullies and wanted to see legal restraints imposed on them."[4] U.S. negotiators argued strongly for a much weaker treaty and a narrower, more constrained commission, arguing that the treaty draft "departed from the established practice of the American government not to agree to submit questions not already at issue to an international tribunal,'' that it ''empowered the Commission to deal with matters under the undisputed control of the United States government,'' and that the commission might ultimately ''become an agency for the creation of international law.''[5]

Secretary Root's letter of June 4, 1908, to British Ambassador Bryce indicates the U.S. attitude:

> The difficulty of the United States in assenting to an agreement that all questions within the broad field described by the Gibbons-Clinton draft [of the treaty] shall be referred for final determination to such a commission as is proposed, is in the main that such questions necessarily involve, not merely questions of fact and of law suitable for the determination of a commission or arbitral tribunal but many questions of policy, of mutual concession and of the give and take which is in so great a number of cases the efficient means of reaching possible settlement of difficult controversies. Such questions of policy, of concession, of discretion, it is impossible for the Government of the United States to commit to any commission under our system of government.[6]

The ultimate result of this debate was a compromise. The United States softened its stand on the commission's scope and authority but won Canadian concession for its view that waters which did not actually straddle the boundary should be excluded, that arbitration in the case of deadlock should not be automatic, and that existing river diversions such as the Chicago Diversion would be exempted. One of the Canadian negotiators' chief desires

> was to see one treaty solve the many problems facing the two countries over the uses of common waters. They . . . strongly believed that treating each case on its own merits, through special temporary commissions, would pit their small nation time and again against more powerful American interests with predictable results. What they felt was needed were fixed and fair principles applied by a permanent commission.[7]

Canada did not achieve its goal but did succeed in getting the United States to agree to a potentially wide-ranging treaty and commission, one which has survived for almost three-quarters of a century. A long-time IJC commissioner, Charles Ross, has noted ''that the United States in 1909 would ever agree to participate in an international institution where its clout would be diluted by equal voting privileges marked a milestone in international relations. It was not an easy decision for the United States.''[8] It is quite possible that this was all that could be achieved and indeed that we cannot expect the United States to go further even now.

The Boundary Waters Treaty

Former Canadian IJC chairman and ambassador to the United States Arnold Heeney, paraphrasing Canadian negotiator Gibbons, noted in a 1969 article that what Canada needed early in the century was a means of direct contact between Ottawa and Washington through a permanent, jointly constituted, unbiased body.[9] The Boundary Waters Treaty achieved this and also established rules by which the boundary waters relationship would be governed.

The principal purpose of the Boundary Waters Treaty of 1909 was to create joint dispute preventative machinery applicable to the use of boundary waters.[10] This joint machinery is, of course, the IJC and its rules of procedure.

The treaty deals with five categories of waters:

1. boundary waters—defined as those along which the international boundary runs, including all bays and inlets of these waters, but not including tributaries to and distributaries from these waters:
2. upstream transboundary waters—waters which are upstream from and flow across the boundary;
3. waters which are tributary to boundary waters;
4. waters which flow from boundary waters; and
5. downstream transboundary waters—waters which are downstream from the boundary, having flowed across the boundary.

The treaty's objectives were to settle all pending disputes along the frontier and to prevent, or to provide for the adjustment and settlement of, similar difficulties in the future. Articles I, II, III, IV, and VIII set forth the principles that were to govern the use, obstruction, and diversion of boundary and transboundary waters. Article VII provided for the creation of the IJC and individual articles covered the four categories of functions that the commission was expected to discharge.[11]

The four categories of functions the new agency was expected to discharge can be summarized as *administrative* (Article VI, which directs the measurement and division of the waters of the Milk and St. Mary rivers); *quasi-judicial* (Articles III, IV, and VIII, which allow for the acceptance of applications for permission to use, divert, or obstruct treaty waters); *arbitral* (Article X, which provides for binding decisions relative to any questions arising between the two countries, giving government unlimited breadth in referring matters to the commission); and *investigative*

(Article IX, which provides for examining and making recommendations on any difference arising along the common boundary).[12]

A contentious issue in the treaty negotiations was the Harmon Doctrine (so-called after former U.S. Attorney General Judson Harmon). The United States has traditionally upheld this doctrine and it is reflected in Article II, which the U.S. negotiators worked hard to win a place for in the treaty.[13] The doctrine entails a significant reservation of sovereignty, since it gives the upstream state exclusive control over the use of all waters on its own side of the line, and it represents the extreme of nationalistic positions.[14] Under the doctrine upstream nations can underprice the cost of development, since the nation undertaking a project only needs to consider costs arising within its own jurisdiction: "such a situation implies that the real economic costs of development will be understated by way of shifting the diseconomies of development onto the downstream riparian."[15]

Why did the United States maintain this position so strongly against Canadian (and British) objections during the treaty negotiations? In 1909 the U.S. economy was more industrially developed than the Canadian, and therefore more likely to develop irrigation and hydro power potential. The United States thus did not want to run the risk of Canada enjoining potential upstream works in the United States and backed Article II, which reduced the claim of downstream interests. The article undermined Canada's position and cost it the common law right to enjoin harmful upstream exploitation.[16] Maxwell Cohen has written,

> The evolution of the "Harmon Doctrine," as initiated by the United States, emphasized the sovereignty of each State to do as it willed with water on its own side of a transboundary river or water-course. The upstream position of the United States in relation to Mexican and Canadian rivers at the turn of the century, crossing into those countries from the United States, and the dry west and south-west perspectives of the United States, made the U.S. very conscious of the need to assert doctrines favorable to the use of waters within their jurisdiction or sovereignty. Thus the "Harmon Doctrine" entered into the treaty regime that has determined fundamental Canadian-U.S. transboundary relations since 1909 namely in Article II of the Boundary Waters Treaty. It is now virtually an irreversible position technically, however moderated the doctrine may have become because of the facts of life both legal and geo-economic.[17]

However, the article does contain compensatory language, since it gives to injured downstream interests rights to legal remedies equivalent to those in effect domestically. This provision thus affords the injured party access to the courts of the injuring party's jurisdiction. It has been asserted, however, that this was an inexpensive concession for the United States to make.

In addition, recourse to the courts of the country making a diversion presents formidable obstacles and little use has been made of it.

The Harmon Doctrine is important because it demonstrates U.S. attitudes toward sovereignty which have characterized much of that nation's history. It shows much foresight in the protection of national interests, and an interested awareness on the part of the U.S. negotiators toward geography and the "lay of the land." It also teaches the lesson that times change and nationalistic theories can outlive their usefulness. The Harmon Doctrine occasionally works in Canada's interests, and will even more so as the Yukon-Alaska frontier becomes developed.

It is difficult to separate description of the treaty from that of the commission, and thus the treatment of the remaining parts of the treaty is incorporated with the institutional treatment of the commission.

The Commission

ORGANIZATION AND FUNCTIONING

The International Joint Commission was established to administer and discharge the purposes of the Boundary Waters Treaty. The commission is composed of three Canadian members who are appointed by the governor general in council (prime minister and cabinet) for fixed terms, and three U.S. members who are appointed by the president with Senate approval and who serve at the president's pleasure. Each section, Canadian and U.S., has a full-time chairman, with the other commissioners serving part-time, and each is served by a small full-time staff of professionals in Ottawa and Washington. These staffs cover disciplines ranging from the traditional engineering and law to the more recent addition of economics, biology, and other disciplines. The commission also administers a Great Lakes Regional Office with a scientific-technical staff whose role is limited to carrying out the water quality research and monitoring mandate of the Great Lakes Water Quality Agreements of 1972 and 1978, a topic discussed in greater depth in a later chapter.

The real work of the commission, however, is carried out by the various boards and groups (investigative boards, pollution surveillance boards, boards of control, Water Quality Board, Science Advisory Board, and two reference groups—Upper Lakes, and Pollution from Land Use). These bodies are composed of career civil servants, mainly engineers and scientists, who are "seconded" on a part-time basis (borrowed, with pay deriving from their home agencies) from a number of federal agencies, commonly including Environment Canada (especially the Inland Waters Directorate), Environmental Protection Agency, U.S. Army Corps of Engineers, U.S. Bureau of Reclamation, and the U.S. Geological Survey (see fig. 1). The boards and groups provide formal reports and recommendations to the six commissioners who, in turn, refer this output to their

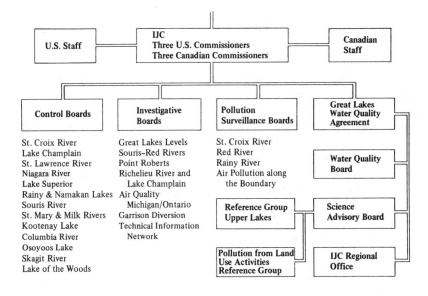

Fig. 1. IJC organizational arrangement and boards (From *70 Years of Accomplishment: Report for Years 1978-1979* [Ottawa and Washington, D.C.: International Joint Commission, 1980].)

Washington and Ottawa professional staffs for further refinement prior to determining their own final recommendations and advice on a matter before them. (The two reference groups, the Science Advisory Board, and the Water Quality Board are limited to work under the Great Lakes Water Quality Agreement and supplement the Great Lakes Regional Office staff at Windsor, Ontario, in advising the commission.)

Each board and its supplementary working committees are composed of an equal number of nationals of each country and serve under a cochairman from each. The boards themselves are appointed by the commission after consultation with the governments and report directly to the commission, not to their governments or their agencies. This system of seconding career bureaucrats has the advantage of making some of the best expertise available to the commission at a less direct cost (the cost being borne by their agencies rather than by the commission). It also has the advantage of leading to greater acceptability of the commission's work by the officials of the agencies represented and therefore by their respective governments. However, there are also disadvantages to this system. Investigative studies take a long time because of inability of board members to work full-time, and because of rapid turnover of membership. The practice of seconding public servants also raises conflict of interest situations.[18] To remedy this, it has been suggested that the IJC employ its own technical personnel to perform an evaluation function and, to a limited extent, this is what takes place in the work of the small IJC staffs. A larger-scale effort, however, would be a major departure, and would have to be paid for by both governments.

The two sections of the commission in Washington and Ottawa operate as a single unit and also as a response unit to the two governments and to governmental and nongovernmental applicants. Cohen has remarked that the commission is unitary and

> the Canadian and U.S. sections are only there for purposes of organization, with their offices in Ottawa and Washington, but they are not separate entities legally, nor are the Canadians or Americans "delegates" . . . from their respective Governments with mandates to behave as delegates under instruction. On the contrary, they are there in their independent capacities to operate as a united Commission.[19]

Since its creation, the commission has concentrated on five major subjects:

1. approval of hydroelectric, flood control, and reclamation and irrigation storage structures which entail flood damage upstream across the frontier;
2. approval of hydroelectric structures and navigation improvements on boundary waters;
3. application for approval of assorted minor river works;
4. water appropriations from boundary and transboundary waters, both intra- and inter-basin; and
5. investigations of various types (for example, lake levels, Great Lakes water pollution).[20]

The commission itself divides its duties into three categories: regulatory, investigative, and surveillance/coordination. Additionally, there is the fourth unutilized function of binding decision. The commission's mandate can also be divided into two broad categories: quasi-judicial/arbitral, and investigative. It is rather more difficult to involve the commission's investigative than its quasi-judicial jurisdiction. Private individuals, corporations, and local governments can get an application before the IJC, but only national governments have the power to initiate an investigation.

MANDATE AND AUTHORITY

The IJC essentially has two broad categories of authority—judicial and investigative. The right to approve or reject applications for projects affecting boundary waters, waters flowing from boundary waters, and transboundary waters below the boundary constitute the commission's judicial authority. This can be considered compulsory jurisdiction granted under Articles III and IV. The commission's binding decisions in these cases are called orders of approval, and are not subject to appeal. Most of the first half-century's work was of this type.

However, the IJC has a broad investigative power granted under Articles IX and X. Article IX confers the power to investigate, study, report, and recommend, but only in an advisory sense. Its findings under this article are not binding.[21] Article X, which is yet to be used,[22] provides for binding arbitration in any matter given to it by the governments. Its investigative powers can be broad, but their terms are dictated by the specific requests of either country. It cannot initiate its own investigative work and thus is seriously constrained as to what it can accomplish, even though the ways in which it could be used are very broad indeed, including virtually any matter in Canadian-U.S. bilateral relations, from trade to border broadcasting.

A specific directive given to the commission by the treaty is the order of preference of water uses to be observed in determining whether to approve an application. Uses for domestic and sanitary purposes are to be considered first, followed by navigation, power, and irrigation, in that order. This reflects a priority ordering that was natural in the early twentieth century but does not represent more contemporary concerns for recreational values, aesthetics, or general pollution problems. Nor does it give electric power generation the high place it might well be given today. There has been no move to change this order of priority, perhaps because the treaty itself would have to be amended, which is a difficult task!

OPERATION

During its first thirty years, the commission's agenda was dominated by applications for water apportionment permits under its quasi-judicial authority granted by Articles III, IV, and VIII.[23] For the most part, these were minor matters having only local or regional significance and raising technical questions requiring engineering knowledge. Rarely did they have serious policy implications or affect more than a small number of people. (Matters pertaining to Niagara Falls, Great Lakes levels, and the St. Lawrence Seaway were an exception to this rule.) In this first half-century, the commission was staffed mainly with engineering expertise, since little else was called for.

However, the past three decades or so have witnessed dramatic changes.[24] As many of the old apportionment disputes were adjudicated, the commission was gradually working itself out of a job (except for the monitoring and surveillance role, which was never-ending). Simultaneously, so many new pollution problems arose in the transborder water and air relationship that the two governments began to increase the number of their requests to the commission to make recommendations under the reference or investigative authority granted under Article IX. Questions of water quality and air quality were arising and were increasingly referred to the commission, not for adjudication but for advice and recommendation. These questions required the expertise of lawyers, economists, and many different types of natural and physical scientists, in addition to engineers. They also often were national in scope, affected many people, and had broad policy implications of a controversial nature. No longer could the commission quietly pronounce upon narrow technical questions of minor impact about which few knew and fewer cared. It was gradually thrust into the public spotlight and became a very different institution with a very different role to play, without any change in its powers or formal mandate.

The commission's new role, to tackle pollution of both water and air and to interest itself in much more abstract quality as well as quantity questions, was founded upon the original language of the Boundary Waters Treaty's Article IV.

> Boundary waters and waters flowing across the boundary shall not be polluted on either side to the injury of health or property on the other.[25]

This is an unusual concept for 1909, but does represent an early mandate upon which most current water quality and some current air quality involvement is based. A high percentage of today's IJC work load occurs in the environmental quality area and is thus predicated upon this clause of Article IV (see fig. 2 for the IJC case load).

Former Canadian Chairman Maxwell Cohen views the commission as having gone through four distinct stages:

(a) the period of shaping the work of the Commission, from 1912 to the beginning of World War II;

(b) the "great works" period of post–World War II [a reference to the St. Lawrence Power and Seaway and Columbia River issues];

(c) the gradual shift away from Orders of Approval to References after 1956 as the principal work of the Commission;

(d) the growing importance of air pollution and water quality problems and the emergence of an increasingly environmental perspective from 1960 onwards.[26]

EVALUATION

Although a rather little known (especially in the United States) and widely misunderstood institution, the International Joint Commission has nevertheless been viewed in differing ways. The range of perspectives is extreme, especially as regards the need for reform, and a small but rich

Workload: 102 Dockets (cases)

By Subject: 81 —water uses, diversions, etc.
                13 —water and/or air quality
                 3 —other border disputes

By Classes: 59 applications
                    —36 construction and maintenance
                       of dams and rivers
                    —15 changes in the levels and flows
                    — 3 diversions
                    — 3 obstructions
                    — 2 remedial works
                41 references
                    —16 regulation of levels and flows
                    —10 water pollution
                    — 4 air pollution
                    — 4 river basin development studies
                    — 3 water apportionment
                    — 4 other (tidal power, Point Roberts, etc.)

Fig. 2. The IJC case load, 1912–75. (From International Joint Commission, Ottawa and Washington, D.C., 1976.)

literature is now developing in this area. An understanding of these various perspectives sheds light on the commission's future potential.

The commission is viewed in some quarters as an outstanding symbol and example of the highest ideal of what can be accomplished in the bilateral relationship of two countries. Some agree it is a great symbol but that its real value has not surpassed the symbolic level. Others believe it has been a good experiment but has worked (or, rather, has only been permitted to work) well below potential. Some believe it has had an outstanding record in certain constrained narrow areas such as common scientific fact-finding, but that it has failed miserably in broader endeavors. A few believe it is nothing more than a pawn of governments designed to rubberstamp or validate decisions already made by others. And, finally, a few believe the institution is a weak attempt at a high ideal that in fact accomplishes nothing and is a cost to society in time and money.

Just as viewpoints differ as to what the commission is or was meant to be, so also views differ greatly as to what it could or should become and the various changes, if any, that are necessary to reach their image of the final product and its role in society. In all of this diversity of view, it can be said there is no strong national affinity to any one viewpoint. The entire range of opinions can be found on each side of the border. Indeed, most agree that one of the real values of the institution is the relative deemphasis of national viewpoints within or surrounding the commission, exemplified in its method of operation, its long history of decision making, as well as in the lack of national consensus in perspectives toward it. Perhaps the only nationally differentiated comment one could make toward the commission is that it is much better known north of the border and better supported financially by the Canadian government (although there is no real evidence to indicate this imbalance has influenced it to favor Canadian positions in any way).

The differing appreciation of the IJC among diplomats and environmentalists illustrates the ambivalence with which the commission has had to learn to live. Diplomatic viewpoints toward the IJC tend to be cautious. Professional diplomats recognize that its long history has for the most part been a very successful one, but at the same time emphasize limits and constraints on its present and future activity. The commission's function is viewed as that of providing an agreed technical data base and nothing more. In this it can be very successful, the more so when it does not have any adjudicatory role. There is a strong attitude among diplomats that the treaty sharply limits its scope (they take a narrow approach to treaty interpretation), and that its powers should not (perhaps cannot) be broadened.

There is feeling in some diplomatic circles that the IJC is losing its spirit of collegiality, partly due to commission membership but more to the heavy work load and the fact that the equity aspects of many of the issues it is called upon to deal with are so much greater than in the past. With such high stakes, collegiality is more difficult to maintain and nationalism to avoid.

Diplomats also often feel that, on Article IX (advisory) references, the IJC role is completed when it makes its report to government and that it has no right or mandate to seek a government response to its recommendations, a point that is disputed by a number of commission members and others. Diplomats agree it is desirable to accept or reject recommendations and to answer the commission when possible, but they make it clear this is often not possible. They also note that technical agencies can in practice accept or reject an IJC recommendation, whether or not governments formally respond. A general complaint of diplomats toward the commission is the time it takes to complete its work, an especially acute problem in recent years with the Poplar water quality and Champlain-Richelieu flooding questions, as we shall see in Part 2. Diplomatic frustration results when the commission delays, for whatever reason, and diplomats argue it would be better for it to admit it cannot conclude the assignment and release it back to them for action. Institutionalization of the bilateral relationship, whether in the form of the IJC or some other entity, does reduce flexibility, whereas ad hoc diplomatic treatment enhances flexibility.

In general the commission stands for institutionalization and stability under set rules of procedure and priorities in contrast to the politically constrained but freer atmosphere of diplomatic negotiation. Its work and findings cannot be controlled by the bilateral negotiators beyond the stage of writing the references, nor can the behavior of sometimes controversial commissioners be controlled. Thus its work, especially if expanded to include initiatory authority or binding recommendations, may be perceived as a threat to freedom of diplomatic negotiation. While the commission can be a useful tool for diplomats, especially on complicated technical issues or those which simply cannot be resolved by normal negotiation, it also presents a threat of competition for turf. Thus the relationship between diplomats and the commission can be a strained one.

Nondiplomatic civil servants who have played the greatest role in IJC boards and in implementing the commission's recommendations are those from Environment Canada in Ottawa and its regions. No U.S. civil servants have had the level of IJC experience of Environment Canada personnel and thus a cross section of these Canadian bureaucratic views is in order. Environment Canada accepts the "seconding" of many of its people to IJC boards and, although this can create temporary staffing problems, in general it gives the department and its thinking a great deal of say in IJC findings. Hence, the department is a strong defender of the role of bureaucrats on commission boards. Having boards with members who are sensitive to government policy is not a bad idea, since it increases both the acceptability and smooth implementation of the commission's recommendations—a great practical advantage. At the same time, however, the fact is acknowledged that board members' views tend to be colored by agency perspectives on how to do things.

Environmental bureaucrats generally exhibit a freer, less structured, and perhaps less cautious attitude than diplomats toward the commission, and sometimes make suggestions diplomats might well regard as unrealistic if not undesirable. For example, the idea of a buffer zone along the border with all water-polluting and air-polluting activities in a designated border strip being under IJC jurisdiction and permit-granting authority has been broached by a few such officials. A certain feeling that IJC solutions to transborder problems are better than diplomatic solutions is evident in environmental circles. A difference of opinion occurs on the subject of whether or not the commission should seek follow-up on its recommendations, with some feeling that "wing-clipping" might ultimately result. However, there is more of a belief that government is obligated to the IJC to either accept or reject their recommendations than is the case among diplomats. Environmental bureaucrats share the belief of diplomats that the IJC does not produce results fast enough but believe part of the problem is that they are given insufficient guidelines.

From a knowledge of the commission and its operation and an awareness of these differing views, one may well deduce that the commission by its internal composition and board leadership is akin to federal environmental agencies and environmental policy bureaucrats. However, the external influences of diplomatic authority over the nature and breadth of most of its work load and the need for the commissioners to make their findings palatable to both governments dictate that the commission must at the same time behave as a diplomatic entity. In a highly complex way, then, it is a composite of both extremes and to be effective must walk on a tightrope between the two.

In summary, it can be said that awareness of and interest in the commission are stronger north of the border; that environmental bureaucrats identify more closely with the commission than do diplomats; and that, among those who are familiar with it, some regard it as a successful symbol to the world while others view it more critically as a pawn in the bilateral relationship. There is good reason to believe, however, that while on paper it is the symbol for its admirers, in practice it may at times be the pawn suspected by its detractors. The two governments have not really permitted it ample latitude to operate to its full potential as provided by the treaty. Until that occurs, it cannot adequately be judged.

COMMISSIONERS' PERSPECTIVES

The International Joint Commission has successfully enabled the two governments to avoid conflict and settle disputes. It has had 75 to 80 percent of its recommendations accepted by government, and these recommendations have stood up over time. Maxwell Cohen attributes the commission's success as deriving from "common fact-finding by technical peers sitting in equal numbers and seeking a commonly accepted solution, under a permanently organized and impartial umbrella."[27] It has achieved this, in Cohen's words, through the "conversion of rhetoric to arithmetic, from noise to numbers."[28] It has enabled the sharing of boundary waters while simultaneously minimizing obstruction to development in the boundary regions. It has exercised, although with some hesitation, a general mandate to alert government to coming problems.[29] It has accomplished these ends with minimal funding support, the latter due to the failure of government to correctly calculate the value to both societies of dispute avoidance. Government has also failed to make use of its arbitration authority under Article X, nor has it taken advantage of the fact that Article IX can be used to get a reference with the support of only one side. Some argue Article X has never been used because it is never needed, while others claim government fears giving such power to the commission. The unilateral use of Article IX would create practical problems (must both sides pay and participate?), and has diplomatic repercussions.

Also contributing to the commission's success is its advisory rather than policy-making or regulatory role. While the institution does require sufficient support to provide independent advice and recommendations necessary to solve problems, it can easily defeat its purpose if it fails to recognize the limits of its role or permits itself to be used by the public to avoid legitimate national policies. Its role has been that of an adviser rather than a participant, in keeping with the treaty's purpose of preventing and avoiding controversy. Its dominant power, therefore, relies upon the force of its technical and moral position rather than upon economic or political imperialism.[30] In the words of former commissioner Charles Ross,

> the proper and correct solution becomes the goal, not the goal sought by the nation with the most clout. To the greatest extent possible, national sovereignty becomes lost in the shuffle.[31]

In focusing on the commission, one must focus on the commission's lifeblood, the boards. Former commissioner Anthony Scott has credited the continued strength of the IJC as stemming from its success in persuading the professionals who serve on the boards to maintain their detachment and give the commission the information that it needs. As much as half the time of the commissioners is spent on the appointment and review of the boards and their work. Scott believes that traditional common fact-finding by expert boards serving a small commission is as well, if not better, designed for today's complex of roles and tasks than it was in the past era of simpler work. A major reform in thinking which Professor Scott, as an economist, advocates is a recognition that there are times when all of the advantages may lie with one country, and the IJC must not hesitate to so report, avoiding the temptation "to distort the question so that some sort of even division of benefits and costs emerges in every case."[32] The remedy, he suggests, is monetary compensation, and we should not be reluctant to use this approach.

More recently, Scott has urged that "marketable private pollution rights might become a vehicle for implementing IJC decisions and recommendations."[33] He regards this technique as more desirable than the other two compensation possibilities—direct cash compensation for costs endured or the present system of requiring equal behavior on both sides, which he finds more costly.

Scott cites three strong advantages to his marketable private pollution rights concept:

(1) it would allow different regions to have different local pollution standards or policies;

(2) similar schemes could be used for water apportionment, air pollution, water pollution, and fisheries sharing; and

(3) it would be flexible for all involved.[34]

This would, of course, give the IJC a wholly new task but may well be a key to enabling it to remain "slim and more taut" in the long run.

Scott also believes in the importance of encouraging local bodies to cooperate without intervention by the commission, and feels the time has now arrived that

> the IJC placed itself at the center of a web, the strands of which give it only remote control over locally-decided and locally-defined issues. Then it would be free to devote its attention to the new problems of complex economic resource-use conflicts and complicated ecological and environmental incompatibilities. . . . (T)he time has come to specialize it more, by narrowing its mission and exploring new instruments for it to use.[35]

NEED FOR A CHANGING ROLE?

Broader concerns which have arisen include questions as to the commission's general mandate, which is often a matter of interpretation. Unanswered questions include: How much will governments tolerate? How far should the IJC go? Does it find the facts and report them as it sees them, or does it mediate? Does government want an objective umpire's judgment or a mediated diplomatic solution? Should boards be continued indefinitely? Should the commission follow up on recommendations and press government for reaction? Should the staff role be merely to explain and transmit to the commission the board reports, or should it more broadly serve as a "think tank," an interdisciplinary team moving beyond the board reports and giving advice to the commission? These are complicated yet important questions that must be answered if the six commissioners and their staffs are to proceed smoothly and effectively in the future.

Some of the more immediate methods that might be instituted to ensure a firm and continuing role for the IJC in Canadian-U.S. environmental relations include a greater willingness on the part of government to turn over controversial issues to the commission, a willingness to use Article X (binding arbitration), and a willingness to show greater confidence in the commissioners it appoints. The commission for its part might speed up its work, press vigorously for the implementation of its recommendations, develop a constituency, and press government for needed procedural reforms. The latter might include consistent joint funding procedures; insisting on more specifically worded references; filing of interim recommendations (to answer government's need for quicker action); placing of more nongovernmental people on IJC boards; developing better public information mechanisms; instituting a feedback mechanism for government to respond to commission recommendations; and working out joint agency guidelines for the commission to use in apportioning upstream and downstream benefits.

In viewing prospects for reform, however, it is wise to accept political reality, for the governments are not likely at this time to increase the commission's jurisdiction or authority. Munton ascribes the reasons as fundamental and not likely to change merely because new parties or personalities come to power.[36] It is likely the commission could not have matured as it has if it had ignored the political milieu surrounding it.[37] Failure to accept its limitations might deprive the commission of an opportunity to play a major role in the future. Munton contends that transboundary environmental issues, especially those which are health-related, will reemerge in the 1980s, and predicts that

> if this pattern unfolds, then there can be little doubt that the governments . . . will come to the IJC for answers and recommendations. The resulting investigations will be inescapably the broadest, most far-reaching in its history. One can hope and trust the Commission will be ready.[38]

With the advent of a new U.S. administration in 1981, the commission entered what may prove to be an era of benign neglect. President Reagan and Prime Minister Trudeau failed for months to fill five vacant positions (of six) on the commission and, for the first time in its history, the IJC was reduced to a membership insufficient for a quorum on either side and thus in practice eliminated as a factor on the scene. All the government rhetoric about the uniqueness and high value of the IJC to the contrary, one must seriously question whether in fact either government believes its own rhetoric when one is confronted with the obviously uncaring attitude of both governments in 1981. The continuing failure of the Canadian government to appoint a permanent chairman further exacerbates an already difficult situation for the commission. What impact this neglect will have on the future of this valuable vehicle for continental amity remains to be seen.

The IJC thus cannot be judged overall but only on a case-by-case basis. There are cases of clear success and clear failure, as described in the case study chapters. However, on any broad overall evaluation, whether the institution has succeeded or failed depends on the level of expectation one brings to the judgment.

A noble experiment, a globally significant innovation, a tangible sign of the good health of Canadian-U.S. relations, or a device to avoid answering larger questions? The reader may decide after seeing the IJC in action in the case studies which follow, and we will return to a consideration of the IJC's role in Part 3.

NOTES

1. Perhaps the best and most comprehensive study to date on the origins and early days of both the treaty and the IJC is N. F. Dreiziger, ''The International Joint Commission of the United States and Canada, 1895–1920: A Study in Canadian-

American Relations" (Ph.D. diss., University of Toronto, 1974). Particularly good among other papers by this same author, who is Professor of History at the Royal Military College, Kingston, Ontario, is "Peace on This Continent: Dreams and Disappointments in Planning for An International Joint Commission Between the United States and Canada" (Paper prepared for the seventieth anniversary of the IJC's establishment, celebrated at the University of Toronto, June, 1979). Other important works include the now classic *Boundary Waters Problems of Canada and the United States* by L. M. Bloomfield and G. F. Fitzgerald (Toronto: Carswell, 1958); *The Regime of Boundary Waters—The Canadian–United States Experience (Lectures to the Hague Academy)* by Maxwell Cohen (Leyden, Netherlands: A. W. Sythoff, 1977) (hereafter cited as *The Hague Lectures*); "Along the Common Frontier: The International Joint Commission" by A. D. P. Heeney *(Behind the Headlines* 26, no. 5, pp. 1–8); "The IJC and Canada–United States Boundary Relations" by F. J. E. Jordan (in *Canadian Perspectives on International Law and Organization*, ed. R. St. J. MacDonald et al. [Toronto: University of Toronto Press, 1974], pp. 522–44); *The Joint Organizations of Canada and the United States* by William R. Willoughby (Toronto: University of Toronto Press, 1979); O. P. Dwivedi, ed., *Resources and the Environment: Policy Perspectives for Canada* (Toronto: McClelland and Stewart, 1980); and "The Work of the IJC" in *External Affairs Bulletin* (Canada Department of External Affairs) 23 (June 1971):208–14. Papers on the IJC by former commissioners Maxwell Cohen, Anthony Scott, Christian Herter, Jr., Matthew Welsh, A. D. P. Heeney, Eugene Weber, Bernard Beaupré, A. G. L. McNaughton, and Charles Ross are available, as are others by Professors Donald Munton, N. F. Dreiziger, William Willoughby, Richard Bilder, Charles B. Bourne, Kim Nossal, and John E. Carroll. Finally, one should not overlook the works of the commission itself, and particularly the annual reports which were begun in the early 1970s. Works on the treaty and IJC are largely Canadian in origin, reflecting the greater Canadian interest in these matters, a bias which is reflected in the citations in this note. Most recently, *The International Joint Commission Seventy Years On*, ed. R. Spencer, J. Kirton, and K. Nossal (Toronto: University of Toronto Centre for International Studies, 1981) is invaluable as an IJC source.

   2. Dreiziger, "Peace on This Continent," p. 7.

   3. Ibid., p. 3.

   4. Ibid., p. 14.

   5. Ibid., p. 15.

   6. Numerical file 1906-10, Department of State, National Archives 5934/25; S. Doc. 118, 85th Congress, 2d Sess., p. 27 (1958).

   7. Dreiziger, "Peace on This Continent," p. 17.

   8. Charles R. Ross, "The International Joint Commission, United States–Canada," in *Entente Cordiale Part II: Bilateral Commission and International Legal Methods of Adjustment*, Proceedings of the 68th Annual Meeting of the American Society of International Law, April, 1974, p. 230.

   9. Heeney, "Along the Common Frontier," p. 5.

   10. Treaty Between the United States and Great Britain relating to Boundary Waters, and Questions Arising between the United States and Canada, January 11, 1909, *Canada Statutes* 1911, chap. 28, art. VII.

   11. Willoughby, *Joint Organizations*, p. 17.

12. Cohen has noted that the result of these articles was "to make it impossible to build a structure, deepen a channel, or in any way interfere with a boundary water on either side by one country if that structure, etc. 'materially' affected the level or flows of the boundary on the other side." Further, "the parties promise . . . that they will not permit construction or maintenance of any remedial or protective works or dams or other obstructions, except with the permission of the International Joint Commission" (*The Hague Lectures*, p. 252). Cohen also remarks that Article III excludes tributaries from its diversion jurisdiction, as per the Harmon Doctrine, but that Article IV, which concerns the raising of levels, by definition must include jurisdiction over tributaries, as long as work on them raises levels at the border. These articles, III and IV, represent the heart of the quasi-judicial, regulatory, licensing power of the commission.

13. International Joint Commission (IJC), *Rules of Procedure and Text of Treaty* (Washington and Ottawa: International Joint Commission, 1965), p. 14. Article II states:

> Each of the High Contracting Parties reserves to itself . . . the exclusive jurisdiction and control over the use and diversion . . . of all waters on its own side of the line which in their natural channels would flow across the boundary or into boundary waters.

14. Ian A. McDougall, "The Development of International Law with Respect to Trans-Boundary Water Resources: Cooperation for Mutual Advantage or Continentalism's Thin Edge of the Wedge?" *Osgoode Hall Law Journal* 9, no. 2 (1971):264.

15. Ibid., p. 266.

16. Ibid., p. 267.

17. Cohen, *The Hague Lectures*, p. 238.

18. One former member of a Great Lakes water quality board subcommittee claims avoidance of conflict of interest is impossible, noting that he was asked by his government to criticize a subcommittee report and recommendation which he himself had written.

19. Cohen, *The Hague Lectures*, p. 257.

20. Ibid., p. 267. *See also* Bloomfield and Fitzgerald, *Boundary Waters Problems*.

21. Article IX technically gives the commission the authority to investigate without the consent of both countries and it has been suggested this could constitute an important qualification of sovereignty; however, in practice, investigations have only gone forward when both countries desired it. Hence, references on the West Coast tanker issue, desired by Canada alone, and the Atikokan power plant, supported only by the United States, did not go forward.

22. Although both governments have been criticized for not using this binding arbitration article, another view of the matter is expressed by R. L. Pentland and W. R. Long, who have remarked, "it is notable that Article X of the Treaty, a provision under which a matter can be submitted by the two governments for binding arbitration, has *never had to be used*" (emphasis added) (in "Boundary Waters Management" [Paper presented to the United Nations Water Conference, Mar del Plata, Argentina, March 14–25, 1977], p. 34. Also cited in Pentland, "Boundary Waters Management," in *Canadian-American Natural Resource Pa-*

pers 1975–1976, ed. John E. Carroll [Durham, N.H.: University of New Hampshire, 1976], pp. 13–22).

23. Willoughby reports that thirty-nine of the fifty cases handled before 1944 were applications for approval under Article VIII and only eleven were references for investigation under Article IX. William R. Willoughby, "The International Joint Commission: Evolving Expectations and Experience, 1909–1979" (Paper presented to the Conference to Commemorate the 70th Anniversary of the International Joint Commission, University of Toronto, June, 1979), p. 6.

24. Between 1944 and March, 1979, there were thirty-five references and twenty applications (Willoughby, "The IJC," p. 6).

25. IJC, *Rules of Procedure,* p. 14.

26. Maxwell Cohen, "Canada and the United States: Dispute Settlement and the International Joint Commission—Can This Experience Be Applied to Law of the Sea Issues?" *Journal of International Law* 8 (Winter 1976):77.

27. Maxwell Cohen, "The Patterns of Settlement—Canada, the United States, and the International Joint Commission" (Remarks to the Conference Board in Canada, Toronto, Ontario, November 7, 1976), p. 12.

28. Maxwell Cohen, International Joint Commission, personal interview, Ottawa, Ontario, June 13, 1977.

29. Keith Henry, International Joint Commission, personal interview, Vancouver, British Columbia, August 5, 1977. Former IJC Commissioner Henry, an advocate of a small commission with a limited role, believes that if a general mandate to watch the border and investigate whatever it wanted were carried too far the commission would soon disappear.

30. Ross, "The International Joint Commission, United States–Canada," p. 236.

31. Ibid.

32. Anthony Scott, "The IJC After Five Years" (Notes for Remarks to a Dinner Meeting of the International Law Association's Committee on International Water Resources Law, Vancouver, British Columbia, July 19, 1977), p. 24.

33. Anthony Scott, "Can the IJC Concept be Strengthened to Deal with Future Western Border Issues?" (Paper presented to the Western Social Sciences Association Annual Meeting, Albuquerque, New Mexico, April, 1980), p. 27. Scott is referring, of course, to creating a market for pollution rights where buying and selling could take place. Only so many rights would be available at any one time.

34. Ibid.

35. Scott, "The IJC After Five Years," pp. 26–27.

36. Don Munton, "The International Joint Commission: Continuing Paradoxes, Future Prospects" (Paper presented to the Conference to Commemorate the Seventieth Anniversary of the International Joint Commission, University of Toronto, June, 1979), p. 20.

37. Don Munton, "The Political Roles of the International Joint Commission" (Paper presented to the Conference to Commemorate the Seventieth Anniversary of the International Joint Commission, University of Toronto, June, 1979), p. 30.

38. Munton, "The International Joint Commission: Continuing Paradoxes, Future Prospects," p. 26.

Part 2
The Problems

CHAPTER 4   Oil and the Coastline

The Eastport oil refinery issue pits the builders of a proposed large U.S. oil refinery to be built at Eastport, Maine, against Canadian fishermen and Canadian and U.S. environmentalists fearful of marine ecological damage from oil spills that are perceived as inevitable. It raises questions concerning rights of free navigation and innocent passage, the impact of large quantities of oil on marine organisms and a highly productive commercial fishery, the inevitability of spills, and how much faith should be put in man's ability to retrieve the oil before its absorption into the natural environment. The West Coast oil tanker issue raises similar questions on the Pacific Coast, although the specific issue here concerns transport of U.S. oil to existing as well as proposed facilities and involves the transport of oil through international waters in such a manner that territorial waters and coastlines of Canada are threatened. The Beaufort Sea issue concerns the impact of hydrocarbon exploration and development on a disputed international boundary, on the Arctic marine ecosystem on and under the ice, and on Canadian threats to U.S. shoreline and Eskimo ecosystems. Serious bilateral boundary and fishery disputes are present as background to all the coastal issues, and linkages, indirect if not direct, are viewed as possible.

With the discovery of a large commercially exploitable deposit of oil in Arctic Alaska in the late 1960s, and particularly with the oil embargo imposed by the Organization of Petroleum Exporting Countries (OPEC) and the emerging energy crisis in the early 1970s, public attention in both the United States and Canada turned toward thoughts of energy security and energy self-sufficiency. Naturally, part of this attention focused on the transport of oil via both sea and pipelines, and on the development of new domestic reserves. Potential domestic sources of oil included untapped frontier reserves in the Arctic and offshore on the outer continental shelves. Any such development of frontier hydrocarbon resources naturally has ramifications on marine ecology, on commercial and recreational fisheries, on coastal zone environments, and, occasionally, on marine mammals and on navigation. Since many of the potentially valuable offshore reserves are in the border regions (see maps 1–3) and since routes of transport from the north and from other parts of the world traverse the territorial waters and exposed coastlines of both countries, there are ample occasions for bilateral environmental issues to arise.

61

This chapter is concerned with the changing and unknown character of the threat to the coastline. The history of Canadian-U.S. relations is full of fisheries disputes, and waterways and fish are part of the stuff of Canadian history, as are energy resources. These are now brought together in a political and emotional atmosphere highly charged by the threat of oil shortages and by vast quantities of new scientific data that are nonetheless inclusive in many important respects. Emotion is also easily inflamed by the apparent deadlines facing decision makers and the bargaining power, which sometimes seems like blackmail, of the oil majors, and confrontations inevitably result. The subject matter of this chapter illustrates the way in which transborder environmental relations have assumed a central role in the bilateral relationship. As in other bilateral marine and coastal disputes, the International Joint Commission is notable for its absence. The two federal governments have not chosen to use it on the coastline or in marine situations, although such use is within the mandate of the commission.

Eastport

"As a firm believer in Murphy's Law (if anything can go wrong, it will), the notion of locating an oil refinery, dependent on the movement of supertankers through Head Harbour Passage, is [I believe] like locating a gunpowder factory next to a blast furnace."[1] These sentiments of the president of the University of New Brunswick have long reflected Canadian feelings on this volatile issue and today increasingly represent the sentiment of American residents of the area as well.

The Pittston Company of New York, a major coal company that is now investing in oil development, has proposed for some years the construction of a large 250,000 barrels/day oil refinery and adjacent marine off-loading terminal at Eastport, Maine, on the Canadian border. Though the refinery would be on U.S. territory, access to it is through Canadian territorial waters. The core of this issue is therefore that the right of access to the refinery site rests with the Canadian government, but that Canada cannot easily deny this right to a close ally. Although vessel size or cargo restrictions can conceivably be implemented against the United States, such arbitrary action could invite retaliation on the basis of unfair discrimination, justifying some caution in Canada's handling of the dispute.

Canada has sought a new interpretation of "innocent passage" (see note 45) and believes there is precedent for government to control the movement of oil (controls on nuclear submarines being one example). In response to a U.S. environmentalist suggestion to turn the matter over to the IJC, Canada responded that this is unnecessary, in that Canada has the authority to deny passage of the oil and has done so. Given that the Canadian position has been so well articulated, there is nothing left to do

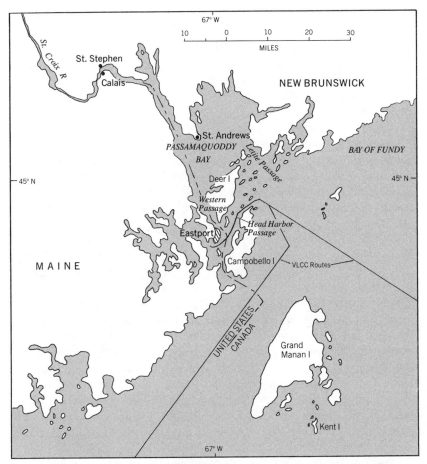

Map 1. East Coast—Oil and Coastline Problem, Maine and New Brunswick. (Adapted from Thomas F. Moffatt, "Beauty in Crisis at Passamaquoddy Bay," *Canadian Geographical Journal* 94, no. 1 [February-March, 1977]: 11.)

but exchange scientific information and wait, with the knowledge that international arbitration is a possibility if the company persists. In the course of time, Canadian fishery and recreational resources in the area are under threat since it is assumed a ship will have an accident sooner or later. The analysis of positive benefits made by the Environmental Protection Agency (EPA) and Pittston are not equal to the disbenefits of the refinery; nevertheless, the thrust of Canadian federal and provincial policy has been to let the United States take care of it by itself.

The U.S. position has not ignored the problems, for the U.S. Environmental Protection Agency's Draft Environmental Impact Statement on the proposed Eastport refinery early recognized that the operation of the refinery represents the commitment of the community, region and State to accept the risk, however small, that an accident could affect all or portions of the fisheries in the area. Thus, the refinery may represent a commitment to an industrial activity which could result in the suppression or even elimination of a renewable resource—fishing.[2]

THE SETTING AND THE PROBLEM

The present proposal and the source of the Canadian-U.S. bilateral controversy dates from the spring of 1973 when the Pittston Company applied to the State of Maine's Board of Environmental Protection (BEP), a permit-granting and licensing agency, for a permit to construct a marine terminal and oil refinery complex with a capacity of 250,000 barrels/day in the city of Eastport, an economically disadvantaged community in easternmost Maine on the New Brunswick border. Eastport contains a sheltered deepwater port, available land at an acceptable price, and an available work force suffering from high unemployment. Until recently, it also bore a positive attitude toward the proposal as an economic boost to the city. The concept involved plans to construct a deepwater supertanker port to receive crude oil from the Middle East, refine this into heating oil and gasoline, and then reship it by smaller coastal vessels to various points in the northeastern United States. The facility would have cost $750 million, by 1979 estimates.[3]

The major bilateral concerns in this issue are environmental in nature. There are also sovereignty aspects, in the form of a boundary dispute, and, indirectly, an economic element, in the form of presently existing excess capacity in Canadian oil refineries that could serve the same market as Eastport.[4] However, the ecological destruction likely to be suffered by inevitable oil spills is the crux of the diplomatic debate. Eastport is a transboundary issue where one nation (the United States) would receive all the benefits while the other nation (Canada) would suffer most of the costs.

In order to serve a refinery at Eastport, it is necessary for large supertankers—known as Very Large Crude Containers (VLCCs) of 250,000 deadweight tonnage (dwt) or more, with a length of over 1,148 feet, a width of 141 feet, and a draft of 66 feet—to negotiate Head Harbor Passage. It is known that a VLCC traveling at twelve knots can stop at best within 11,500 feet and at six knots within 4,000 feet,[5] creating a potential threat of accident and spillage. Head Harbor Passage is the link between Cobscook Bay and the Bay of Fundy, a distance of 7.4 miles. Separating Deer and Campobello islands, this necessary access route is only 1,617 feet wide at its narrowest and has an average width of 3,120 feet and a minimum depth of 75 feet. Swift tidal currents of as much as five knots and large whirlpools characterize the local waters. Also characteristic is an exceptionally heavy fog regime. Visibility is less than half a mile for 34 percent of the time in July, 28 percent in August, and 20 percent in June in the Bay of Fundy region, including the area of the proposed refinery site.[6] In the winter, when fog is less persistent, strong northwesterly winds provide a docking hazard.[7] Given that a VLCC has limited maneuverability, a serious threat of accident and resultant oil spillage appears to exist.

> The strength of these currents would necessitate very tight scheduling for transit of the Head Harbor Passage, and very short time slots or "windows" would be needed for safe movement of VLCCs to the berths which they must reach at slack water. In this scheduling there is no time to allow for emergencies.[8]

Slack water (i.e., the period between the change of tides when the water is not moving) lasts for only thirty minutes every six hours. Four tanker captains and pilots out of six interviewed in an unpublished study conducted for the Pittston Company noted that the route is more difficult than any now being used by supertankers, while a fifth said the route is impossible and a sixth said it would be difficult without tugs.[9]

Even so, the U.S. Coast Guard gave qualified approval that the passage could be safely navigated by 250,000 dwt vessels, but that precise vessel dimensions, maneuverability, speed, and other factors would have to be known for final determination of navigability.[10] It should also be noted that the industry does have a good record in serving its refineries at St. John and Point Tupper, Canadian Maritime ports with somewhat similar conditions. Gulf Oil has brought in 200 million barrels in ten years, with spills totaling less than 700 barrels.

All indications point to the need for heavy expenditures and meticulous organization if spills are to be avoided. The requirement includes a very expensive, highly sophisticated shore-based navigation system employing the latest technology in communications and surveillance; allegedly fail-safe backup systems; heavy-duty escort tugs; pilots to be picked up from shore; and coordination and approval of all decision makers on ship and shore prior to entry of a vessel into the passage. Reliance on any of these aids may be misguided and unrealistic given the high cost, and in any event the best available technology in this area is unproven given the limited real world experience characteristic of most of it.

A Canadian Ministry of Transport study found that there were eleven categories of serious navigability problems with the Eastport proposal: tidal currents, adverse meteorological conditions, uncertainty of depths, channel width and configuration, lack of a safe anchorage, adverse weather conditions, possibility of berthing and unberthing problems, security of vessels moored at berths, pollution containment and cleanup difficulties, lack of technical data, and overall navigability. This study, upon which much of the Canadian negotiating position is based, concluded that "the risk of pollution remains high and is environmentally unacceptable."[11]

The Canadian government is not alone in raising navigational concerns in this matter, for the U.S. Army Corps of Engineers confirmed Canadian fears by acknowledging that "its [Eastport Harbor's] approaches are winding; its currents extremely difficult to judge, and the area has the highest number of fog days along the coast." Further, "Highly sophisticated navigational controls would have to be installed if traffic in this harbor were ever to become heavy."[12]

The nature and characteristics of Head Harbor Passage and the navigational problems of supertankers resultant therefrom constitute the greatest single concern in the Eastport issue as well as the horn of a diplomatic dilemma. Any serious accident in navigation by a loaded supertanker would undoubtedly result in the entry of some oil, and perhaps a great quantity, into the marine and shoreline ecosystems. The introduction of such oil is largely viewed as damaging, and spills, including catastrophic large-scale spills, are viewed as inevitable. Thus, the precise impact of such spills on regional natural resources is of major importance.

It is alleged that chronic spills would occur at the rate of twenty to eighty barrels per year and that a major spill of at least 500 tons would occur every six years. There appears to be no known method for containing such spilled oil in currents of over two knots, while currents in this passage reach five knots.[13] Trites has speculated on a 500-ton spill once every eight years at Eastport.[14] The same study has projected a rapid dispersement of the oil, with both shorelines of the passage and many of the Cobscook Bay beaches contaminated within the first twelve hours. Such a dispersement would easily outstrip the capacity of booms designed to contain it. Spilled oil is projected to reach the extremes of Cobscook and Passamaquoddy bays within one week and to circulate throughout the Bay of Fundy and Gulf of Maine systems from western Nova Scotia to Cape Cod within six weeks.[15] Transport Canada's assessment found that containment of a spill anywhere within the Eastport terminal system would be impracticable.[16] Containment would necessitate the use of dispersants and acceptance of their ecological ramifications. Finally, reduced visibility not only hampers navigation and thus increases the threat of spills but significantly hampers cleanup as well.

There is a difference of opinion in the literature on the effect of spilled oil on life forms, although some effects are obvious. Lighter oils dissolve more rapidly but are more toxic and are widely believed to be carcinogenic or mutagenic to certain organisms and in certain concentrations. Heavier residual oils which do not dissolve as readily do mechanical damage to organisms and habitats. Trites concludes that the overall effects of all types of oil to living things can be devastating,[17] as does Dr. David Scarratt in a more recent work.[18]

Potential damage to the fisheries includes not only the direct impact on the stocks but also damage to plankton, critical components of the food chain of commercially valuable species. It also entails damage to nets, weirs, and vessels, and especially damage to the area's numerous herring weirs from spills of heavy oils which would readily attach. Trites estimates that the total cost of replacing all 225 weirs in Charlotte County, New Brunswick, would amount to over $1.1 million,[19] and a major spill could force replacement of all of them. Any spill that sets back the fishery for even a few years could undermine the economics of the industry; it may not require a major spill to do this.

The herring fishery has been long established in this area of New Brunswick and Maine, and the industry in Canadian waters has expanded dramatically during the past decade. Because almost all the catch now goes to direct human consumption, its unit value has increased 2.5 times since 1975 and now totals over $200 million per annum.

The invertebrate fishery, composed largely of lobsters, was valued at over $66 million in 1978. This is also an important area for live lobster storage in tidal pounds that are vulnerable to floating oil borne in on the tides. These pounds have a capacity of 8.5 million pounds, and Trites estimates that two million pounds are at risk at any one time.[20]

Harmful impacts to the clam, scallop, mussel, and baitworm industries are also feared, as are impacts to winter flounder and other groundfish, and in particular to salmon. Commercially valuable Irish moss, dulse, and other marine plants could also be harmed.

All or most of these vulnerable fisheries are of greater significance to Canada than to the United States, and therein lies an important reason for Eastport to be a bilateral issue—most of the benefits accrue to one country, and most of the costs to another. Whether one considers present employment patterns, capital investment in the fishery, percent of public equity invested, disruption of a well-established life-style and culture for the inhabitants of coastal communities, or the overall role and importance of fishery resource development as a whole to a nation, Canada has much more at stake. In addition, Canadian Maritimers have expectations of benefits from extension of the two-hundred-mile limit, and major investments have resulted.

GOVERNMENT AND THE ENVIRONMENTALISTS

In order to begin construction, the Pittston Company must obtain permits from both the Maine Board of Environmental Protection and Region I (New England) of the U.S. Environmental Protection Agency. Maine issued its approval of the Pittston application in June, 1975, subject to several conditions, the most significant of which was a stipulation that Pittston must "secure Canada's approval both for tanker transit through Canadian waters and for the construction of navigation aids."[21] It is questionable if a state government can constitutionally allow a permit it issues to be contingent upon a decision of a foreign government. Thus far, this has not been tested in court, but it could conceivably be found unconstitutional, given the preeminent role of the federal government in foreign relations.

At the federal level, the required EPA National Pollutant Discharge Elimination System (NPDES) permit has been issued, although two other agencies, the National Marine Fisheries Service and the U.S. Fish and Wildlife Service, have advised EPA that the risk associated with tanker traffic in Head Harbor Passage is "unacceptable."[22] An earlier air quality permit granted by EPA has now expired. The U.S. Fish and Wildlife Service (FWS) attempted to have the Passamaquoddy Bay region declared a critical habitat for the northern bald eagle, an endangered species known to breed there.[23] The Fish and Wildlife Service noted that the Eastport area is "the only location in the northeastern United States where the eagle population has begun to recover,"[24] but their effort to block the issuance of the federal permit failed a court challenge.

The province of New Brunswick has not taken a vocal role in the Eastport issue but is on record as opposed. In its only formal statement, in December, 1973, the province recognized most of the common values endangered by the proposal, expressed particular concern over loss of tourism and recreational values, and worried that "strong economic pressures would develop for relaxing environmental navigation standards from those proposed initially." It concluded that economic pressure would result in more and more tankers sailing the passage under increasingly difficult conditions with increasing risk of accident.[25] New Brunswick found that the benefits of the project do not justify the risks and that "it is the concern for our marine resources and no other reason that we oppose the Eastport Terminal and Refinery proposal."[26] The government's Environmental Council of New Brunswick also finds the risk to the New Brunswick coastline and Bay of Fundy waters unacceptable.[27]

The Eastport issue also has an air quality component, in that emissions of sulfur dioxide (SO_2) and other pollutants will likely be borne downwind to Campobello and Deer islands, and perhaps Grand Manan, although there is likely to be no impact on the New Brunswick mainland. This is an issue that has been pursued with vigor by environmentalists in Eastport who are opposed to the refinery. There is also a chance of increased acid rain over the Bay of Fundy and Nova Scotia. However, the New Brunswick Environment Ministry noted in 1978 that the air quality issue would never stir up people in the province, but the fishery and marine oil spill issues would.[28] The ministry remarked at that time that while it was strongly opposed to the transport of oil through provincial waters near Eastport, its opposition was based on concern for fisheries and recreational resources. The province was not worried about air pollution or opposed to a refinery per se. In response to U.S. charges that the province opposed an Eastport refinery because it hoped to have the refinery built in the province, the ministry responded that New Brunswick had substantial excess refining capacity and thus would not want to attract this additional refinery. As to local concerns, it was noted that the voters in the St. Andrews area across from Eastport are inactive on the issue (though the mayor of St. Andrews is opposed). New Brunswick also reiterated in 1978 that it would support federal strategies and follow the federal lead in dealing with this bilateral dispute.[29] The provincial Environment Ministry has never wavered in its opposition based on marine pollution, but there has never been any official concern over air quality.[30]

In the mid-1970s, in response to the perceived threat to the Eastport and adjacent Lubec, Maine, environments, a small group of citizen environmentalists formed the Friends of Eastport (FOE), an organization committed to stopping the refinery proposal. FOE systematically fought, through data gathering, testifying, and lobbying, every step of the Maine, and particularly the U.S. EPA, permit-issuing processes. In contrast to concerned Canadians, FOE took a stronger interest in air pollution than in the threat of oil spills. However, it recognized a common cause with some Canadian officials and to a certain extent collaborated on strategy with them.[31] FOE was established as a single-issue organization that "will dissolve when the refinery proposal at Eastport is officially dead" and whose purpose is to "educate people to what we have so that they can compare this to what Pittston offers"[32] and then to decide among the alternatives. Although most of this group's work has been domestic, the organization has successfully raised the issue of air pollution on Campobello Island, New Brunswick. Part of this island is an international peace park whose air quality is designated Class 1—pristine[33]—under the U.S. Clean Air Act Amendments of 1977 and thus cannot legally be deteriorated. The EPA has said the air quality would not be eroded. FOE rejects this claim, and has brought litigation. There is no Canadian or New Brunswick interest in the matter, except as it might contribute to stopping the refinery. FOE has also warned Nova Scotia of an acid rain threat from the refinery and maintains regular contact with Fisheries and Oceans Canada on joint data analysis and strategy planning.

The government of Canada has been on record since June, 1973, that "the risks inherent in the transport of a large volume of pollutants through Head Harbour Passage would be unacceptable, and that the Government was therefore opposed to such transport in these difficult waters," as expressed in a formal diplomatic note from the Canadian embassy to the U.S. State Department.[34] This position was reiterated in a letter to the Pittston Company in July, 1976, which noted, "in view of the well established Canadian position, it will not be possible for the Canadian Government to enter into . . . agreements, extend . . . approval, or grant . . . permits," and warned "regulations to give effect to our position will be issued, should this prove necessary."[35]

Canadian diplomats in early 1977 were stressing that Eastport was a very serious bilateral issue and that Canada would not accept oil going through Canadian waters to reach Eastport. They considered this matter to be a serious sovereignty issue and expressed the feeling that both governments preferred to let the issue lie. They also expressed Canada's hope that Pittston would not go to the State Department and complain that it had not been given a fair hearing, a move that could create problems for the Canadian-U.S. relationship.[36]

By summer, 1977, Canada believed that the company wanted to isolate the problem of Canadian opposition and overcome all other issues first, then pressure the State Department to negotiate around the Canadian opposition. Canada also believed that diplomats on both sides wanted to avoid sovereignty issues and allowed for the fact that the whole issue might wind up in third-party arbitration at the International Court of Justice.

By the summer of 1978, so many domestic hurdles had arisen to challenge the refinery proponents that Canada's position became more one of watching and waiting. One Canadian official on the scene remarked that if oil were shipped, Canada would invoke its Shipping Act to put controls and conditions on the movement of oil which could be moved through the passage, thereby making the project nonviable. Specifically, the strategy would be to permit tankers to carry just enough oil for local Eastport area consumption to pass, so that Canada could not be portrayed as denying Americans needed oil supplies.[37] This tactic would, of course, preclude the refinery and insure that no tankers, with or without oil, would ever pass through these waters.

There is, however, an additional factor which Canada must be concerned about, and that is the potential for the Eastport issue to complicate an already complicated and difficult East Coast fisheries dispute. One Canadian government scientist estimates the capital investment in the fishery in the Bay of Fundy to be $100 million per year (current value) for the herring industry, with a potential herring and mackerel combined value of half a billion dollars.[38] The oil refinery is not worth nearly as much, but this single project is believed to have the potential to wipe out the whole fishery. Both New Brunswick and Nova Scotia have a big stake in such a fishery loss. The fear has developed of a trade-off, with Canada dropping opposition to the refinery in return for U.S. concessions on fishery stocks.[39] Such a trade-off could create serious future problems.

In late 1979 Canada's minister of External Affairs responded to U.S. Senator Edmund Muskie's concerns about overall pollution of the Campobello Island environment if the refinery should be built, providing Canada with another opportunity to restate its long-held position.[40] Shortly thereafter, the prime minister wrote, ''my Government regards the large-scale transit of toxic pollutants through Head Harbour Passage as unacceptable, and is prepared if necessary to regulate to prevent it.''[41] He continued:

It does not appear to me that the promoters of this project, or responsible officials and political figures in the U.S., should have any reason to be uncertain about the views of the Canadian Government on this issue. . . . (S)hould the Pittston Company again approach the Canadian Government for . . . permission [to transport oil], our response will be as clear and firm as that which was given in 1976.[42]

The U.S. EPA's withdrawal of its former objections to the Eastport refinery in spring, 1981, undoubtedly contributed to the dramatic action of the Canadian Government in February, 1982. In that month, Canada issued new oil tanker regulations by amending the Canada Shipping Act so as to limit to 5,000 metric tons the quantity of crude oil or oil product which may be carried in tankers in the waters of Head Harbor Passage. This action prohibits the volume and type of tanker traffic comtemplated for the Eastport refinery, and in fact represents about the quantity of oil needed to serve already established needs of a local nature. The action was carried out expressly to protect the environmental quality and fishery resources of local waters. The action also strengthens Canada's claim to sovereignty over these waters. The U.S. State Department immediately rejected Canada's claim to jurisdiction over Head Harbor Passage implied in the issuance of such regulations and said it would file a formal protest with the Canadian government. Thus, in 1982 the Eastport dispute was elevated to a higher order of magnitude.

SUMMARY

Basic to the Canadian-U.S. bilateral conflict over Eastport is, of course, the passage of U.S. and other non-Canadian tankers carrying U.S.-destined oil at considerable risk through Canadian waters to a U.S. refinery and U.S. markets, thereby endangering Canadian property and resources in the process. The disagreement centers "over both the legal status of Head Harbour Passage and the degree of control Canada can exercise over ships passing through it."[43]

The United States recognizes these waters as being within the territorial sea of Canada, but insists that they constitute an international strait under customary international law. Canada refers to these simply as "Canadian waters" and has not been more specific, but opponents of the project in the United States contend that the passage is not territorial sea but internal waters. It would appear that this viewpoint is strengthened by the 1910 treaty that settled the Canadian-U.S. boundary from the mouth of the St. Croix River to the Bay of Fundy and which is said to have created internal waters for each state on its side of the line.[44]

Considering the passage an international strait permits the United States, according to international legal practice, to claim the right of innocent passage.[45] With such a claim, it is the U.S. position that, "while Canada may enact reasonable regulations which the U.S. must obey, these regulations may not have the effect of prohibiting passage."[46] By this rule a 5,000-ton limit on pollutants (which Canada has suggested) would be viewed as unreasonable and beyond Canada's legal authority.

Supporters of the Canadian position argue that if these be internal waters, then Canada enjoys full territorial sovereignty and thus is not obligated to recognize U.S. rights of passage.[47] Alternatively, if this be an international strait, then it might be argued that under international law a coastal state has considerable latitude in deciding what constitutes "innocent passage." One viewpoint holds that the navigational risks and ecological vulnerability of this area make tanker traffic here "noninnocent."[48] If the Eastport refinery receives final go-ahead in the United States, these legal arguments might well be subsumed by political questions of who can bring how much influence to bear and at what price. However, it is not inconceivable that legal resolution through third-party arbitration might be necessary first if required navigational aids on Canadian territory are to be installed to facilitate passage. Hence, political resolution could well force a legal decision.

As time passes, the people of Eastport, once supporters of the refinery, are souring, viewing it as a threat to their hope for a rejuvenated fishery. The entire Eastport affair may ultimately go down in history as an exercise in futility, but not without making its mark in the long history of Canadian-U.S. relations.[49]

The West Coast Oil Tankers Issue

The West Coast situation involves a much more heavily populated area (see map 2) and presents more of a broad transport-related problem. However, the idea of new refinery construction in Washington State requiring U.S. oil transport near (but not through) Canadian waters bears similarities to Eastport. Both issues constitute perceived U.S.-originating environmental threats to Canadian waters and shorelines. However, whereas no benefits to Canada are offered by Eastport, Canada may be in line for some West Coast benefits (depending on the outcome of prolonged siting decisions). Thus the Canadian diplomatic position on the Pacific is less clear and perhaps more equivocal than has been the case on the Atlantic.

VESSEL TRAFFIC MANAGEMENT

On December 19, 1979, a decade of bilateral planning and three years of diplomatic negotiation culminated in the signing of a Canada-U.S. Vessel Traffic Management Agreement. This agreement provides for the establishment of vessel traffic routes and a system of traffic management which asks vessel masters to report information to, and receive information and advice

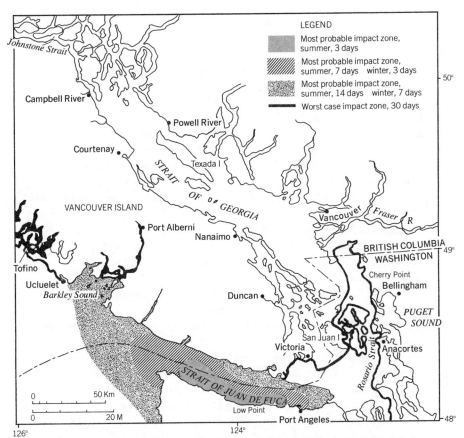

Map 2. West Coast—Oil and Coastline Problem Using Low Point, Washington, as the Site of the Projected Oil Spill, Showing the Probable Impact Areas. (Adapted from W. H. Wolferstan, *Oil Tanker Traffic: Assessing the Risks for the Southern Coast of British Columbia*, APD Bulletin 9 [British Columbia, Ministry of Environment, 1980].)

from, designated vessel traffic management (VTM) centers being established in the Washington and British Columbia coastal regions. The agreement relies on voluntary compliance, and it is described by the Canada Department of External Affairs as

> the latest in a series of cooperative measures taken over the past six years by Canada and the United States to support efficient navigation, to protect life and property, and to reduce as much as possible the environmental risk presented by tanker traffic in Juan de Fuca [Strait] and Puget Sound. [The accord] does not prejudge separate decisions as to whether, or in what numbers and sizes and by what routes, tankers should carry crude oil from Alaska to U.S. markets. Rather,

it serves to put in place the safest possible traffic management system.[50]

The advent of a vessel traffic control system in West Coast waters in January, 1974, was occasioned by the U.S. government's decision to transport Alaskan oil via tanker to U.S. West Coast ports and to the eastern United States via the Panama Canal. A series of ship-ferry collisions and near-misses near Vancouver heightened public interest in a control system. Taking a page from aviation, the Canadian Coast Guard established a vessel traffic routing scheme and a vessel movement reporting system on the West Coast. Three zones were established from Vancouver to Prince Rupert.[51] The system covered all vessels over 65.6 feet in length and excluded all pleasure boats as well as most fishing vessels. Traffic lanes were established for inbound and outbound traffic, with separation zones in between. Thus the traffic control program, headquartered in Vancouver, operates much like air traffic control at a major airport, though on a voluntary rather than a mandatory basis.

Canada has shown substantial pride in this very sophisticated and advanced, as well as expensive, system for reducing accidents in coastal waters and in the mid-1970s claimed high voluntary compliance. Given this pride and good record, it was natural for Canadian diplomats, when pressured by the public outcry in British Columbia over U.S. oil shipments in coastal waters, to press the United States to institute a similar system for U.S. waters and for the overlapping zone of disputed waters as well. Indeed, the system itself could not be completed or made mandatory without a negotiated bilateral agreement, if for no other reason than the necessary overlapping of traffic lanes in the Strait of Juan de Fuca.

Transport Canada and External Affairs both believed in 1976–77 that such a negotiated agreement with mandatory compliance was precisely what was needed to defuse this bilateral issue. Some officials optimistically believed that an agreement could be signed before the end of 1977, in time to govern the movement of the first oil tankers from Alaska. The only question remaining at that time was one of liability and compensation: given that U.S. oil moving to U.S. ports would threaten Canadian resources, how was Canada to be compensated for damages? There was also a recognized need for a cooperative research effort on oil spill effects and control, a joint oil spill management plan, and a belief that Canada should be looking at alternate ports in both countries so as to discourage the United States from relying on Puget Sound deliveries. Among alternative sites in the United States, the Canadian and British Columbian governments have consistently favored Port Angeles on Juan de Fuca over Cherry Point on Puget Sound. British Columbia supported the VTM concept with the proviso that it be made mandatory, a condition that remains unfulfilled.

Canada's hopes for an early VTM agreement were dashed for a number of reasons. The U.S. Navy objected to VTM as an undesirable precedent that might weaken the country's negotiating positions at the Law of the Sea Conference, given that it has consistently supported an international nation-states regime promoting freedom of the seas and open, unrestricted passage in all international straits. This is a predictable position to be expected from a major maritime power which in its own interests advocates freedom of movement for all military traffic around the world. The U.S. Navy assumes the Strait of Juan de Fuca to be international, but Canada does not. Further, the navy wanted an exemption for all military vessels in any negotiated agreement. However, the sticking point for a time was the required acceptance by the United States of more stringent Canadian safety regulations, present and future. An answer to this issue could be a reciprocal waiver of regulations, but British Columbia opposed this because of the associated increase in the chance of an accident and spill. These differences were enough to delay signing of the agreement for over two years, until the end of 1979. U.S. Navy objections were ultimately overcome and military vessels must comply with the agreement's provisions except when compliance would interfere with defense operations. On the other side of the issue, Scott and Le Marquand note that

> Canada's credibility . . . has been reduced by [its] decisions to transport its own oil out through the same waters. . . . Furthermore, British Columbia may have to receive its future oil through the same waters if Alberta can no longer supply the British Columbia market. This realization that crude oil transportation is a *common problem* has led Ottawa to concentrate on making the Straits of Georgia and Juan de Fuca tanker route as safe as possible. [Emphasis added][52]

OIL PORT SITING

Because of increased environmentalist attention to the linkage of oil port location to the West Coast oil pollution issue and because of a serious Canadian proposal to build an oil superport for Alaska and overseas oil at Kitimat in northern British Columbia, the focal point of the whole controversy shifted in the later 1970s to the matter of the location of an oil off-loading and transshipment terminal. A spirited debate ensued between competing interests representing several locations among the many alternatives. This debate did not assume an official bilateral character, since neither federal government established a diplomatic position. The Canadian government did set up a formal inquiry, however, at first limited to the Kitimat proposal and then broadened to consider all possible sites. The inquiry examined three main proposals: Trans-Mountain Pipelines and Atlantic Richfield's application to construct a terminal at Cherry Point, Washington;

Kitimat Pipelines' proposal to construct a terminal at Kitimat, British Columbia; and Northern Tier Pipelines' proposal to construct a terminal at Port Angeles, Washington. Major points brought out by this inquiry were that the two U.S. projects (Cherry Point and Port Angeles) have a direct and significant impact on Canada, for their location will determine the vessel traffic routes. Also, the U.S. projects, if completed, eliminate the need for Kitimat and, vice versa, mean that impacts on Canada are much more than just environmental.

However, the inquiry was short-lived and did not complete its work. It was terminated allegedly because a need no longer existed for a Canadian port. Ottawa felt that this work was largely premature, especially in view of the fact that U.S. decision making was moving along quite independently.[53] Environmentalists early recognized that some site, however undesirable, must be chosen, and were adamant on both sides of the border that there must be a single site to serve both countries. There was also an international environmentalist consensus that the site should be on the Olympic Peninsula of Washington at some point from Port Angeles west. The rationale for this was the desire to keep oil tankers out of Puget Sound and away from Vancouver. A British Columbia–commissioned study found that

> Environmental resources at stake are estimated to be five times less (at Port Angeles) than in Puget Sound, twenty times less than in Georgia Strait, and possibly three times less than at Kitimat.[54]

CANADIAN GOVERNMENT CONCERNS

The Canadian government has demonstrated great interest in the concept of risk assessment along both coastlines. Studies were carried out on the Atlantic Coast to demonstrate (among other things) the high risk associated with locating an oil refinery at Eastport, Maine. In May, 1977, Ottawa issued a detailed interagency report under Ministry of Transport (Coast Guard) auspices but involving a number of agencies.[55] The interagency group formed the Terminal Policy (TERMPOL) Coordinating Committee for the Kitimat Proposal, and included in their TERMPOL Assessment detailed consideration of ship-caused pollution, tanker design and construction, manning, tanker equipment, operating procedures, composition of tanker fleets, the issue of flags of convenience, control of tankers calling at Kitimat, the chartering of tankers, and claims for pollution damage. The committee also inventoried the physiography, terminal facilities, and aids to navigation, and made recommendations on contingency planning, environmental and socioeconomic assessment, and optimal routing.

The United States has not devoted such attention to coastal oil spill risk assessment. The scope and quality of this work leaves no doubt that Ottawa is serious and highly committed to optimal decision making in this area. It appears that the Canadian federal commitment has been of a higher order than that to be found in Washington and bespeaks Canada's capability to develop a stronger, better-founded diplomatic position than the United States at such future time as this becomes necessary. The committee made major recommendations, including three sets of specific routes into Kitimat and a fully implemented mandatory vessel traffic management system. However, the fruits of this study have not yet been reaped because of the withdrawal of the applicant (a consortium called Kitimat Pipeline Ltd.), as well as a decision of the Canadian cabinet, upon National Energy Board recommendation, that owing to lower anticipated demand and improved supply from oil sand "it simply was not necessary to discuss the relative merits of new import facility sites on the two coasts."[56] Thus, the Canadian concept of an international oil port at Kitimat, broadened with the short-lived 1977 Thompson Oil Ports Inquiry's widened mandate to include all possibilities, came to an end in early 1978 with Prime Minister Trudeau's remark, "if it is just for the Americans my off-hand attitude is why don't they build a port on their west coast if they want to bring oil for themselves."[57]

In late 1977, Senator Warren Magnusson of Washington succeeded in achieving passage by the U.S. Senate of an amendment to the Marine Mammal Protection Act which effectively barred oil-carrying supertankers from Puget Sound after a similar attempt by the state was overturned in the U.S. Supreme Court on the grounds of lack of state jurisdiction. This amendment has two effects: it forced the focusing of greater attention on the building of a terminal at Port Angeles or elsewhere on the Olympic Peninsula and barred Cherry Point or other Puget Sound sites from competing; and it has likely inhibited refinery construction and resulted in the movement of a large number of smaller tankers into the sound for off-loading at Cherry Point. This latter consequence is environmentally somewhat questionable. With the unexpected defeat of Senator Magnusson in the 1980 elections, attention turned to the distinct possibility of a repeal of the amendment and the resultant return of supertankers to the sound as well as possible construction of additional refineries. This may create an entirely new terminal siting situation, and the U.S. Coast Guard has announced that it will now conduct navigation maneuverability tests in the sound.

There have been no significant spills since the various bilateral arrangements were made, and as a consequence public interest in this whole question somewhat dissipated toward the end of the decade.

CONTINUING CONCERNS

Although the West Coast oil tanker issue is more quiescent at the outset of the 1980s than it was a decade earlier, there are at least four remaining concerns that continue to keep this issue active bilaterally: the Northern Tier and Trans-Mountain pipeline projects, salmon, weakness of the VTM, and eventual Asian oil imports.

Both pipeline projects will significantly increase tanker traffic in West Coast waters. However, the Northern Tier project holds no benefits for Canada, only potential harm, and has never been supported by Canada. The Trans-Mountain project, on the other hand, would provide Canada with economic benefits shared with the United States. A spill occurring with this system in place would have much less bilateral diplomatic significance than would a spill in the all-U.S. Northern Tier system. From a diplomatic perspective, therefore, Trans-Mountain is the more desirable alternative, representing less threat of an international incident.

Officials in Ottawa believe that a lucrative salmon fishery is threatened and could be wiped out with only two or three accidents. Ottawa is now spending $150 million over ten to twelve years to rejuvenate the industry and double its value, and hence the salmon issue is not a small one.

The Vessel Traffic Management Agreement is viewed as insufficient by many and is suffering from lack of compliance. While it sets the rules on traffic lanes it does not resolve the problem of the volume and nature of the substances being moved. There is also widespread belief in both countries that compliance is much lower than hoped for. Mandatory VTM appears to be quite a few years away, and the complete installation of the necessary navigational equipment to operate and enforce a mandatory system is still not complete.[58]

Vessels are also coming closer to the coast than they should, to reduce travel time and save fuel, a situation that could easily lead to a diplomatic incident. (British Columbia has not been quick to raise noncompliance concerns, perhaps because of the question of its own jurisdiction in the territorial sea.)[59]

There is increased interest in importing Indonesian and other Asian oil into North America, and a new tanker terminal here might well receive such oil, especially after Alaskan supplies diminish. Although U.S. law requires that oil moving between U.S. ports be carried in U.S. ships, Asian oil would be under no such requirement and would likely be moved in older, less-safe "flag of convenience" vessels which would heighten the chance of accident. Some see required tug escorts as an answer, while others believe Canada should mandate a Canadian pilot and Canadian-approved ship as a prerequisite for entry into these waters, in concert with the VTM system.

The need for institutional arrangements in this area is inevitable to meet the conditions of the future.[60] The VTM agreement already has an institutional aspect, in that it establishes a joint coordinating group consisting of two Coast Guard officers from each country. However, this is less than what supporters of further institutionalization envision.

In addition to Maine–New Brunswick and Washington–British Columbia, a third Canadian-U.S. maritime boundary, that of British

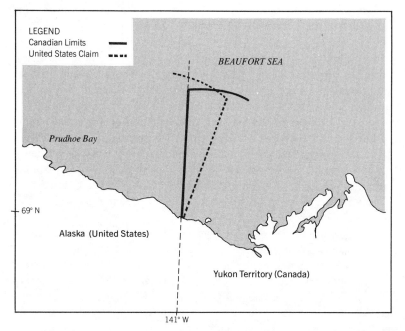

LEGEND
Canadian Limits ▬
United States Claim ▬ ▬ ▬

BEAUFORT SEA

Prudhoe Bay

69° N

Alaska (United States)

Yukon Territory (Canada)

141° W

Map 3. Beaufort Sea—Oil and Coastline Problem. (Redrawn with the permission of the Canadian Hydrographic Service.)

Columbia–Alaska, is in dispute, but has to date occasioned no major bilateral environmental dissension. A fourth boundary, however, that of the Alaska-Yukon boundary in the Beaufort Sea, has had a prominent place in the annals of Canadian-U.S. environmental relations, and will be described in the following section (see map 3).

The Beaufort Sea Issue

The Arctic ice pack has been described as the most significant surface area of the globe for it controls the temperature of much of the Northern Hemisphere. Its continued existence in unspoiled form is vital to all mankind. The single most important threat to the Arctic at this time is the threat of a large oil spill . . . [which] would destroy effectively the primary source of food for Eskimos and carnivorous wildlife throughout an area of thousands of square miles. . . . Because of the minute rate of hydrocarbon decomposition in frigid areas, the presence of any such oil must be regarded as permanent. The disastrous consequences which its presence would have on marine plankton, upon the process of oxygenation in Arctic North America, and upon other natural and vital processes of the biosphere, are incalculable in their extent.[61]

The terrestrial oil and gas discoveries at Prudhoe Bay in Arctic Alaska in the late 1960s and early 1970s spurred exploration interest in offshore outer continental shelf geologic formations in the Beaufort Sea in both Canadian and U.S. waters. Canada moved first. At the mouth of the Mackenzie Delta the drilling season is very short and techniques are largely untried and very expensive, but the determination of Canadian entrepreneurs coupled with that of the federal government was sufficient to launch a serious effort as early as 1974, based on preliminary exploration in the late 1960s. The United States has moved at a much slower pace in the Beaufort Sea, but it appears certain a U.S. program will go forward.

The Beaufort Sea issue is in many ways one of the more intriguing to emerge in recent years. The issue provides diplomacy with more than the usual number of challenges and opportunities, albeit on a low key. Beaufort has not fully developed to challenge diplomats (and may never, for reasons described later), but the potential problem provides a good microcosm of a bilateral environmental issue and essentially defines what constitutes a transboundary environmental conflict.

What really has made the Beaufort Sea an interesting bilateral problem is the complex of "ifs" surrounding the response of the two governments to it. On the one hand, if the Canadian program goes ahead successfully without a major accident, U.S. entrepreneurs stand to gain much of great value from Canadian knowledge and experience relevant to the United States' own Beaufort drilling program. This will undoubtedly save the United States much money and heartache. Canada paves the way and the United States benefits.

On the other hand, if Canadian drilling results in a major spill which either damages U.S. territory or, equally importantly, destroys marine mammals (which are the focus of much attention today), polar bears, and waterfowl, and if this is well publicized (as it assuredly would be) by U.S. media, then the chances of a U.S. program getting off the ground would be set back for years, if not permanently cancelled. This is probably the real reason why U.S. diplomats have exerted some pressure on their Canadian counterparts to enforce stricter safety rules, although significant environmental concerns have probably played a role as well. Canada's risk, of course, is to its own environment and to the harm a sizable spill might bring to Canadian-U.S. relations. However, Canada is running a lesser diplomatic risk. As the U.S. program moves forward and begins to catch up, the bilateral nature of the issue is proportionally reduced. The same amount of damage is, of course, still possible, but the threat of a Canadian accident blocking the development of a U.S. program will disappear, and, further, each nation will have a more equal ability to harm each other in Beaufort, thus neutralizing diplomatic positions.

Hence, the Beaufort Sea debate boils down to the dangers of a blowout or major spill and the threat this may hold not only to the coastal and marine ecosystem but also to its possible effect on the policy climate for drilling, especially in the United States, and the opportunity for the U.S. business community to gain much expertise and knowledge at the expense of Canada. The United States has much to gain or lose.

It has been regarded as quite natural that the United States, which has a significant stake in the future of the Arctic, would wish to follow closely the important Canadian exploration program to insure that its own interests are not in some way jeopardized. Equally, now that the U.S. Beaufort program is moving forward, Canadians will wish to follow closely U.S. exploration and drilling progress with a similar view to protecting the integrity of the Canadian environment. Each country realizes that the other is also concerned to preserve the delicate Arctic ecosystem and that they have a common interest in cooperation in order to maximize the protection given to their shared environment. As an example, a joint oil spill contingency program has been developed by which the resources of both countries would be deployed to clean up any spill involving waters under the jurisdiction of either. Information exchanges have been carried out on a number of occasions at both the cabinet and subcabinet level on many different aspects of the Beaufort Sea issue.[62]

THE SETTING AND THE PROBLEM

A lengthy and much-publicized debate in the Canadian cabinet preceded the granting of initial exploration approval to Dome Petroleum, Ltd. in 1974. This approval was contingent on the provision of an adequate backup drilling system in case a relief well had to be drilled and agreement to undertake comprehensive environmental assessment programs, among other conditions. With progress made in these areas, in early 1976 the federal cabinet approved a new expanded phase of Beaufort Sea exploration by Canadian Marine Drilling, Ltd. (CANMAR), a subsidiary of Dome. This company

> has initiated a unique exploratory drilling program . . . using ice-strengthened drill ships and a growing fleet of supply boats, ice-breakers and ancillary equipment. It is the costliest drilling in the world. Its program, hampered by an extremely short drilling season, has produced nine wells so far, with optimists predicting commercial production by 1985.[63]

The open water season in the Beaufort Sea is a scant four months, with landfast ice extending to depths of over sixty-five feet in winter. Dome, with its CANMAR subsidiary, is the only operator and faces the risk of discharging oil and gas under the pack ice if a blowout should occur near the end of the open water drilling period.[64]

In response to a government proviso, Dome, Imperial Oil, and others spent millions of dollars and carried out one of the most extensive Arctic environmental research programs ever attempted, ranging from baseline studies of flora and fauna and oil spill effects on sea birds and marine mammals to engineering and physical environmental problems, including ice movement, ocean current research, and oil spill cleanup.[65] The comprehensive nature of these studies, as well as the great expenditure incurred, has been made necessary by the fact that there are few areas of the world that pose more unusual problems for offshore drilling.

Oil extracted offshore in the Beaufort would have to be collected and stored prior to transport south.[66] The most likely method of transporting it is by pipeline and tanker. (Research is proceeding to develop a large Class 10 icebreaker so as to facilitate transport by tanker.) Ocean-floor feeder lines would move the oil from subsea wells to shore or offshore storage installations. It is in this phase of the process where the challenges of the elements pose a particular threat, for the lines must be buried at a depth sufficient to protect them from ice keels that gouge the sea bottom to water depths nearing 150 feet. Another threat to these collector lines, and also the well heads, are ice islands and fragments that drift through the area destroying everything they encounter. They are not easily controlled. The weather here is extremely hostile, with low temperatures, high winds, and severe storms, all of which mandate the development of a whole new Arctic technology. Instability of sea floor sediments, sea-ice features, and particularly movement of the polar pack ice and the presence of a thick permafrost layer beneath the sea floor round out the great challenges facing hydrocarbon resource development in this far northern environment.

Industry has thus far met the challenge and has a good record. The very nature of these challenges helps set the stage for diplomatic concern, however, for the cost of overcoming such challenges is sufficiently great to require government subsidy of one type or another and therefore a high level of government participation and commitment. It is also based on an emerging technology which not only entails higher risks than that which is conventional but also holds great opportunity value for those in other nations who are not currently involved in this kind of endeavor but soon will be. In other words, the knowledge and experience gained have a greater-than-usual value.[67]

The most severe ecological impact would result from a blowout, particularly one under the ice (which likely could not be controlled) as it moved toward shore. Oil-soaked waterfowl and pelagic and shore birds, suffocated marine mammals who could not get to the surface to breathe, direct or indirect loss of fish and bottom-dwelling (benthic) organisms, including shellfish, from overexposure to oil, and reductions in the number of polar bears, are all possible in a worst-case scenario. Such blowouts could be caused by human error, equipment failures, or environmental adversities, and under some circumstances, such as in storm conditions or oil under the ice, there is little that can be done to prevent damage. Such a blowout is currently possible in the exploratory stage, as it would be during any commercialization stage.

The principal environmentalist organization concerned about Beaufort Sea drilling (the ecological, not the diplomatic aspects) in either country is the Ottawa-based Canadian Arctic Resources Committee (CARC). There is no equivalent organization in the United States, nor does any other Canadian group approach CARC's level of involvement. Through its newsletter, *Northern Perspectives,* CARC has raised public interest and galvanized citizen concern over Beaufort Sea drilling programs, though its message has been largely restricted to the Canadian audience. Its concerns appear to center philosophically around the long-term effects of drilling on the region's environment.

The Canada-U.S. Environment Committee, composed of environmentalist group leaders from both countries, has included the Beaufort Sea issue on its agenda and has specifically called for postponement of the U.S. lease sale off Alaska and delay in further Dome Petroleum drilling until more information on risk to the ecosystem and technology more appropriate to Arctic conditions become available, and effective spill containment and cleanup techniques are put in place. Environmentalists also call for a cooperative Canadian-U.S. bilateral program of long-range studies, with particular emphasis on whales and other marine mammals, and for the two governments to establish an international board of independent scientific and local experts to oversee the planning and implementation of research in the Beaufort Sea region. They also suggest that the IJC be asked to coordinate the establishment of such a board. Finally, they advocate designating a major portion of the Beaufort Sea as an international marine sanctuary, which would effectively preclude hydrocarbon development activity over the area so designated.

The decision by the two governments of whether to explore further, or to commercialize this resource, while tempered by environmental and social considerations, must also be made on the grounds of need for the energy and suitability of the alternatives.

DIPLOMATIC PROBLEMS

The Beaufort issue entered the diplomatic agenda about 1976, at the time of increased Canadian exploration, when the U.S. government requested consultations concerning measures being taken to avoid environmental damage resulting from a blowout or other disaster in the Beaufort Sea. Since that time yearly meetings have been held between Canadian and U.S. officials, during which the preceding year's experience has been reviewed and the coming year's drilling program and environmental control measures explained in detail. In addition, full ongoing operational reports have been supplied to the U.S. government during the drilling season and detailed briefings have been arranged with respect to specific incidents.

Complicating this issue somewhat has been a jurisdictional difference of opinion within the Canadian bureaucracy between the Department of Indian and Northern Affairs (DINA) and the Department of the Environment (DOE). DINA's orientation is toward economic development and drilling, and at the same time exercises lead jurisdiction in the North. DOE has greater environmental interest and capability, leading to disagreement with DINA over the drilling program and its environmental ramifications. DINA approves of the drilling program, while DOE has substantial reservation. With one agency in support of drilling and the other opposed, External Affairs is caught in the middle.[68] The U.S. State Department does not want the drilling program scrapped, since the United States can learn much from it, but is nevertheless concerned that the driller may be too lax. Forcing the company to exercise increased caution within the regulations has been a challenge for External Affairs owing both to the domestic agency differences and the U.S. State Department.

A further diplomatic problem has been the subject of liability. Canada's request to drillers for $50 million liability has been accepted as a requirement under the Arctic Waters Pollution Prevention Act, but this does not apply west of the 141st meridian of longitude (that is, the United States) and U.S. citizens are excluded from being liable to claim from the Canadian government. Ottawa has responded by requiring $40 million insurance, plus a special $10 million penal bond to cover U.S. citizens. The United States wants reciprocal compensation arrangements in addition, and Ottawa is moving in this direction.

Dome is on the verge of commercialization and passing the point of exploration. Suspicion has also been expressed that the company is on the verge of a very large find, or perhaps has already made such a find. The environmental threat, while not a major issue at the moment, is potentially explosive and is a "time bomb" type of situation.[69] There appears to be more concern over the issue among federal opposition members of the House of Commons than can be found in other quarters, aside from the citizens group CARC. However, since the United States is now moving ahead with its own drilling plan, it might be said Beaufort is becoming less of a bilateral issue per se than it has been, although an accident on the more advanced Canadian side which harms either U.S. interests or common property resources would still have diplomatic repercussions.

One major element of the Beaufort Sea question that has not been addressed is the boundary dispute and related sovereignty issue. This dispute is not close to settlement, but has not affected resource development thus far. However, the disputed area is sizable and is thought to hold great hydrocarbon wealth. Hence, this factor may well emerge as the central issue in the bilateral debate.

In summary, this issue evolved as a bilateral problem because one nation proceeded on a potentially threatening though perhaps necessary course, while the other watched and waited. It can be said that the United States can both win and lose diplomatically no matter which position it takes. Canada faces a risk of lesser proportions. The ideal, perhaps, is for the sovereignty dispute to be settled, by third party arbitration if necessary, and for the drilling program to continue without serious accident. Vigilance and luck will play a role, as will numerous factors involved in Canadian-U.S. diplomatic relations. Only time (a few more drilling seasons) will tell if a "time bomb" is ticking or a rich new industry is being born.

Georges Bank Hydrocarbons

The East Coast fisheries dispute and the attempt to implement a bilateral treaty to provide for joint management of increasingly valuable fish stocks has drawn much national attention in both countries to fishery relations. It has also focused attention on the Gulf of Maine–Bay of Fundy–Georges Bank ecosystem in recent years. The complicated and frustrating fisheries dispute itself is not in the purview of this book, but to the extent that this debate relates to development of hydrocarbon reserves in this area it must be addressed in this chapter.

Because of extension of sovereignty two-hundred-miles seaward with the unilateral declaration of offshore economic zones, a boundary dispute developed in this area as it did elsewhere in the coastal border regions. The boundary dispute has exacerbated an already existing fisheries dispute and has energy ramifications as well, since hydrocarbon interests in both countries have sought exploration leases in anticipation of a find of seabed oil and particularly natural gas. The fishery itself is one of the world's richest, and the hydrocarbon reserves may well be significant, setting the stage for a possible fish versus oil and gas dispute, a marine environmentalist versus hydrocarbon entrepreneur dispute, superimposed on a basic bilateral sovereignty dispute between the two nations. A supplemental issue is transport of imported oil via local shipping lanes, which could conceivably impact across-the-border fisheries in either direction.

The United States and Canada are not close to a solution on the boundary dispute, a matter which both nations agreed at the end of 1981 to refer to the International Court of Justice at The Hague. (Professor Maxwell

Cohen, author of the foreward to this book, was nominated by the government of Canada and accepted by the court as the Canadian judge [ad hoc].) Prior to this referral it was conceivable that the fishery dispute could have been solved without reaching a border-sovereignty solution. The same could never be said, however, for the hydrocarbon aspect, for leases can be let and royalties collected from one nation or the other, but not both. Oil and gas are not mobile, as in the case of fish, and are not renewable or subject to stock management; hence the border must be settled before the subsea reserves of this area can be extracted.

The Canadian position on boundary line demarcation is the same as that found in Article 6 of the Geneva Convention of 1958, which states, "Where the same continental shelf is adjacent to the territories of two or more states whose coasts are opposite each other" and in the absence of agreement, "the boundary is the median line, every point of the baselines from which the breadth of the territorial sea of each State is measured."[70] There is a diplomatic dispute as to whether the two nations are opposite or adjacent, however, and the United States does not recognize the median line as governing. Instead, the United States argues that the deep channels separating Georges Bank from the Nova Scotia shore are the natural dividing line, giving the United States sovereignty over all of Georges Bank, and thus of any hydrocarbons that may exist there. The Carter administration and the Canadian government signed a treaty committing the two governments to resolve this disputed maritime boundary by third-party means. However, the latter treaty has been ratified by the U.S. Senate but the companion fisheries treaty may well never become effective. The Reagan administration consideration of the fisheries and boundary issues has paved the way for movement of the issue to The Hague, and both nations now await the outcome of the International Court's decision.

The direct concern of this volume is, of course, the diplomatic problems that may result from U.S. hydrocarbon extraction or transport which may damage the Canadian fishery or New Brunswick–Nova Scotia shorelines or from the potential damage done by Canadian drilling on the U.S. fishery or New England coast. The fishery dispute over resource allocation has received the lion's share of media attention, the energy and sovereignty aspects have probably received less attention than they deserve, and the bilateral environmental aspect has been largely ignored. Perhaps the real significance of this issue, however, is that since the boundary question has not been settled, the treaty has now triggered third-party adjudication (the International Court) that will force a boundary settlement and thus the allocation of hydrocarbon reserves. This in turn could set in motion a similarly arbitrated settlement of the other three disputed boundaries or alternatively could spur the two countries to reach a bilateral agreement,[71] and for this reason alone may justify both national attention and a place in this chapter.[72]

Other Coastal Issues

There are other bilateral coastal issues that have not been mentioned, basically because they are minor at the present time, although they may create future problems. These include potential oil and gas drilling off the British Columbia–Washington State coast and, though less likely, off the British Columbia–Alaska coastal boundary. Another is the question of Point Roberts, Washington, a peninsula of U.S. territory south of Vancouver that is completely isolated by land from the rest of the United States, and dependent on Canada for most essential services. Point Roberts is a source of discord between Canada and the United States over any number of issues. Both federal governments referred this issue to the IJC in an unusual reference in the early 1970s. The commission recommended establishment of a coastal national park and marine sanctuary there, a concept not well received by either side. The Point Roberts issue is now quiescent, but may reemerge at any time.

　　Near Point Roberts is another potential problem as well as an opportunity. Roberts Bank, Canadian territory, is the site of a coal-handling port through which western Canadian coal is exported to the Far East. The Canadian and British Columbia governments have recently agreed to significantly expand this facility, which could entail pollution ramifications for nearby U.S. waters (not from coal but from the large ship traffic and infrastructural development). Likewise, this expansion could represent an opportunity, in that the United States also appears to need a west coast coal export terminal. If certain hurdles could be overcome, it is conceivable that both governments could operate jointly in this area. The British Columbia–Alaska coastal border has been the most quiescent of the four, but salmon problems have been experienced, and may be impacted positively or negatively by future hydroelectric dam development in British Columbia or increased gold placer mining and dredging in either nation. Finally, any large increase of oil movements into the Portland (Maine)-to-Montreal pipeline, especially as a result of a Canadian-U.S. east-west oil swap, could conceivably increase the chances of oil spills in the Gulf of Maine fishing grounds (although such expansion would at least initially be welcomed by Portland for employment reasons).

　　As with problems in the continental interior, these coastal problems are varied and serious and may well call for basic changes in approach, including new institutional techniques, a subject to be explored later in this book. In any event, settlement of the boundary disputes may well become a prerequisite for any final settlement of these other questions.

Conclusion

Canadian-U.S. bilateral relations in the marine environment focus on hydrocarbon extraction and transportation and on the well-publicized problems of commercial fisheries management and allocation. The various fisheries disputes are outside of the scope of this book, except as they may impact on, or be impacted by, environmental quality disputes. However, oil in the environment, to the extent it becomes a bilateral matter, is well within the scope of the book and dominates coastal and marine concerns. All three coastlines (Atlantic, Pacific, Arctic) have experienced diplomatic problems over oil extraction and transport, with the Eastport and West Coast tankers issues being most significant. All four coastal borders (Atlantic, two Pacific, and Arctic) are characterized by disputed boundaries and sovereignty questions, created or worsened by the unilateral extension of maritime jurisdiction and creation of offshore economic zones by both nations in 1977. (These unilateral decisions also exacerbate fisheries disputes.)

All the foregoing marine and coastal debates thus involve questions of energy policy, pit energy interests against fishery and environmental interests (whether justified or not), and have been handled by diplomats on a purely ad hoc basis. No Boundary Waters Treaty of 1909, no International Joint Commission, no rules of procedure or ordered processes have been available to assist in solving or alleviating these problems. The thus far good record of oil exploration, extraction, and transport has helped avoid problems. Likewise, some problems disappear as bilateral problems when each country reaches the point of representing an equal threat to the other. The present ad hoc system is thus untested, and it is debatable as to whether it is capable of solving real (rather than hypothetical) problems along these coastal margins, should such arise.

In contrast, the next several chapters deal with issues in which the Boundary Waters Treaty and the IJC have long played a role. The most traditional of these is flooding at the border.

NOTES

1. "U.S. Refinery Threatens Fishery and Research," *University Perspectives* (University of New Brunswick), December 6, 1976, p. 6.

2. Ibid.

3. Allen L. Springer, "Licensing the Eastport Refinery: Adequate Recognition of Canadian Environmental Interest?" (Paper presented to the 5th Biennial Meeting of the Association of Canadian Studies in the United States, Washington, D.C., September 29, 1979).

4. Ron Trites, ed., *Summary of Physical, Biological, Socio-Economic and Other Factors Relevant to Potential Oil Spills in the Passamaquoddy Region of the Bay of Fundy*, Fisheries and Marine Service, Technical Report No. 28 (Ottawa, 1974), pp. 5–33 (hereafter cited as *Oil Spill Factors*).

5. Thomas F. Moffatt, "Beauty in Crisis at Passamaquoddy Bay," *Canadian Geographical Journal* 94, no. 1 (February-March, 1977):9, map, p. 11.

6. Trites, *Oil Spill Factors*, p. 52.

7. Ibid., p. 41.

8. Moffatt, "Beauty in Crisis," p. 12.

9. Ibid.

10. U.S. Environmental Protection Agency (EPA), *Final Environmental Impact Statement for the Construction of a 250,000 Barrel/Day Oil Refinery and Marine Terminal—Eastport, Maine* (Boston: U.S. EPA, 1978), pp. v–16.

11. Canada Ministry of Transport, *Eastport Ship Terminal System: Accessibility and Ship Safety Preliminary Analysis and Assessment* (Ottawa: Transport Canada, 1976), pp. 3–8.

12. U.S. Army Corps of Engineers, *Atlantic Coast Deepwater Facilities Study: An Interim Report* (Washington, D.C.: U.S. Army Corps of Engineers, 1973), p. 51.

13. Moffatt, "Beauty in Crisis," p. 13.

14. Trites, *Oil Spill Factors*, p. 149.

15. Ibid., pp. 63–101.

16. Canada Ministry of Transport, *Eastport Ship Terminal*, p. 4.

17. Trites, *Oil Spill Factors*, pp. 141–58.

18. David J. Scarratt, ed., *Evaluation of Recent Data Relative to Potential Oil Spills in the Passamaquoddy Area*, Fisheries and Marine Service, Technical Report No. 901 (Ottawa, 1979), pp. 91–96.

19. Trites, *Oil Spill Factors*, p. 151.

20. Ibid., p. 119.

21. State of Maine, Board of Environmental Protection, Order 4, June, 1975.

22. U.S. Environmental Protection Agency (EPA), *Final Environmental Impact Statement: Proposed Issuance of Federal Permits to the Pittston Company of New York for the Construction of a 250,000 Barrel/Day Oil Refinery and Maine Terminal—Eastport, Maine* (Boston: U.S. EPA, 1978), 2:iv–2.

23. "Exemption Process Stayed as Agencies Reinitiate Consultation on Maine Refinery," *Endangered Species Technical Bulletin* 4, no. 3 (March, 1979).

24. "Eagle Peril Bars Maine Refinery," *New York Times*, January 18, 1979, p. B-8.

25. Brief by the Province of New Brunswick to the Board of Environmental Protection, State of Maine, Public Hearing of Application by the Pittston Company, December, 1973, p. 9.

26. Ibid., p. 3.

27. Margaret Taylor, Environmental Council of New Brunswick, personal correspondence, September 14, 1978.

28. Dr. Owen Washbourn, Director of Environmental Affairs, New Brunswick Ministry of the Environment, personal interview, Fredericton, New Brunswick, August 22, 1978.

29. Ibid.

30. Brian Barnes, Deputy Minister of the Environment, Province of New Brunswick, personal interview, Fredericton, New Brunswick, July 18, 1980.

31. Dallas (Doc) Hodgins, President, Friends of Eastport, personal interview, Lubec, Maine, August 20, 1978.

32. D. Hodgins, memo to Friends of Eastport staff, September 8, 1978.

33. The U.S. federal classification system for air quality, established by the

Federal Clean Air Act of 1970, consists of three classes: 1 (pristine), 2 (moderate pollution levels), and 3 (degraded, generally with much industry). There are active proposals in Congress to abolish this classification system.

34. Canadian Government, Note No. 206 from the Canadian embassy to the U.S. Department of State, June 7, 1973.

35. Letter from Mr. H. B. Robinson, Under-Secretary of State for External Affairs, to Mr. A. F. Kaulakis, Vice President, Pittston Company, December 1, 1976.

36. Ralph Gordon, Environmental Counsellor, Canadian Embassy, personal interview, Washington, D.C., February 25 and May 10, 1977, and Sean Brady, External Affairs, personal interview, Ottawa, May, 1977.

37. David J. Scarratt, Fisheries and Oceans Canada, personal interview, Marine Biological Laboratory, St. Andrews, New Brunswick, August 21, 1978.

38. T. Derrick Iles, Fisheries and Oceans Canada, personal interview, Marine Biological Laboratory, St. Andrews, New Brunswick, August 21, 1978.

39. Canadian Broadcasting Corporation, "Our Troubled Waters," television documentary, 1978. This would involve giving Canada a greater share of the stocks.

40. Flora MacDonald, Minister of State for External Affairs, personal correspondence to Senator Edmund Muskie, Ottawa, November 7, 1979.

41. Joe Clark, Prime Minister of Canada, personal correspondence to Senator R. A. Corbett, M. P., Ottawa, November 20, 1979.

42. Ibid.

43. Springer, "Licensing the Eastport Refinery," p. 4.

44. Horace A. Hildreth and Brice M. Clagett, Brief on International Law for the Coastal Resource Action Committee in the State of Maine, Board of Environmental Protection, in the Matter of Pittston Oil Refinery, pp. 1–5.

45. "The right of innocent passage is the major exception to the general rule that sovereign nations have absolute jurisdiction over their territorial waters" (David Larson, *Major Issues of the Law of the Sea* [Durham, N.H.: University of New Hampshire, Department of Political Science, 1976]). This principle once had its application strictly in the military context but has evolved more broadly over the years. The right of innocent passage through the territorial sea includes traversing that sea without entering internal waters as long as such passage is not "prejudicial to the peace, good order, and security of the coastal State" (Henkin et al., *International Law: Cases and Materials* [St. Paul, Minn.: West Publishing Co., 1980]). Further, the coastal state may not hamper innocent passage, but may take necessary steps to prevent noninnocent passage, traditionally defined as that threatening the security of the coastal state. Latest draft conventions of the Law of the Sea Conferences, which technically are not law, state that a coastal state may adopt laws and regulations with respect to the conservation of living resources and the preservation of the environment, including the prevention, reduction, and control of pollution. This broadening of the term is not yet tested, however, and its force remains to be demonstrated.

46. Springer, "Licensing the Eastport Refinery," p. 5.

47. Hildreth and Clagett, Brief on International Law, p. 11.

48. Barnaby J. Feder, "A Legal Regime for the Arctic," *Ecology Law Quarterly* 6 (1978):808.

49. The Canada Department of Fisheries and Oceans in December 1979

issued a report entitled *Evaluation of Recent Data Relative to Potential Oil Spills in the Passamaquoddy Area* (Fisheries and Marine Service Technical Report No. 901), edited by David Scarratt, which further substantiates the material presented in this chapter and is a compendium of fourteen papers detailing as many different types of oil spill effect in the Passamaquoddy region.

50. *External Affairs News Bulletin* 8, no. 5, January 30, 1980, p. 4.

51. Transport Canada, Vessel Traffic Management System, Western Region, Vancouver, B.C., n.d. p. 6. The international system was ordained by an *Exchange of Notes Constituting an Agreement on Vessel Traffic Management for the Juan de Fuca Region,* Ottawa, December 19, 1979, and its Annex *Agreement for a Cooperative Vessel Traffic Management System for the Juan de Fuca Region.*

52. David Le Marquand and Anthony D. Scott, "Canada–United States Environmental Relations," *Canada–United States Relations, Proceedings of the Academy of Political Science* 32, no. 2 (1976):157.

53. W. R. D. Sewell and N. A. Swainson, "West Coast Oil Pollution Policies: Canadian Responses to Risk Assessment," in *Resources and the Environment: Policy Perspectives for Canada,* ed. O. P. Dwivedi (Toronto: McClelland and Stewart, 1980), pp. 216, 233.

54. British Columbia Land Use Committee Secretariat, *Preliminary Comparison of Kitimat and Port Angeles Tanker Routes* (Victoria, British Columbia: Land Use Committee, 1976).

55. TERMPOL Coordinating Committee for the Kitimat Proposal, *TERMPOL Assessment of the Navigational and Environmental Risks Associated with the Proposal of Kitimat Pipeline, Ltd. to Construct a Marine Oil Terminal at Kitimat, B.C.* (Ottawa: TERMPOL, 1972).

56. Sewell and Swainson, "West Coast Oil Pollution Policies," p. 234.

57. *Vancouver Sun,* February 24, 1978. Sewell and Swainson, in their article on "West Coast Oil Pollution Policies" provide a clear, concise history of this whole matter, and also note the little-publicized fact that in early 1972 the Canadian government had tried unsuccessfully to get the U.S. government to agree to an IJC reference assessing the risks associated with tanker traffic on Canadian and U.S. waters. One can only wonder what impact IJC involvement would have had on the future course of events had the U.S. government been willing to support such a reference. It has been suggested that this U.S. unwillingness led to Canada's 1978 unwillingness to refer another issue, that of the Atikokan power plant, to the IJC and that perhaps both should have gone to the commission for recommendation.

58. Cliff Stainsby, Director, Scientific Pollution and Environmental Control Society (SPEC), personal interview, Vancouver, British Columbia, December 19, 1980.

59. British Columbia and Ottawa have been involved in a lengthy and complicated dispute over jurisdiction on the continental shelf.

60. This viewpoint is also supported by Sewell and Swainson, in "West Coast Oil Pollution Policies," p. 241–42.

61. House of Commons Debates, *Official Report,* April 15, 1970. Thus spoke Prime Minister Pierre Trudeau during the debate on the Arctic Waters Pollution Prevention Act of 1970.

62. Based on interviews with Canadian diplomats.

63. Edgar J. Dosman, "Arctic Seas: Environmental Policy and Natural

Resource Development," in *Resources and the Environment: Policy Perspectives for Canada*, ed. O. P. Dwivedi (Toronto: McClelland and Stewart, 1980), p. 202.

64. Ibid., p. 209.

65. Mary Collins Consultant, Ltd., *Social and Economic Aspects of Dome/ CANMAR's Beaufort Sea Project* (Calgary, Alberta: Mary Collins Consultant, Ltd., 1977); F. F. Slaney and Co., Ltd., *The 1976 White Whale Monitoring Program, Mackenzie Estuary, Northwest Territories* (Calgary, Alberta: F. F. Slaney and Co., for Imperial Oil Co., 1977). *See also* A. Milne, *Oil, Ice and Climatic Change* (Beaufort Sea Project, Institute of Ocean Sciences, Canada Department of Fisheries and Oceans, Sidney, British Columbia); idem, *Offshore Drilling in the Beaufort Sea: A Preliminary Assessment* (Beaufort Sea Project, Institute of Ocean Sciences, Canada Department of Fisheries and Oceans, Sidney, British Columbia), and the following DINA reports: *Annex I: Environmental Review of Beaufort Sea Offshore Drilling (1977); Annex II: Assessment of the Social, Cultural and Economic Impact of Dome/CANMAR Drilling Activities (1977); Annex III: Technical Assessment of Drilling Operations by CANMAR in 1977, Beaufort Sea (1978)*.

66. For further background in the areas of outer continental shelf hydrocarbon extraction under Arctic and other northern environments, see annual reports and fact sheets of Dome Petroleum, Ltd., Toronto; also Kosh, White, et al., *Energy Under the Oceans* (Norman: Oklahoma University of Oklahoma Press, 1973); Douglas Pimlott et al., *Oil Under the Ice* (Ottawa: Canadian Arctic Resources Committee, 1976); and The Technology and Assessment Group, Science and Public Policy Program, *North Sea Oil and Gas* (Norman: University of Oklahoma Press, 1973).

67. The social and environmental impacts of a full-scale commercial drilling program are diverse. While this book concentrates on transboundary environmental issues, certain social impacts, especially on native peoples, cannot be ignored. The people to be most directly affected are the native peoples, the Inuit of two settlements, Inuvik and Tuktoyaktuk, and the nomadic Inuit who still maintain a subsistence hunting and fishing society along this coast and on these waters. Economic impacts on these people of an essentially positive nature include establishment of new business opportunities and expansion of existing business, infusion of substantial public dollars into the region's economy, and, of course, direct employment on the project itself. The availability of new goods and services, including expanded transportation, are added benefits. Naturally these benefits can become disbenefits depending on their impact on the social structure of the native societies and how they are received and used by the native people. The most obvious disbenefits aside from social disruption are impacts on the populations of fish, mammals, and birds upon which the native peoples depend for their subsistence.

68. Based on interviews with Canadian diplomats.

69. Tom McMillan, Progressive Conservative environment critic in the federal House of Commons, is an especially strong advocate of this view.

70. *Convention on Fishing and Conservation of the Living Resources of the High Seas*, done at Geneva, 29 April, 1958, TTAS 5578, 449 UNTS 311 (New York: United Nations, 1958).

71. George C. Van Roggen, Senate of Canada, interview, Ottawa, October 9, 1980, and Vancouver, December 22, 1980.

72. For further information see Hal Mills, *The Strategy for Georges Bank: A Case Study in Canadian-American Diplomatic Relations* (Halifax, Nova Scotia: Dalhousie University Ocean Studies Program, 1981). See also John E. Carroll, "Georges Bank: A Canadian-American Challenge." Paper presented to the American Society for Environmental Education's Technological Conference on Georges Bank Hydrocarbon Exploration and Development, Nantucket, Massachusetts, April 27–30, 1982. (A modification of this paper appeared in *Offshore: The Journal of Ocean Business* 42, no. 5 [May, 1982]:364–71.)

CHAPTER 5

Flooding across the Border

The High Ross Dam–Skagit River issue involves the rights of a U.S. utility to raise the height of an existing hydro dam to generate additional electricity at the cost of inundating a valley forest in Canada. The issue is one of the validity of a contract signed by both sides permitting this development and raises the question of the impact of changing values and attitudes on a signed contract as well as on an International Joint Commission order of approval of long standing. This matter also pits the value of increased electricity (and the environmental and economic costs of alternative forms of electrical generation) against the largely intangible values of a unique valley and its natural ecosystem. Champlain-Richelieu pits quantifiable Canadian property damage without flood control on Quebec's Richelieu River against what appears to be extensive but less quantifiable ecological damage to Lake Champlain and its wetlands and shorelines from a flood control dam on that river, if one were to be built. It illustrates the high values Vermonters place on the lake; their determination to protect it, and the ineffectiveness of the IJC when there is division along national lines.

Transborder flooding differs from coastal oil issues in that territorial sovereignty is not at stake and there is little or no pollution element. It also involves a wholly different set of nondiplomatic actors, a strong role for provincial and state governments, and the introduction of a new actor, the International Joint Commission (IJC). This problem was one of the earliest identifiable types of transboundary environmental issues to arise and therefore one of the first to be dealt with shortly after the turn of the century. It was prominently addressed throughout the negotiations for the Boundary Waters Treaty, ultimately formed a critical part of it, and has represented a significant part of the work of the IJC.

We have seen that U.S. insistence on the application of the Harmon Doctrine meant that a major aspect of flooding is excluded by giving the upstream country exclusive control of all waters on its own side of the line which flow across the boundary or into boundary waters (Article II, Boundary Waters Treaty). This means that the application of the treaty and its instrument, the IJC, were excluded from domestic alteration which, while affecting water levels upstream (on one side of the border), also affected water levels downstream (across the border). However, any obstruction on the downstream side of the border that causes water to back

upstream across the border, whether resulting in flooding or not, is governed by the treaty and thus within the jurisdiction of the commission.

Flooding of the type governed by Article II is typically caused by
downstream flood control (as proposed at Champlain-Richelieu), hydro
power production (as at Skagit and potentially at Dickey-Lincoln), and
irrigation and drainage works of various kinds (as at the Pembilier Dam
and Buffalo–Au Marais sites in North Dakota). Flooding not governed by
Article II can be caused by upstream water diversions, upstream consumptive uses for thermal power development (as at the Poplar plant in Saskatchewan), upstream flood control (as at the Roseau and Red river valleys
in Minnesota and North Dakota), and in the Great Lakes (as with excessive
diversion into Lake Superior at Long Lac–Ogoki, Ontario).

Border flooding disputes are numerous, although many of them are
minor. Collectively these disputes absorb much of the time of diplomats and some of the time of the IJC and can damage bilateral relations.
Major border flooding issues include the Skagit–High Ross Dam in
Washington–British Columbia and the Champlain–Richelieu in Vermont-
Quebec. The Skagit–High Ross Dam debate, smoldering at least since
1941 with frequent outbursts in the past four decades, is the best known
and the most representative.

Skagit–High Ross Dam

The Skagit–High Ross Dam issue in British Columbia and Washington is
one of the oldest and most enduring, as well as one of the best known,
transboundary environmental issues. The background and history of this
issue is surprisingly complicated, especially since the actual stakes have
only regional significance, although the issue does have national connotations north of the border.[1] While Skagit is not quite as much of a cause
célèbre as Garrison, it has been well publicized in Canada for over a
decade and, like Garrison, stands as a symbol of alleged insensitive and
uncaring U.S. attitudes toward Canada. The analogy ends there, however,
for the stakes in Skagit are considerably lower and the area affected much
smaller. Professor Arleigh Laycock of the University of Alberta once
referred to this issue as a "tempest in a teapot,"[2] which was in some ways
accurate, but nevertheless public perceptions and concerns dictate that it be
treated much more seriously. The issues at Skagit involve, in varying
degrees: (1) federal-provincial differences; (2) limits to IJC powers and
flaws in IJC procedures; (3) the importance of citizen involvement; and (4)
long-term planning (Seattle) vs. political vacillations (British Columbia).
Each of these elements plays a role in the unfolding story of the Skagit-
High Ross controversy.

Although the Skagit flooding issue is an old one, its lack of resolution is continuing to create bilateral difficulties. It also brings to the fore

new problems and dimensions not demonstrated in other flooding issues: the validity of an old IJC order not acted upon for many years; challenges to the procedural correctness of the commission; the impacts of changing values and attitudes on a signed contract; and the clash between inexpensive energy and the intangible values of wilderness. Hence, it is by no means a dead issue, and its outcome will undoubtedly influence other similar conflicts which are likely to develop in the future.

In its work on the Skagit issue, the IJC identified a significant contemporary point to be used by dam opponents, for it

> Attached considerable importance to "social and option values," including the value of a wilderness valley to the city dweller who never intends to visit, but simply enjoys knowing that the valley remains in its natural state.[3]

The Skagit–High Ross Dam is a large hydro power dam on the Skagit River which currently backs up water causing inundation of a small acreage of Canadian territory. The current dispute centers around the proposed raising of the dam to generate more electricity, at the cost of further inundation of Canadian (as well as U.S.) acreage. In comparison to this emphasis on the consequences of flooding, less attention has been given in the public debate to some of the basic arguments which have been advanced in favor of the dam. Most basic is the need for the power itself.

The Seattle City Light's service area is one of substantial population and economic growth. The area is also losing some of its less expensive hydro power supplies from the federal Bonneville Project, on which it has depended for over half of this century. Further, hydro is economically very desirable, as most Canadians can well attest, and is viewed even by environmentalists as one of the least environmentally damaging sources of electricity available today. It is renewable, and in this instance no new powerhouse or transmission lines will be required. Also basic to the Skagit case, the public utility responsible has incorporated this dam and river development into its power-generation plans over many years, and has worked diligently to obtain all necessary land, rights, rights-of-way, and permits. It has invested substantial sums of money, always with the understanding it could move forward and rely on this power source when needed and, as per its 1967 contract, has been making agreed upon payments on schedule. Finally, the city did not force British Columbia to acquiesce to an undesirable situation. British Columbia brought its problems upon itself. The record stands clear on these matters.

The Skagit River rises in southern British Columbia and flows south into Washington State, emptying into Puget Sound between Seattle and Bellingham. The river has good hydroelectric power capacity, early recognized by Seattle City Light, which, with great foresight, obtained power rights to the river and began construction of a series of downstream hydro dams in 1927. More recently the utility constructed an upstream dam, Ross Dam, the first to have an effect on Canada. Ross Dam is thirty miles south of the border, but has created Ross Lake, a reservoir that backs up across the border linearly for one mile and inundates 520 acres of British Columbia. This reservoir is now Seattle's primary storage facility for hydroelectricity, and the Canadian acreage underlying its northern end was purchased by the utility in 1929. In 1941 the city applied to the IJC to raise the reservoir's level to such an extent as to back water across the border to 1,725 feet (it is 130 feet in depth at the boundary). The IJC approved the application in 1942 subject to compensation by the city to the provincial government and affected private interests and

> provided that the water level at the International Boundary would not be raised unless and until there was a binding agreement between the City of Seattle and the Province of British Columbia providing for indemnification for any injury resulting in British Columbia.[4]

Provincial enabling legislation (Skagit Valley Lands Act) was passed in 1947 authorizing the flooding, subject to compensation. From 1954 to 1966 the U.S. city and Canadian province entered into annual agreements permitting the flooding that has occurred to date. In 1967 the city and province agreed to permit the city to flood for ninety-nine years the full Canadian portion of the Ross Reservoir to an elevation of 1,725 feet in return for a cash compensation of $34,566 per year. Following this agreement, flooding to 1,602.5 feet, the current level, commenced. (Present flooding extends approximately one linear mile into Canada, whereas permitted flooding to 1725 feet would extend ten or eleven miles into Canada, covering 5,200 acres.) In late 1970, Seattle filed an application with the U.S. Federal Power Commission for an amendment to its 1927 license to construct the final phase (that is, to raise the elevation) of Ross Dam in order to generate additional power and to utilize its rights to flood additional Canadian acreage in the process. In the same year, the British Columbian government announced plans to establish a new provincial park and recreation area in the Skagit Valley to take advantage of the water provided by the proposed dam.

Attitudes toward boundary area resource decisions had changed during the 1960s. These were exemplified by the feeling that the Columbia River Treaty represented a sellout of British Columbian interests. Intensified Canadian nationalism, heightened concern for environmental quality, and the attention focused on North Cascades National Park all contributed to this change.[5]

HISTORICAL OPPOSITION TO THE HIGH ROSS DAM

British Columbian opposition first surfaced in late 1969, initially involving the low compensation figure,[6] but soon broadening into environmental quality concerns.[7] The activist ROSS Committee (Run Out Skagit Spoilers) was formed in Vancouver. Public pressure grew to such an extent that in 1970 federal environment minister Davis was able to persuade the United States to refer the problem back to the IJC, this time as a nonbinding Article IX reference, but with two provisos insisted upon by the United States: that the IJC could not rescind the 1942 order and that the study could only cover Canadian territory.

In turning to the IJC, the Canadian federal government was able to avoid, at least temporarily, the bind it found itself in during 1969–70 facing, on the one hand, an apparently legal contract between Seattle and British Columbia and the earlier British Columbian approval of the dam, and, on the other, an increasing crescendo of environmentalist voices condemning the project and urging its revocation. The decision to ask for an IJC study was a compromise, given the U.S. conditions. However, environmentalists, initially satisfied with the mandate given to the IJC, were incensed to learn that, regardless of its findings, the commission could not come up with recommendations inconsistent with its 1942 order. When the findings of the IJC study were released, however, environmentalists were satisfied with the increased credibility given to their position.

The ROSS Committee and other environmentalists are concerned with several interrelated threats, including the flooding of a unique, and one of the last, flat-bottomed valleys in British Columbia, the loss of river fishing, canoeing, and camping along ten miles of river, and the loss of wildlife habitat. The aesthetic issue associated with necessary lake drawdown is a contributing concern, given the general ugliness of extensive mud flats and stumps.[8]

The IJC study was completed in late 1971, and although it did not make recommendations, it did conclude the economic consequences of the flooding would not be great but that severe environmental and ecological consequences would result in the Upper Skagit Valley.[9]

Few were satisfied however, and the opposition parties in the British Columbia legislature criticized the commission for not doing precisely what it was prohibited from doing, namely, recommending termination of the project. The federal minister of the environment took the study as further evidence that the project should not proceed, while his colleagues in External Affairs soon raised a different concern, that of insufficient payment. On January 9, 1973, the Canadian government sent a diplomatic note to the United States requesting an opening of discussions on this subject. In this note, External Affairs Minister Mitchell Sharp said "the outstanding issue at the moment is not the question of the flooding but the question of the cost and this is what we are discussing at present."[10]

There was hope in British Columbia during 1973 that the project would die and that the only outstanding problem would be one of compensation for the surrender of the rights. Ottawa, recognizing U.S. rights, took the position that British Columbia should manage any compensation arrangements without direct federal involvement, leading to later accusations from the Victoria government that it had been abandoned by Ottawa. The federal government was also concerned over the implications for bilateral relations if Canada were to breach unilaterally a contract it considered legally binding. At the same time, the government was disappointed that the United States was not willing to allow a reopening of the original 1942 IJC order and realized the United States was clearly reluctant to give up Seattle's rights under that order.

THE NEGOTIATIONS

Throughout the 1970s the International Joint Commission favored negotiation between the utility and the province in the hope that the latter would be able to piece together some type of monetary compensation and power package that would persuade the utility to alter its plans and give up its rights to flood Canadian acreage. The stance of both federal governments was to observe from a distance, hoping the affected parties could manage it in their own way.

It has been suggested that Seattle City Light's persistence in protecting its rights derived not so much from a desire to raise the dam as it did from a desire to use this leverage to extract the best possible power trade with British Columbia from British Columbia Hydro's Seven Mile Dam. It does appear that the utility has always wanted energy rather than monetary compensation. On the other side of the border, the British Columbian premier's office argued strongly in 1977 that the crux of the issue was not the economic but the political repercussions. The environmental aspect even took second place to these concerns over the political repercussions of a "sellout" to the United States. The British Columbian environment ministry, however, maintained that the environmental aspects of a loss of the valley was the root of the province's concern. The basic long-standing problem appears to be the legal strength of the U.S. position and the embarrassment Ottawa experiences in trying to defend British Columbia, compounded further by serious Victoria-Ottawa difficulties in other areas (such as natural gas) which spill over into the Skagit problem.

The negotiations between the Seattle utility and British Columbia, which the two federal governments and the IJC put so much faith in, dragged on for several years in the later 1970s and were not successful. British Columbia initially offered cheap power (that is, cheaper than would be available from High Ross Dam), but not in large enough quantities to satisfy Seattle. An early offer (1977) was one of additional power from raising Seven Mile Dam, providing less base power than Skagit, but at a lot cheaper cost. The province also offered additional peaking power from Mica Dam for up to twelve years, but on a returnable basis if British Columbia needed it. This the utility firmly refused.[11] A further British

Columbian offer of electricity was made, part of which would come from the new Seven Mile Dam on the Pend d'Oreille River where "B.C. Hydro spent an extra $10 million to raise the dam 15 feet more than needed to create a reservoir back to the U.S. border."[12] According to Ken Farquharson of the ROSS Committee,

> the British Columbia proposal is that the additional 19 MW of firm energy and 41 MW of peak power provided by this extra output be sent to Seattle, in return for payment of the additional $10 million cost to raise the Seven Mile Dam. This compares with 37 MW of firm energy and 234 MW of peak capacity that Seattle would obtain from raising the High Ross Dam at a cost of $120 million.[13]

British Columbia Hydro could raise the High Seven Mile Dam and flood across the border, selling this extra electricity to Seattle, providing half the energy to be received from High Ross Dam. The canyon to be flooded is a steep one that would have no significant ecological effect from flooding, and the site on the British Columbia–Idaho border would provide cheap energy. Seattle has not shown interest,[14] likely because the offer was inadequate.

The British Columbian government considered offering to trade off its post-1984 right under the Columbia River Treaty to divert the Kootenay River (a diversion which would reduce power generation on downstream dams in the United States) in return for Seattle City Light's surrendering its right to raise High Ross Dam. However, the informal U.S. response was one of disinterest.[15]

With no response from Seattle City Light after several years of offers, a frustrated British Columbia government terminated the rather one-sided negotiations in mid-1980 and asked the IJC to reopen the order of 1942. Thus the continuing efforts of the IJC, supported by both federal governments, to get the province and the utility to reach agreement had failed, and the whole matter was thrown back to the bitter period of the early 1970s, precisely what the IJC had feared as it watched the negotiations flounder during the latter half of the decade.

THE INTERNATIONAL JOINT COMMISSION'S ROLE

Does the IJC have jurisdiction to reinvolve itself, given that it had essentially signed off in 1942? Both the U.S. and Canadian governments appear to agree that it still has jurisdiction, as it itself maintains. However, precisely what does that jurisdiction consist of? Or, more broadly, can we legitimately reexamine and change established past decisions, given the evolution of environmental values and knowledge?

On August 14, 1980, British Columbia officially requested that the IJC exercise continuing jurisdiction,[16] thus terminating negotiations and returning the matter to the commission. The province had concluded that dispute resolution by direct discussions could not be successful.[17] In this request, the province asked the commission to annul or rescind the 1942 order of approval. British Columbia contended that the original and required 1942 hearing was contrary to the treaty and commission rules by being conducted by less than a majority of the commission.

The province contends further that the conditions surrounding the 1942 decision were those of wartime emergency and national security and that no works were ever constructed during this emergency period, therefore nullifying the order. It also notes that environmental factors were not given consideration and that, in 1953, the city raised the level of the river and flooded Canadian territory without Canadian approval and therefore in violation of the treaty (though with later Canadian acquiescence). Finally, and perhaps most significantly, the provincial request states that the 1942 order violated Article VIII of the treaty since it wrongfully delegated to the city and province matters within the commission's own exclusive jurisdiction (the protection and indemnity of Canadian interests). British Columbia also contended that the 1967 agreement on compensation is null because it usurps the commission's jurisdiction by providing for additional flooding in Canada, because it agrees to submit differences to a board of arbitration which has no jurisdiction over such matters, and because the commission never adopted or approved its provisions.

An IJC official acknowledged that it was unusual for the commission to reopen such an old case,[18] but it finally agreed to do so in November, 1974, in a landmark decision. Much later, External Affairs Minister Mark MacGuigan refused to support the British Columbian position initially, saying the province had agreed to be compensated for the project and now must try to negotiate its way out of it—a position consistent with the official Ottawa stance in recent years. Former Environment Minister John Fraser, a Vancouver MP, was angry at the lack of federal support in the dispute and accused the external affairs minister of "trying to cover up on the messy and seamy side of Canadian arrangements."[19] He referred to Canadian diplomats as being fearful of challenging an IJC ruling lest the United States challenge other rulings favorable to Canada.

At the invitation of the IJC, the ROSS Committee quickly wrote a draft statement in response supporting the province, and urging that the negotiating positions of the province and the city should be researched, with the research results to be used as a basis for solution.[20] The ROSS Committee also wants assurance that any alternative power supply provided Seattle by the province in lieu of High Ross Dam be provided at a price that accounts for the social and environmental costs of generating that power. The committee complained that the IJC has not given Canadians the protection to which they are entitled.[21] It also implied that since the 1942 order was a response to the wartime emergency it would become null and void if not used by the end of that emergency, a position not otherwise supported in the record. Finally, the committee argued that it is immoral for Seattle to take advantage of Canadians today by insisting on rights to an agreement made for mutual protection in 1942, given that only justification is their own current self-interest.

There are various interpretations of the current dilemma. Perhaps the strongest legal argument raised in the British Columbian request is that the Boundary Waters Treaty requires that the IJC make proper compensation and that it cannot delegate this task to others. Acceptance of this argument would entail invalidation of the 1967 compensation agreement, which resulted from the allegedly illegal delegation. One view holds that the commission does have the right to reopen the order and to vary it as it wishes.[22] A question is also raised as to the legality of Seattle's negotiation with a foreign government, with or without IJC mandate.[23] It is also contended that the IJC's failure to approve the 1967 agreement is a legal flaw. But since all concerned accepted it at the time, it may well not be the flaw asserted.

An argument can be made that Skagit is a political rather than a legal issue. Viewed in this light, while the province's legal grounds may be weak, the political costs to the United States in soured bilateral relations might be the real issue.

To what extent was Seattle blind to Canadian repercussions of the project? Given such possible blindness, it may have been a big mistake to turn the Skagit negotiations over to British Columbia and Seattle City Light, especially if the two cannot relate to each other.

Some cost has already been incurred by the IJC, for there is anger in the province that the commission did not act to protect Canadian interests when it permitted the province in 1967 to sign an unfair agreement. Nor did the commission, it is alleged, protect British Columbian interests when Seattle violated the agreement and illegally flooded Canadian territory in 1953. The commission is also criticized for inconsistency between the Skagit and Columbia issues in not considering downstream benefits in Skagit but making them a big issue in the Columbia case.[24]

Some British Columbian interests believe that the commission should prohibit the dam but order compensation paid to Seattle City Light for actual costs incurred. Others (including the ROSS Committee) believe the commission should force Seattle to accept British Columbia's offer, perhaps through British Columbia Hydro's raising the Seven Mile Dam and exporting this additional energy to Seattle. Support is strong for mandatory IJC arbitration based on the province's proposals and also, should such fail, for new public hearings.[25]

Not all British Columbians support these views, however. Some note that the existing Ross Dam has created the environment which others now want to protect from the High Ross Dam and fear Canada's recent behavior borders on breach of an international agreement with the United States.[26] Still others believe the politicians and media have made this into the major problem that it has become.[27]

REVIEW AND CONCLUSIONS

The Washington Department of Ecology has become convinced that the bitterness of the Skagit debate is now spilling over into other areas, having delayed talks on possible joint efforts to control flooding on the Similkameen River. The state believes that, since British Columbia has been burned by both the Columbia River Treaty and the High Ross Dam issue, it will not consider any other U.S.-induced flooding of its territory, no matter how justified the reason.[28]

There appears to be a consensus in this long and bitter bilateral debate that, if Seattle City Light is deprived of its perceived legal right to raise the dam, then British Columbia owes the utility compensation because of the province's past practice and earlier acceptance. There also appears to be consensus that the IJC erred in its 1942 decision, though whether sufficiently so to invalidate the order and agreement is debatable, especially in light of British Columbia's long acceptance of the situation. There is also no disagreement that the arrangement results in a very substantial annual savings to Seattle (now estimated at over $4 million annually). However, the various legal accusations by British Columbia (and especially those of commission absenteeism and changing social and environmental values) are weak reeds upon which to build a case. (The question of IJC authority to delegate the crafting of a compensation agreement is another matter.) Finally, all would agree on the precise environmental effects of raising a dam and inundating the valley bottom, though sharp disagreement ensues over whether such action is justified. This is especially true when weighed against the economic and environmental advantages of reliance on this particular type of energy source. Agreement also breaks down on the various motives of Seattle and British Columbia on this matter and what each ultimately hopes to attain.

Be that as it may, the Skagit–High Ross Dam issue has taken on national diplomatic connotations in Canada, a fact that must be recognized by the U.S. diplomats. The issue is not of the magnitude of acid rain, Great Lakes water quality, or even the Garrison Diversion issue, but neither is it a purely local Canadian concern. Thus, it can cause harm, and in this instance perhaps the body most vulnerable is the International Joint Commission. If the IJC suffers a loss of public image, deservedly or undeservedly, then both governments must be concerned, for both lose in the long run. However the matter is resolved, Seattle City Light's rights must be protected. Past practice has supported the utility's claim to those rights, and procedural errors of the commission or feelings in British Columbia do not lessen them. Perhaps the final challenge for diplomats will be to balance the need for this electric power against the costs in Canadian-U.S. relations, while protecting the rights of those to whom they have legitimately been granted and at the same time doing all possible to support the International Joint Commission. The stakes appear higher than one dam and one small valley might at first glance suggest.

The Skagit–High Ross Dam issue achieved another milestone with the issuance on April 28, 1982, of a long-awaited Supplemental Order by the IJC (see appendix 6 for text). This new order recognized Seattle's clear legal claim to a right to raise the dam, and found that the British Columbian request does not constitute sufficient grounds to persuade the commission to grant the province relief from Seattle's proposed action. However, it did find that the Skagit Valley should not be flooded beyond its current level, as long as compensation in the form of money, energy, or other means is made to the city of Seattle for the loss of its valuable right and the electricity that right represents. Interestingly, the IJC announced the formation of an unusual four-member committee, giving the province and the city the right to name two of the members, such committee to oversee the negotiation of an acceptable agreement and report back to the commission. This decision represents the long-recognized desire of the two federal governments to let the province and the city work the matter out themselves, as well as the desire of the U.S. administration to let state and local government play a greater role in decision making. It is to be hoped that this technique will succeed and that a negotiated settlement will be in hand soon.

Champlain-Richelieu

Professor Maxwell Cohen, who chaired the Canadian section of the IJC during much of the history of the Champlain-Richelieu dispute, has written:

> [t]he balance to be drawn between environmental protection through leaving nature alone, on the U.S. side, and flood control structures for protection on the Canadian side, has presented a mix of arithmetic and emotions unresolved to this day.

Approaching the root of the problem, he continues,

> there is very little common cause on this issue since the environmental values are measured so dearly on the U.S. side while the priority of feeling for flood control often has drowned out the more modest environmental voice in Quebec.[29]

The Champlain-Richelieu issue to which he was referring involves a Canadian flood control structure that would alter water levels on U.S.

territory. Champlain-Richelieu offers insight into nationalistically polarized attitudes and demonstrates a case where basic thinking and perceptions differ greatly on both sides of the border. It is similar to Skagit in that
it represents a failure of the IJC to come to grips with a transboundary
flooding question. Also as with Skagit, it is an issue that may never be
satisfactorily resolved for both technical reasons and uncompromising attitudes without recourse to forces external to the regions involved.

At Champlain-Richelieu, riparian flooding, a normal and formerly
acceptable event to area farmers (given valuable silt deposition) became a
hindrance with the development of waterfront housing. An international
problem soon resulted.

Lake Champlain is a large and scenic lake of major recreational
importance bordered by Vermont, New York, and a very small portion of
Quebec. The lake drains northward into the Richelieu River, the entirety of
which is in Quebec. The lake basin includes a large area of adjacent
Vermont, much of which is a floodplain originally formed by the lake.
Vermont's richest agricultural land (and most of the state's flat land), its
largest city and university community, its only shoreline, islands, and most
of its wetlands are all on, dependent upon, or in some way associated with,
Lake Champlain, a fact that has had much to do with Vermont's strong role
in this issue. New York has a small city and college community, Plattsburgh, on the lake, but for the most part this region of New York is
sparsely populated and an economic and political backwater of the state.
Quebec's portion of the lake is small and inconsequential to the province.
The Richelieu River, entirely in Quebec, drains a very rich agricultural
district, flowing past (and threatening in times of flood) a major regional
urban center, St. Jean-d'Iberville. Recently, the riverfront has become host
to small-lot waterfront housing subdivisions for second-home owners from
Montreal. The United States has no stake in the river. These geographical
facts set the stage for this issue.

The origin of bilateral involvement in the issue was the 1937 approval by the International Joint Commission for construction of water
control works in the Richelieu River in Quebec to control flooding of lands
adjacent to Lake Champlain and the river. This Canadian request for flood
relief and the IJC acquiescence to the request were unopposed. The Fryers
Island Dam, constructed just downstream from St. Jean on the Richelieu
River, resulted from this project, but the dredging and diking which would
have made the project operative were not completed because of the onset of
World War II. Thus, the purpose of this original project was never
achieved.[30]

Many dry years ensued, and serious problems did not develop until a
period of high lake water occurred as part of Lake Champlain's normal
fluctuation cycle in the early 1970s. Just prior to this high-water period,
extensive conversion of Richelieu Valley farmlands had begun, with new
second-home housing subdivisions being carved out of riparian farmland in

the immediate floodplain of the river. The high-water cycles of Lake Champlain and the Richelieu River flooding that results from this only significant contributor to its water supply are a natural phenomenon. The normal lake level range is five feet, although it can be greater depending on precipitation extremes. General high-water damages in the river include damages to permanent and seasonal residences and recreational facilities and some shoreline erosion and reduction of agricultural production by delays in spring planting. Low-water damages can also result, including reduced access to recreational facilities, effects on water supply intakes, the aesthetic problem of rotting vegetation, and low-flow-induced water quality problems.

The more housing development in the floodplain in years of low water, the more property damage that was bound to result when high waters returned. Real estate entrepreneurs, perhaps unsuspecting themselves, sold the land in dry years to unsuspecting buyers. These buyers began to feel the effects of high water in the early 1970s and called upon their provincial and federal governments to protect their interests. Hence, Ottawa asked Washington to join in a reference to the IJC. In March, 1973, the commission received from government a reference

> to investigate and report upon the feasibility and desirability of regulation of the Richelieu River . . . for the purpose of alleviating extreme water conditions in the Richelieu River and in Lake Champlain.[31]

The commission was specifically requested to make recommendations on the desirability of operating the 1937 Fryers Island project or alternative works to alleviate the extreme water conditions, taking into consideration water supply and sanitation, recreation, navigation, and environmental factors, including fish and wildlife, wetland reclamation, and other beneficial purposes. It was asked to recommend the most practicable and economically feasible system of regulatory works and methods to alleviate the extreme water conditions, and the costs, benefits, and impacts of each alternative. All this it was requested to do in one year. As the decade of the 1980s began some seven years later, the commission had still not completed its deliberations.

The IJC established its International Champlain-Richelieu Engineering Board in 1973. The board concluded its work in late 1974 and found that regulation of the lake for flood control on the river could be accomplished, but board members differed sharply on the environmental effects of such regulations. This board concluded that the environmental acceptability of the project could not be determined without further studies. After holding public hearings, the commission recommended to the two governments than an intensive study be undertaken "to determine the environmental, physical and economic effects of regulations in both countries."[32] This was accepted and a second board, the International Champlain-Richelieu Board, was established and reported to the commission in mid-1978.

In January, 1976, after conclusion of the first set of IJC studies but prior to the beginning of the second board's work, the government of Canada, supported by Quebec, applied to the IJC for permission to dredge the river, thereby reducing peak water levels during flood, and to construct a fixed crest weir to maintain levels during low water.[33] The application noted that Canadian and Quebec government studies indicate this weir and dredging arrangement "offer the best possibility of rendering a measure of flood control benefit at minimum cost and with minimum environmental disruption."[34] The commission did not grant permission for this work, deferring to its new International Champlain-Richelieu Board and to its own ultimate decision on the whole issue.

The commission and its two boards have concluded that regulation of some sort is desirable. But before treating the findings of the second board, which the commission spent such effort deliberating over, investigation of the role of a variety of actors who have succeeded in polarizing this issue is in order.

ENVIRONMENTAL ORGANIZATIONS

The local Quebec reaction to the IJC recommendations toward lake level regulation were positive, while the New York, and particularly the Vermont, reaction was quite the opposite. This polarization annoyed and even angered affected Quebec residents, who viewed the U.S. reaction as selfish environmental elitism which valued birds and fish over property damage.

Given geographical realities, it is not surprising that Vermonters, with a much greater dependence upon Lake Champlain (over 80 percent of Vermont residents rely on the lake in one way or another), should react much more strongly than New Yorkers to the proposal to alter the lake's levels. Thus, the principal citizen environmentalist opposition formed in Vermont. The Lake Champlain Committee worked consistently through the 1970s. The committee has continually opposed any artificial change in lake levels, insisted upon full and detailed environmental impact statements prior to any actions affecting the levels, and opposed completion of the Fryers Island Dam. It has insisted, as have other environmental groups and the state of Vermont, that strict floodplain zoning in Quebec's Richelieu Valley is the answer. Originally a regional environmental organization fighting water pollution in the lake, the committee with its intense lobbying has now moved into the forefront of this international issue.

The Champlain Coalition has come together expressly to find an alternative to lake-level manipulation for flood control. The coalition is dedicated to floodplain land use management in the basin and believes the

IJC has given insufficient attention to floodplain zoning as an alternative.[35] New York environmentalists have generally supported these two Vermont-based organizations, but have not seen fit to form a group of their own. Quebec environmentalist groups are not involved in the issue.

On the national and international level, the Canada-U.S. Environment Committee (CUSEC) has ascribed sufficient bilateral significance to this issue to find that dam construction will lead to even further floodplain development and harm wildlife habitat, and has therefore urged the IJC and the Canadian and Quebec governments to seek nonstructural alternatives, specifically including zoning and flood insurance, as the solution.[36] It should be noted that one Quebec organization (Société pour vaincre la pollution), fourteen other Canadian groups, and four U.S. groups endorsed this position.

Movement was made in the direction desired by environmentalists when the Quebec and Canadian governments signed a federal-provincial agreement designed to reduce future floodplain development. The agreement provided $5 million to identify and map flood risk areas, prohibited federal or provincial spending in these areas, and encouraged zoning where possible.[37] However, although it may well reduce or discourage future development on the Richelieu floodplain, it cannot affect in any way development which is already in place.

VERMONT

Vermont has been in the forefront of adamant opposition to any structural controls on the Richelieu River that would in any way affect the levels of Lake Champlain, and has stated officially that it cannot abide by the findings of the IJC,[38] finding its documentation "misleading and inconclusive."[39] (Such blanket refusal may be a precedent.) Vermont's principal concerns revolve around lake level criteria needed to provide spawning habitat for northern pike, an important recreational fishery. The state has a basic technical disagreement with the IJC on this matter. First, it disagrees with the commission's findings that the more rapid evolution of wet-meadow ecological succession will provide sufficient habitat for northern pike spawning requirements. Second, the state contends the total acreage loss itself is unknown because of an inadequate data base. Vermont's third technical concern is in the area of net benefits, wherein it questions the whole justification of this project. Put simply, it contends that the costs ($1.5 million average annual project cost) outweigh the annual damages (which it believes to be greater than $1.39 million). The state also contends the cost-benefit analysis does not include lost revenue from the destruction of critical wetlands.[40]

The governor of Vermont, arguing that 22 percent of the state's wetlands would be lost, together with associated waterfowl, fish, and furbearers, has also expressed doubts about IJC findings and suggests that the commission might be undervaluing the basin's wetlands.[41] He complains of incomplete consideration of the true costs to Vermont, although he does not question the claimed benefits in Quebec. He advocates a detailed wetlands mapping and monitoring program, more research and a closer look at nonstructural alternatives in Quebec, including floodplain zoning. Finally, he wants consideration to be given to the purchase of development rights and the relocation and evacuation of existing development on the floodplain, and concludes:

> The extraordinary importance of Lake Champlain to the people of Vermont needs no special explanation to the Board. Our forefathers and their children have found that this lake speaks to them in a virtually sacred fashion of the true meaning of Vermont. Whatever we can do to preserve that feeling is our debt and our duty to the future citizens of Vermont.[42]

Hence, Vermont's strong concerns focus on damage to wetlands habitat and associated recreation values, and its general disagreement with the technical data developed by the commission. Left unsaid are concerns over shoreline erosion, flooding of private property, particularly on Grand Isle and other Vermont islands in the lake, and possible navigation concerns in periods of low water. In all, Vermont's position has been clear, consistent, and absolutely uncompromising. The issue is well known and polarized statewide, and the political leadership has widespread support for its position, not only from such environmental groups as the Lake Champlain Committee but also from the important University of Vermont community at Burlington and from the state's press as well. Compromise in Vermont appears unlikely and the U.S. State Department recognizes this and has generally supported the state's position and interests.

NEW YORK

Like Vermont, New York places a high value on the wetlands to be lost with lake level regulation. New York echoes Vermont's concerns about IJC failure to consider the value of lost wetlands, noting that New York's loss would be 11,400 acres of wetlands representing more than 22 percent of the total loss occurring under high-water conditions.[43] New York's own Freshwater Wetlands Law requires the state to place a high level of protection on these environments and to avoid altering them when their full values are unknown and when there are reasonable alternatives. Thus, New York cannot condone this Champlain Basin wetland destruction and be consistent with its own statute. New York goes a step further than Vermont in questioning the negative effects of flooding across the border in Quebec, noting the type of flood damage sustained in Quebec is not the catastrophic life-threatening type associated with fast rivers and currents. For this reason and because most damages occur to grounds rather than structures, New York believes that property damage figures from flooding in Quebec are exaggerated, further reducing justification for a control structure (since by definition the protection benefits would be inflated as well). New York also calls for additional and more refined information including the addition of replacement cost figures (acquisition and overhead) for wetlands lost through water level regulation before it can take a final position on manipulation of lake levels.[44] New York joins Vermont and area environmentalists in advocating immediate implementation of nonstructural flood control measures in both countries and demands that the IJC withhold recommendation on any structural measures until a U.S.-type environmental impact statement, showing all reasonable alternatives, is prepared.

QUEBEC

Quebec's position is tempered with the reality of existing flooding and property damage. Immediate political pressure within the province changes with the rise and fall of lake level cycles. Quebec finds difficulty in understanding the vigor of U.S., and particularly Vermont, opposition. There is a tendency in Quebec to oversimplify and view the U.S. reaction as favoring fish and ducks over people, just as there is a tendency for oversimplification by U.S. environmentalists and state officials who advocate floodplain zoning as a solution to the problem. Neither position is helpful to the beseiged property owners of the Richelieu Valley who, by commonly held public values in both countries, deserve public protection and consideration of some sort. Hence, Quebec's position has been understandably to protect its people, and there is no reason why the province should be concerned about wetlands or wildlife habitat in the U.S. Champlain Basin (although a small amount of its own shoreline would also be affected).

Another aspect of Quebec's position in this issue may lie in the nature of the people affected. Those who have purchased second homes on the Richelieu's floodplain are essentially middle-class French-speaking Montrealers who, for the most part, are purchasing seasonal homes (or their own year-round homes) for the first time. It is inevitable that comparisons might be made between these people and those more affluent English-speaking Quebecers who earlier built seasonal homes along the shores of Lake Memphremagog, a lake west of Richelieu, likewise on the Vermont border and likewise threatened by conditions (in this case, pollution) from the U.S. side. In the Lake Memphremagog case, the Canadian federal government moved quickly to protect the interests of those Quebec Canadians who were threatened from across the border. Although the question is not entirely a fair one, some may ask, will Ottawa move just as quickly and strongly to protect middle-class French interests as it did to protect more affluent English interests? Hence, the broader issues of French-English and Quebec-Ottawa relations are playing, or may play, a role here.

THE FEDERAL POSITION

Washington must protect Vermont–New York interests from enforced lake level alterations from downstream regulation and Ottawa must protect Quebec rights to implement river regulation. The basic federal response on both sides has been to put the issue in the hands of the IJC where it has remained for most of the past decade. The federal role has been virtually nonexistent and remained so until the commission disposed of the issue.

IJC FINDINGS

The final recommendations of the last International Champlain-Richelieu Board, those which Vermont has so vigorously opposed and which the commission vexed over for two years, include:

- a refusal to recommend a nonstructural alternative by itself, believing only 20 percent of the damage could be eliminated in this way; and recommendations
- to implement a combined structural/nonstructural alternative to reduce flood damages while maintaining the seasonal rhythms of lake levels and protection of the lake and river ecosystem;
- to construct a new gated structure near St. Jean and adoption of a regulatory scheme;
- to implement a flood forecasting and warning system and floodplain zoning regulations;
- to share equally all capital construction and operational costs; and
- to conduct additional environmental studies and monitoring during the first ten years of operation; with any permanent board of control established to contain representatives of area environmental control and management agencies in its membership.[45]

The long two-year commission deliberation, coupled with the fact that both governments turned the matter over to the commission eight years ago, has not been lost sight of by the diplomats: Canadian diplomats have experienced particular problems with Quebec over the long delay, and some feel it would have been better for the IJC to admit it cannot handle the question and thus turn it back to government for action. As it is, with jurisdiction turned over to the IJC, the governments could do nothing, Richelieu Valley residents remained threatened, and those concerned with the lake and its levels remained burdened with uncertainty. Thus, regardless of the wisdom of its final findings, the IJC's image has been tarnished in this case by what many perceive as inordinate delay and the costs in uncertainty which that delay has caused.

In January, 1981, after so many years of study and investigation, the IJC, operating with one vacancy and under the tangible threat of three or four more vacancies, submitted its final conclusions and recommendations on the last day of the Carter administration. The commission stated its difficulties in balancing the interactions between the environmental, physical, and economic aspects of this complex issue.[46] It concluded that the gated water control structure at St. Jean was economically feasible, in that damage reduction outweighed the costs. However, it questioned whether the benefit-cost analysis used truly reflects the social gains and losses to be expected from regulation. Also questioned was the lack of attention given to intangible future options—social preservation values that the report concludes are worthy of separate attention. The commission sided with Vermont's strong arguments for wetlands and wildlife habitat protection and added a role for aesthetic values, noting that these resources were significant in the Champlain-Richelieu system and legal protection for them was consistent with U.S. policy. It also concluded that flow management schemes under the gated structure plan are capable of accommodating environmental criteria.

In its final conclusions, the IJC largely accepted the board's recommendations. However, the commission concluded, perhaps surprisingly, that it should be the job of the two governments to determine the desirability of control works rather than have the IJC make that determination. The governments thus are given the exceedingly difficult task of weighing the intangible socioeconomic values and environmental criteria.

The commission thus came close to concluding this thorny case, then wound up by handing it back to government and failing to recommend its own solution. Its final recommendation reads:

> Although the Commission has concluded that it is technically feasible to operate a gated structure at St. Jean that accommodates the proposed environmental criteria, the Commission was unable to determine the desirability of the gated structure and therefore is unable to make recommendations regarding the regulation of Lake Champlain and the Richelieu River.[47]

The International Joint Commission has clearly made a tangible contribution by overseeing a technical board's conduct of a good study which has helped both peoples, and particularly Vermonters and Quebeckers, to gain a greater understanding of the Lake Champlain–Richelieu system and of the complexities of solving this difficult problem. The commission's work also instilled into Vermonters a greater interest in and love for "their" lake. However, by any reasonable standard, the commission failed in its basic task and, unfortunately, further discredited itself by delaying its "nondecision" for an excessively long time, thus preventing diplomatic negotiators from tackling the issue. It is experiences such as Champlain-Richelieu that lend credence to the negative criticism that is sometimes directed at the IJC, namely, that too much of its work takes up excessive time and too many of its findings are indecisive and inconclusive.

THE DIFFICULTY OF A SOLUTION

Many have despaired of a solution, given Quebec's adamant desire to protect its flood-beleaguered citizens and Vermont's unwillingness to alter its position on lake level regulation which seems to hold no benefits for it. Maintenance of the status quo, through deadlock or for whatever reason, benefits Vermont and New York at the expense of Quebec, since this means no structural regulatory measures, no lake level manipulation, and no flood protection in the high-water cycles. To achieve its goals, Quebec needs a break in the current long-standing impasse, while U.S. interests are protected, albeit weakly, by its continuation.

An answer may lie in some type of U.S. payment to the afflicted Quebec property owners, however indirectly contrived, to enable them to vacate the floodplain and then institute floodplain zoning, converting the land to some use which is not harmed by periodic flooding. It is not irrational to suggest that those benefiting most from this result, Vermont and New York lake users and shoreline interests, should play a special role in raising the funds to enable this U.S. payment to be made. In any event, the major polarization which has developed in the region has reduced IJC effectiveness and may well negate diplomatic opportunities now that the matter is being returned by the IJC to the diplomats. As a result, it may be necessary to create a special body to adjudicate this specific issue.[48] Such a body would reduce surprise and dissipate much of the adversary atmosphere of the present. Ultimately it might be converted into a planning and regulatory authority. Seasoned diplomats might well abhor the proliferation of such lower-level joint institutions on the scene of international relations. Yet, where other lines of communication have failed and polarization has developed, is there an alternative to a more locally based institution? The alternatives may well be as limited as those available to Vermonters, New Yorkers, and Quebeckers adversely affected by the Champlain-Richelieu problem.

Conclusion

The transborder flooding issues are gradually becoming less frequent as appropriate sites for water control are less available; thus water quality is overtaking water quantity on diplomatic and IJC agendas. An exception to this generalization is North America's renewed interest in hydroelectricity and the fact that many previously uneconomic generating sites are now becoming economic as a result of the rising costs of fossil fuel alternatives (for example, the High Ross Dam's quadrupling of its annual value to Seattle in only half a decade). A further exception is the coming development of the Alaska–British Columbia–Yukon border regions, an area with much hydro potential, and the new industrialization of Alberta with its likely impacts on Montana. Hence, agendas will continue to feature some flooding questions, and Article II of the Boundary Waters Treaty will continue to be invoked where appropriate, even if less frequently than in the past.

The experience of the Skagit–High Ross and Champlain-Richelieu transborder flooding issues has taught us that minor flooding issues affecting small geographical areas can easily become national causes célèbres and have impacts on the bilateral environmental relationship well out of proportion to their size, impacts which are long-lasting and possibly unresolvable. The experience also teaches that the IJC does not always succeed, even in this, its most traditional area of transborder activity. The IJC deserves blame in the Skagit–High Ross case for basic errors in judgment and in Champlain-Richelieu for unwillingness (or inability) to face differences squarely and dispose of those differences in a timely manner. The lessons learned from these experiences are important, for similar flooding issues will continue to arise in the future. The tool is there to resolve them, but it must be used well.

Where the tool of the IJC has been used, and rather well, is in the area of Great Lakes levels. It is these levels and the issue of Great Lakes water quality that are the subject of chapter 6.

NOTES

1. For detailed background on the Skagit River, see *Of Man, Time and a River: The Skagit River, How Should It be Used?* ed. by R. L. De Lorme, Occasional Paper #10, Center for Pacific Northwest Studies, Western Washington State College, Bellingham, Washington, 1977.

2. Arleigh Laycock, Department of Geography, University of Alberta, personal interview, Edmonton, Alberta, July 11, 1977.

3. Thomas L. Perry, Jr., "The Skagit Valley Controversy: A Case History in Environment Politics," in *Managing Canada's Renewable Resources,* ed. Ralph R. Krueger and Bruce Mitchell (Toronto: Methuen, 1977), p. 249.

4. International Joint Commission (IJC), *Environmental and Ecological Consequences in Canada of Raising Ross Lake in the Skagit Valley to Elevation 1725* (Washington and Ottawa: IJC, 1971), p. 52.

5. William M. Ross and Marion E. Marts, "The High Ross Dam Project: Environmental Decisions and Changing Environmental Attitudes," *Canadian Geographer* 19, no. 3 (Autumn, 1975):222.

6. Environmentalist Thomas Perry writes: "it was obvious to most that the province had been skinned once again. The yearly rental of seven dollars per acre was less than the Skagit's value as a Christmas tree farm, and $35,000 compared pitifully with the $1 million Seattle expected to save each year by raising Ross Dam instead of choosing an alternative." Perry continues, "if the Columbia Treaty was a bad deal for Canada, the Skagit settlement signed by British Columbia Minister Ray Williston on January 10, 1967 was a dead giveaway. For an annual rental of $34,566 (or its equivalent in power), clearing of the flood basin, and replacement

of an existing road, Seattle Light gained the right to flood 5,716 acres of prime Canadian land and to save itself $1 million per year on the cost of building alternative electric generating capacity'' (''The Skagit Valley Controversy,'' pp. 245–46).

7. David Brousson, personal interview, British Columbia Institute of Technology, Burnaby, British Columbia, December 18, 1980. David Brousson, then a provincial legislator and a long-involved environmentalist, states that the issue started with a single letter to him from a constituent which catapulted into a national issue because the ''right mixture of people and places interacted at the right time.'' Brousson was the needed key provincial politician, and his constituents in the Vancouver area were represented by the soon-to-be first minister of the environment in the Trudeau cabinet, Jack Davis. Brousson and Davis working together were able to interest the prime minister, and it was not long before media coverage was extensive.

8. Ross and Marts, ''The High Ross Dam Project,'' p. 226.

The ecological impact . . . will not be critical in the sense of destroying unique ecological systems. The real question is whether steps have been taken . . . to protect areas with ecological characteristics similar to the Upper Skagit or whether they too will succumb to incremental demands for development. This broader question . . . is the real concern of opponents and is at the heart of their antagonism to raising Ross Dam.

9. IJC, *Environmental and Ecological Consequences.* This, of course, is rather obvious, in view of the fact that alteration impact by inundation is total and results in a completely different aquatic rather than terrestrial and aquatic environment.

10. Canadian Embassy, Note to U.S. Department of State, January 9, 1973.

11. Based on interviews with Canadian diplomats and British Columbian officials in mid-1977.

12. ''B.C. Deal on High Ross in Trouble,'' *Seattle Post-Intelligencer,* September 10, 1979.

13. Ibid.

14. Robin Rounds, Assistant Head, Power and Special Projects, British Columbia Environment Ministry, personal interview, Victoria, British Columbia, December 15, 1980.

15. Kenneth Farquharson, Secretary, ROSS Committee, personal interview, Vancouver, British Columbia, December 17, 1980, and Charles Nash, Vice President for Corporate Affairs, British Columbia Hydro, personal interview, Vancouver, British Columbia, November 26, 1980. British Columbia Hydro has been caught in the middle of much of the negotiation. The crown corporation has no formal involvement in the matter but would necessarily have to provide any power to Seattle offered by British Columbia in negotiation. British Columbia Hydro is philosophically opposed to its own government's position on Skagit, being a consistent supporter of the dam, and believes the province has entered into a legally binding commitment which it must honor.

16. Allan Williams, attorney general of British Columbia, letter to David Chance, secretary, Canadian section, IJC, August 14, 1980.

17. International Joint Commission, "In the Matter of the Application of the City of Seattle for Authority to Raise the Water Level of the Skagit River Approximately 130 Feet at the International Boundary Between the United States and Canada, Request in the Application," August 14, 1980, p.3.

18. Robert Sheppard, "Ottawa Won't Aid B.C. to Appeal Skagit Case," *Toronto Globe and Mail,* October 22, 1980.

19. Ibid.

20. ROSS Committee, *Draft Statement in Response* (Vancouver, British Columbia: ROSS Committee, 1980), p. 10.

21. The response states "There is a strong perception in Canada . . . supported by two unanimous votes of the House of Commons and the direct argument of the Province to the Commission, that the 1942 Order and 1967 Agreement impose an unacceptable loss on the people of Canada and B.C. . . . (T)he public perceives that the Commission has failed in its responsibility to achieve a fair solution, and has allowed the situation to proceed to the point where Canadians feel aggrieved by the one-sided nature of the proposal" (ROSS Committee, *Statement in Response,* p. 4-4).

22. Prof. Charles Bourne, University of British Columbia, personal interview, Vancouver, British Columbia, November 27, 1980.

23. Roger Leed, attorney, personal interview, Seattle, Washington, December 2, 1980. Canadian diplomats also raise the legal question of whether the IJC had a right to ask Seattle to negotiate with a foreign government over an international matter. Interestingly, British Columbia has never raised this issue in its request, and one may speculate as to whether it wishes to avoid this important question for fear of giving Ottawa another reason to assert federal rights at the expense of the province.

24. Farquarhson interview.

25. David M. Brousson, P. Eng., Statement of Response to the Request of the Province of British Columbia, December 11, 1980, p. 4.

26. Parzival Copes, Director of Canadian Studies, Simon Fraser University, personal interview, Burnaby, British Columbia, December 11, 1980.

27. Neil Swainson, Department of Political Science, University of Victoria, personal interview, Victoria, British Columbia, December 16, 1980.

28. John Spencer, Acting Director, Washington Department of Ecology, personal interview, Olympia, Washington, December 4, 1980.

29. Maxwell Cohen, "Transboundary Environmental Attitudes and Policy—Some Canadian Perspectives" (Paper presented to the Harvard Center for International Affairs, Harvard University, Cambridge, Massachusetts, October 21, 1980), p. 37.

30. International Champlain-Richelieu Board, *Regulation of Lake Champlain and the Upper Richelieu River: Report to the International Joint Commission by the International Champlain-Richelieu Board* (Ottawa and Washington: IJC, 1978), p. 1.

31. Ibid., p. 75.

32. Ibid., p. 1.

33. International Joint Commission, press release, Ottawa and Washington, January 22, 1976.

34. Ibid. By mandate of the Boundary Waters Treaty of 1909, such application must be made to the IJC. Article IV states that neither country "will . . . permit the construction . . . of any remedial or protective works or any dam or other obstruction in waters flowing from boundary waters or in waters at a lower level than the boundary in rivers flowing across the boundary the effect of which is to raise the natural level of waters on the side of the boundary unless the construction or maintenance thereof is approved" by the IJC (see treaty in appendix).

35. The Champlain Coalition, *Information Sheet*, Burlington, Vermont, n.d. The coalition is composed of a group of existing organizations, including the Sierra Club.

36. Canada-U.S. Environment Committee, List of Resolutions, Third Meeting, Ottawa, March 16–17, 1977, p. 4.

37. Frederick W. Stetson, "Vermont Environmentalists Welcome Reduced Flood Plain Development," *Burlington Free Press*, Burlington, Vermont, October 10, 1976.

38. Dr. Brendan Whittaker, Secretary of Environmental Conservation, State of Vermont, personal interview, Montpelier, Vermont, June 25, 1980.

39. Vermont Agency of Environmental Conservation, *A Case Against the Regulation of Lake Champlain* (Montpelier: Vermont Agency of Environmental Conservation, 1980), p. 1.

40. Ibid., p. 4.

41. Richard Snelling, Governor of Vermont, Remarks to the International Joint Commission Hearing, September 26, 1978, pp. 5–6.

42. Ibid., p. 13.

43. Peter A. A. Berle, Commissioner of Environmental Conservation, State of New York, Statement of New York State to the International Joint Commission Public Hearing at Plattsburgh, New York, June 7, 1978, p. 3.

44. Ibid., p. 5.

45. International Champlain-Richelieu Board, *Regulation of Lake Champlain*, pp. 3–4.

46. International Joint Commission (IJC), *Regulation of the Richelieu River and Lake Champlain: An IJC Report to the Governments of Canada and the United States* (Washington and Ottawa: IJC, 1981), p. 11.

47. Ibid., p. 24.

48. Justin Brande and Mark Lapping, "Exchanging Information Across Boundaries: The Richelieu-Champlain Experience," *Canadian Water Resources Journal* 4, no. 4 (Fall, 1979):49. Brande and Lapping have written, "we must strive to develop . . . a recognized, continuing institution as will allow us at the outset at least to discuss our interests, ideas and plans on a regular basis. This could be done by treaty, compact and/or simple agreement."

CHAPTER 6

Issues in the Great Lakes

There are two basic and quite different bilateral Great Lakes issues: lake levels and water quality.

The central bilateral issue in Great Lakes levels is the diplomatic or IJC adjudication of a number of divergent interests, some of which have a stake in higher levels all or part of the time, and others of which similarly have a stake in lower levels. Adjudication of external interests whose actions impact the level of the lakes through diversion into or out of the system for various reasons is a further critical element.

In the water quality area, decades of diplomatic and IJC attention and two Great Lakes Water Quality Agreements have still failed to come to grips with the central question of national rights to assimilative capacity of the lakes, and whether such rights should be based on a formula of current population and industrial demand (favored by the United States) or of geography (favored by Canada). Further questions of the funding and construction of sewage treatment plants, bans on phosphorus, levels of treatment necessary, contribution of nonpoint sources, and control of long-range transport of airborne pollutants are also central to the bilateral quality debate on the lakes.

The Canadian-U.S. border consists of 3,145 miles of land and 2,381 miles of water. Most of the latter is in the Great Lakes–St. Lawrence System. The Great Lakes system, the world's largest freshwater inland lakes system, means much to the people and national economies of both the United States and Canada, the basin which it occupies being home to one-third of the Canadian population and one-quarter of the U.S. population. Four of the lakes—Superior, Huron, Erie, and Ontario—and their connecting rivers—the St. Marys, St. Clair, Detroit, Niagara, and part of the St. Lawrence—form the international border from the heart of the continent almost to the Atlantic coast. Thus, there is a high probability of many varied transboundary environmental issues arising along this aquatic boundary. Indeed, the number of these issues, the intensity with which they are pursued, and the economic stakes associated with them could justify a book themselves or dominate this book, if space permitted. One chapter cannot do these matters justice, but it can serve to:

- introduce the Great Lakes as an international basin and the setting for numerous bilateral conflicts as well as opportunities;

- describe an important aspect of the work of the International Joint Commission, historically and at present;
- present the lakes issues in their context as a continuing item on the Canadian-U.S. diplomatic agenda;
- express concerns of industry, using steel as an example, and of environmentalists; and
- give the reader an understanding of the lessons that can be learned from the Great Lakes international experience.

The bilateral Great Lakes issues can be broadly divided into two categories: lake levels and water quality. A third type of issue, that of winter navigation, is a lesser issue, but one that may emerge in the 1980s. (A U.S. proposal to open the lakes to all-season navigation through ice control would have a variety of complicated economic and environmental effects. Its bilateral implications remain potential and may never develop—hence, they are not treated here.) Air quality and acid rain in the Great Lakes will be treated in chapters 10 and 11. The oldest issues are in the area of lake levels, man's impact on those levels, and his unsuccessful attempts to manipulate them to his benefit. These questions will be treated first.[1]

Great Lakes Levels

The earliest of the Great Lakes bilateral issues to emerge was that of lake levels, in recognition of the fact that there is competition for use of the lakes' water, both among users and between the two countries. This section describes the competing uses by category (shoreline property owners, shipping, hydroelectric power, and fish and wildlife). It then introduces the role of the IJC in levels and diversions, describes the 1976 IJC report on lake level regulation, and discusses the Chicago Diversion as a bilateral issue.

Four of the five Great Lakes form the international boundary and, together with their connecting channels and the St. Lawrence River, constitute a shared resource in which fluctuations in one body of water are likely to have repercussions throughout the system. Although Lake Michigan is wholly within the United States and has been expressly excluded (by the United States) from international status, level changes in that lake are intimately related to the rest of the system.[2]

Even if the lakes were not international, there would be competition among user groups. Such competition ignores international boundaries. A survey of the specialized requirements of various user groups leads inevitably to the conclusion that all categories of users in the basin cannot be satisfied at all times. This is especially true on the three unregulated lakes, Huron, Michigan, and Erie. However, there is a consensus that lake level regulation can produce favorable conditions for all uses.[3]

With such competition it is natural that concern should arise over consumption. Canada is becoming concerned about increasing consumptive uses (those uses which represent a net withdrawal of water) of the lakes, most of which is by the United States. For a nation so heavily dependent on the lake levels for hydro power generation, this is understandable. As will be discussed in the following section, there are also water quality ramifications from consumptive use. The population ratio in the boundary water basins is three Americans to every one Canadian. Does this mean that consumptive use of water (or rights to pollute, as discussed later) should be split three to one based on the population ratio or fifty-fifty based on the geographical allocation of the basin? The United States naturally prefers equity (on a population basis, benefiting that country) while Canada prefers equality (on a geographical basis, benefiting Canada). A real political problem results.

INTRODUCTION

The Great Lakes system accounts for a sixth of the world's total freshwater supply. (See map 4 for detail.) Water levels in the system are regulated by natural hydrologic forces with the maximum system outflow ever recorded being less than two and one-half times the minimum recorded outflow.[4] Nevertheless, the fluctuation that does occur has significant impact. This impact was most noticeable during the high periods of the 1950s and early 1970s, which coincided with an extended period of high precipitation, and during the low periods of the 1930s and mid-1960s, which coincided with an extended drought.

In spite of this high degree of natural self-regulation, the U.S. and Canadian governments have attempted to control the outflows of two of the lakes—Superior and Ontario. The former was accomplished by the construction of compensating works at the mouth of the St. Marys River and at Sault Ste. Marie, and the latter by the Moses-Saunders Generating Station and Dam and the adjacent Long Sault Dam.

In addition, the natural lake levels are affected by two major diversions. One is into Lake Superior from Canada (Long Lac, dating from 1939, and Ogoki, since 1943), accounting for 5,400 cubic feet per second (cfs), or 7 percent of the average outflow of the lake. The other is a major diversion from Lake Michigan, the Chicago Diversion, dating from 1848 and accounting currently for 3,200 cfs. The Canadian diversion into the lakes is used for hydroelectric generation at sites upstream on the Albany River before draining into Lake Superior, whereas the U.S. diversion out of the lakes is primarily to dilute sewage effluent from Chicago while carrying it south into the Illinois-Mississippi river system, and secondarily to enhance navigation and to provide a small amount of hydro power. Fluctuations which result from these diversions affect the three unregulated lakes, Huron, Michigan, and Erie, while lakes Superior and Ontario are

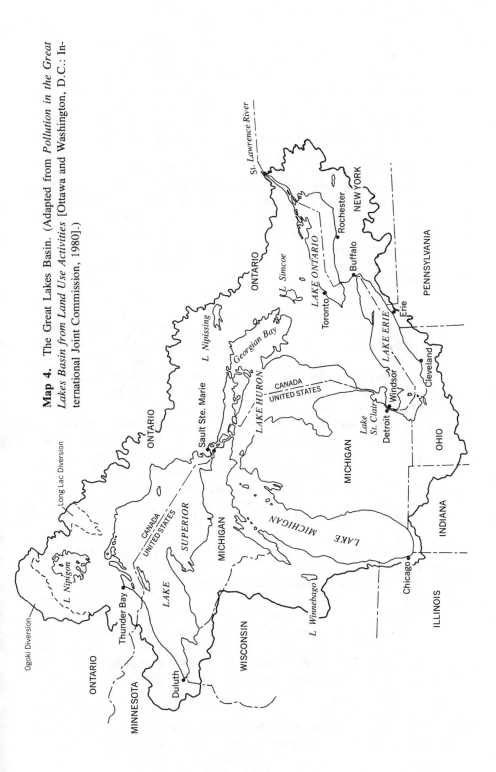

Map 4. The Great Lakes Basin. (Adapted from *Pollution in the Great Lakes Basin from Land Use Activities* [Ottawa and Washington, D.C.: International Joint Commission, 1980].)

unaffected as their regulation plans account for the diversions. A small number of other diversions have only negligible impact.

Consumption of Great Lakes water constitutes a diversion of sorts. The principal consumptive uses in order of magnitude are municipal and rural uses (1,285 cfs); and irrigation (145 cfs). The consumptive use total was 2,770 cfs in 1977, and the IJC estimates that this will increase to 6,000 cfs by the year 2000 and to 12,000 cfs by the year 2030.[5]

COMPETING USES FOR GREAT LAKES WATER

Competition for Great Lakes water does not divide along national lines— except for dilution capacity, which will be discussed in the water quality section—but rather by user categories, with transborder alliances being common today. There are four principal user categories, each defining optimum lake levels in a different way: shoreline property owners, navigation/transportation interests, hydroelectric companies, and fish and wildlife interests.

Shoreline Property Owners

Vulnerable shoreline property includes homes and cottages, port facilities, marinas, recreational developments, and various industrial and municipal facilities, all of which suffer from excessively high water (storm surges, flooding, erosion) and some of which suffered from low water (unaesthetic mudflats, loss of navigational access for recreational boating). Shoreline property damage constitutes the single greatest economic problem resulting from water level fluctuations.[6] While both U.S. and Canadian interests suffer, the U.S. shoreline is more intensively developed and also, in many places, lower in elevation; it is thus subject to greater economic damage. The shoreline interests are best served by maintenance of neither exceptionally high nor exceptionally low water but rather by stabilization of the levels and maintenance of a medium level.

Shipping

Each year over 100 billion ton-miles of waterborne freight destined for domestic and foreign markets passes through the Great Lakes system, including a third of all Canadian ship cargoes.[7] These navigation/transportation interests, both U.S. and Canadian, are best served by high water, which allows ships to transport heavier cargoes, reduces the frequency of trips, and saves on costs. Although commercial navigation is of great importance to both countries, a greater share of Canada's total shipping occurs on the lakes than is the case for the United States. Hence, the system is relatively more important to Canada.

Hydroelectric Power

As U.S. residents sometimes need to be reminded, the Great Lakes are a significant source of hydroelectric power. No less than 8,000 Mw of generating capacity are vitally affected by Great Lakes levels, including 3,200 megawatts in the United States and 4,800 megawatts in Canada.[8] (New York is likely to increase its take from Niagara Falls soon, thus raising the U.S. total hydroelectricity demand from the system, but this will still be surpassed by Canadian demand.) Given the higher Canadian figure as well as a relatively greater national dependency on the system, Canada is more vulnerable to the loss of this generating capacity or any portion thereof. In common with shipping interests, the hydroelectric companies are interested in maintaining high lake levels to insure minimum flows necessary to meet demands for power. Unlike shipping interests, however, hydro power needs are for high winter levels, while shippers' needs for high water are in summer.

Fish and Wildlife

Any drainage basin as large or as ecologically diverse as the Great Lakes will contain habitats for many species of fish and wildlife; many species and their habitats are affected by lake level fluctuations and by associated exposure or draining of wetlands. Since the requirements for individual species vary considerably, different fish and wildlife interests require a different definition of optimum lake levels. Levels management, therefore, must reconcile complex competing demands to satisfy fish and wildlife interests, demands which take no account of the international border.

THE IJC, WATER LEVELS, AND DIVERSIONS

Management of the water levels in the Great Lakes is carried out under the provisions of the Boundary Waters Treaty of 1909, and for this reason the histories of Great Lakes levels issues and the IJC are intertwined over many years. The treaty formally established, in Articles II through V, specific injunctions against unilateral diversions and obstructions of boundary waters, as discussed in chapter 3, ensuring the IJC a steady and continuing portion of its work load.

The International Joint Commission has been involved with the regulation of water levels and flows since the very first year of its operation in 1912. Questions that have been referred to it include:

- Detroit River and St. Marys River flow questions (1912 and 1913, respectively);
- St. Clair River flows (1916);
- Great Lakes–St. Lawrence hydroelectric power questions (1918), and navigation (1920);
- Niagara River navigation (1925)
- Niagara Falls preservation of natural beauty (1950);
- Lake Ontario levels and St. Lawrence power (1952);
- Niagara Falls remedial work (1961) and shoal removal (1963);
- Lake Erie and the Niagara River ice booms (1964);

- general Great Lakes levels (1964);
- American Falls on the Niagara River (1967);
- Raisin River power diversion from the St. Lawrence (1968);
- emergency regulation of Lake Superior (1973); and
- Lake Erie regulation, Great Lakes diversions and consumptive uses studies, and Great Lakes Levels Advisory Board activities (1977).[9]

The history of IJC Great Lakes levels activity is thus a catalog of all bilateral activity in this area. The commission's Great Lakes Levels Advisory Board, together with its boards of control on lakes Superior and Ontario, the St. Lawrence River, and the Niagara River, its study boards on Lake Erie, and its boards on the subject of diversions and consumptive uses are the principal international institutional mechanisms operating in this area on the lakes today.[10]

IJC activity in this field began to intensify with high water levels of the early 1950s as a result of the inevitable complaints of shore erosion and flooding damages and the consequent political pressure to ameliorate the situation.[11] By the summer of 1952 the two governments had agreed to an Article IX reference directing the commission to study the problem of reducing the extreme flood stages on Lake Ontario and complementary applications under Article VIII for approval of the construction of hydro power works in the St. Lawrence River. The commission established the International Lake Ontario Board of Engineers to conduct studies. It then approved construction of the hydro power works with certain provisions, including the right to establish another board—the International St. Lawrence River Board of Control—to advise the commission regarding the maintenance of an optimum level for the waters of Lake Ontario and the St. Lawrence River. These boards' studies led the commission to develop an operating and flow regulation plan for the St. Lawrence, the outlet of Lake Ontario. By 1960 it could boast that that year marked

> the first occasion on which man has taken charge of the flows of the great St. Lawrence River and has redistributed those flows from a state of nature.[12]

During the early 1960s, excessively low lake levels began to occur, and once again demand for remedial action was heard. Michigan sought diversion of more water into the system, and U.S. IJC chairman Teno Roncalio wrote to President Lyndon Johnson suggesting that the time had come

> to encourage the reversing of rivers in order to contribute to the storage capabilities of the Great Lakes instead of continuing tapping water from the Great Lakes.[13]

Canada originally rejected the inclusion of diversion of Canadian waters southward in the reference, viewing this as a domestic question, but ultimately agreed to referring it to the commission for future study. On the other hand, Canada was willing to allow the IJC to study levels questions immediately. U.S. authorities were so eager for such studies to begin that they agreed to include Lake Michigan (previously off limits) and the Chicago Diversion in the terms of reference.[14] In the face of such eagerness on both sides, a further IJC lake levels reference was forwarded in late 1964, simply directing the commission to study and recommend what could be done to regulate the levels so as to reduce the extremes. To carry out this mandate, the commission established its International Great Lakes Board and various working committees.

The history of lake levels manipulation and IJC involvement in this area up to 1964 seemed to prove that boundary water levels in the lakes can be controlled, but only with international cooperation. At that time the further conclusion was drawn that such control can best be achieved through the IJC and its boards and that the two federal governments must be alert to insure that vested interests do not obtain privileged concessions harmful to other equally deserving lake users.

By the time of the release of the IJC report *Further Regulation of the Great Lakes* in June, 1976, the levels of the lakes were returning to normal and the public demand was quieting down accordingly. Inevitably, however, the furor would rise again as the lakes rose or fell. The basic message of the 1976 study was that the public had better be willing to accept the consequences, for the IJC found that lake level fluctuations were primarily the result of natural forces and that only a minimal impact was caused by man's activity. The report furthermore concluded that only limited reductions in levels changes were practicable, owing to the high degree of natural regulation, and it warned that major reductions in the fluctuations in one lake would cause wider variations in outflows among all the lakes. To control such variations would require regulatory works and remedial measures, and the commission maintained that possibly serious upstream and downstream impacts would result from these control activities.[15] This would appear to contradict the earlier conclusion in 1964 that the levels could be controlled by international cooperation alone. We now know, based on the more recent research, that nature stands in the way.

In its 1976 report,

> the Commission concluded that protection from high and low water levels cannot be achieved from lake regulation alone. Protection is best derived from systematic management using all of the tools available. These tools include not only lake regulation but also encompass careful planning of residential, recreational and industrial activities along the shoreline to assure wise use of vulnerable areas.

To assist in the expansion of knowledge to reduce fluctuations, the Commission recommended that it be authorized to study and determine environmental and other effects of limited regulation of Lake Erie; the effects of existing or new diversions into or out of the Great Lakes Basin; the effect of future consumptive use of water on Great Lakes levels and flows.[16]

Thus, once again it can be seen that environmental diplomacy is a moving target. Diplomats must contend with a constantly evolving data base. In the case of lake levels, what was thought of as possible in the 1960s, given only international cooperation, was shown to be impossible by the new knowledge of the 1970s, regardless of international cooperation.

The governmental response to this IJC finding was to issue two new references. The first directed the commission to determine the possibility for, and effects of, limited regulation of Lake Erie. The second reference called for examination of the effects of diversions as well as the effects of existing and foreseeable patterns of consumptive use on levels and flows. This mandate led to the establishment of two currently active study boards: the International Lake Erie Regulation Study Board and the International Great Lakes Diversions and Consumptive Uses Study Board.

In keeping with the commission's recent findings in other flood situations (Roseau River, Champlain-Richelieu), the 1976 lake levels report concluded that land use planning was needed to ameliorate the effects of extreme water levels. This nonstructural approach represents a departure from the commission's traditional structural water control approaches of earlier years and is a tangible sign of the commission's response to modern-day environmental sensitivities. The governments of both countries have concurred that shoreline planning is central to the long-term reduction of impacts and have endorsed the concept of land use regulation.[17]

The Chicago Diversion

The Chicago Diversion issue has been a particularly thorny aspect of the Great Lakes levels debate for many years and emerged in force during the 1950s. The original diversion was established in 1848 without Canadian (or British) consultation. During the 1950s Canada expressed considerable concern over a series of U.S. proposals to increase the diversion of water from Lake Michigan at Chicago. Although not an international boundary water and, therefore, not under the jurisdiction of the Boundary Waters Treaty, Lake Michigan's diversion into the Mississippi drainage basin unilaterally by the United States has been protested against by Canada as it results in decreased hydroelectric power generation downstream in the lakes (affecting both nations but Canada more so) and increased navigation problems. Canadians were supported in their opposition by U.S. power and navigation interests working against Chicago municipal interests. Proponents of increased diversions at Chicago have asserted that increased Lake Michigan water is needed to dilute municipal pollution in the Illinois waterway, to restore fish populations and recreational values, and, most importantly, to provide increased economic gains to downstream shipping and power generation. Diversion proponents also claimed that Canadian interests should not be considered, since Lake Michigan was a national rather than an international body of water and since Canadians had earlier made unilateral diversions into the system at Long Lac and Ogoki.[18] Diversion proponents had considerable clout but were unable to succeed in four consecutive sessions of Congress, at least partly owing to the sustained and concerted efforts of the Canadian government.[19]

Canadian concerns continue over loss of hydroelectric generation at Niagara Falls and in the St. Lawrence River as a result of the Chicago Diversion. The country's current negotiating position makes clear its expectations that compensation will be paid by the United States for all losses incurred by Canadian hydroelectric companies as a result of any increased diversion at Chicago.[20] In a diplomatic note of October 8, 1976, the Canadian embassy advised the U.S. State Department that if a proposed demonstration project involving increased diversions were carried out unilaterally,

> there will be significant adverse economic and environmental consequences affecting particularly Ontario and Quebec. Not the least of these would be a much greater consumption of fossil fuels with serious economic and environmental costs, to replace power lost at Niagara, Cornwall and on the Canadian Section of the St. Lawrence. Very large compensation costs could be involved and *Canada expects that the U.S. would provide such compensation* (emphasis added). . . . By this Note the Government of Canada wishes to ensure that the United States Government is fully aware of Canada's longstanding opposition to unilateral increases in diversions from the Great Lakes system.[21]

Proposals to increase the Chicago Diversion involve such high stakes that they will be a recurrent theme at various times in the future, as predictably as will future natural fluctuations of high and low water conditions in the lakes themselves.

U.S. diplomats and their Canadian peers recognize the direct relationship between Chicago water diversions and the generation of hydroelectric power at Niagara Falls and elsewhere downstream and accept the premise that U.S. unilateral action at Chicago without Canadian consultation would fly in the face of U.S. efforts at negotiation. Such unilateral action, while never brought to fruition, has been attempted on a number of occasions. They also recognize that Canada has, at some cost, foregone possible diversion into Lake Superior at Long Lac–Ogoki for the purpose of power production when the lakes are high, so as to aid U.S. property interests suffering from high water. Hence the United States is indebted to Canada. It is not, therefore, in the United States' self-interest to upset Canadians by depriving them of power by pumping additional water through the Chicago Diversion Canal. It is indeed ironic that Canada may well have to burn more coal at Nanticoke, further polluting New York and Pennsylvania air, or build the Atikokan coal-fired power plant over U.S. objections, if the United States deprives it of needed power through the Chicago Diversion!

The problem for these same U.S. diplomats, however, has been communicating this message to the U.S. Congress and particularly the Illinois and downstream Mississippi Basin congressional delegations, which have shown little sensitivity to Canadian matters in the past. Indeed, they are sufficiently removed geographically to have little reason to be knowledgeable about Canada, and political interests and competing concerns often preclude sensitivity to Canadian needs.

During periods of lower lake levels the political situation is reversed. Demand for increased flow through the Chicago Diversion diminishes since the water is simply not available to supply the engineering works. In these periods U.S. pressure builds to encourage Canada to increase its water inflow at Long Lac–Ogoki to prevent the levels from becoming too low so as to avoid damage to power and navigation interests.

In late 1980 the IJC's Lake Erie Regulation Board concluded against attempting to control the level of Lake Erie, saying it would do more harm than good; it would be better to live with the occasional damage caused by high water rather than draining more water out of the lake. The board estimated that lowering the lake would save about $58 million in erosion and flood damage and expose recreational beaches worth $51 million. However, there would be a $156 million increase in navigational costs to commercial shipping and pleasure boats and a $10 million loss of hydroelectric power production. (These figures are not universally accepted outside the board.) There would also be further environmental costs, namely, wetlands destruction and water quality deterioration. This current Lake Erie case points out the classic trade-offs associated with artificially regulating the lakes for any purpose.

Conclusion

In periods such as the late 1970s–early 1980s when Great Lakes levels are close to average, bilateral disputes are nearly nonexistent. The occasional attempt to divert additional water out of the system via the Chicago Diversion Canal, usually initiated in the U.S. Congress, provides the occasional exception. As lake levels rise or fall, however, the response of various interests as well as IJC adjudicatory activity and research endeavors are predictable. Bilaterally, management of this cyclic problem is what is called for, and the commission generally performs its task well.

Although the actors have been somewhat different, both in terms of those affected and those representing the governments in the work of negotiation, and although the history has been shorter, there is a linkage between Great Lakes water levels and water quality. It is to the complex subject of international pollution and international pollution control cooperation on the lakes that the next section addresses itself.[22]

Great Lakes Water Quality

Few subjects in the history of Canadian-U.S. environmental relations have achieved the level of complexity or the notoriety of the Great Lakes water quality issue. This issue persisted throughout the 1960s and 1970s in various manifestations for one or more of the following reasons:

- the number of people resident in the basin who are affected;
- the major industries affected and the magnitude of monetary considerations at stake;
- the signing of two historic bilateral agreements (the first well publicized, the second distinguised by diplomatic and bureaucratic secrecy); and
- the significant expansion (and regionalization) of the work of the International Joint Commission.

This issue continues to confound the talents of the two nations' best diplomats. A case could well be made, however, that the work of the IJC and the two agreements which it has been charged with implementing are the strongest commitments yet made to orderly cooperation in the environmental field as opposed to "expedient ad hocracy," thereby making the experience particularly important. It is only with the advent of the acid rain and toxic substances concerns of the 1980s that the place of the Great Lakes water quality issue may be challenged on the bilateral agenda.

Great Lakes water quality constitutes a subject of vast dimensions which would be highly complex purely as a domestic issue. As an international issue it may be the most complex topic dealt with in this book, with the possible exception of acid rain. After detailing early concerns, this section examines the two agreements of 1972 and 1978, describing the differences in the two nations' approaches to pollution control, evolving pollution control standards as they apply to the lakes, and various perspectives on the agreements—those of the steel industry, of environmentalists, and of an academic seminar. It concludes with an explanation and evaluation of the final 1978 document, which is now in effect and is used as a model to the world.

EARLY CONCERNS

Bilateral concern about the quality of water in the Great Lakes was first raised indirectly only three years after the Boundary Waters Treaty was signed (1912) and had already had a long-standing history by mid-century.

During the 1950s and early 1960s, pollution worsened as the water quality in the system continued to deteriorate with increasing population and industrial discharge. The consequences of multiple and fragmented jurisdictions were becoming increasingly apparent, and there was growing recognition that the problem could be solved only by a concerted and coordinated effort at all governmental levels. In 1964, faced with continued deterioration of lakes Erie and Ontario and their depository, the international section of the St. Lawrence River, the two governments forwarded a reference to the commission, requesting it to determine whether the lakes and connecting channels were polluted and asking for remedial recommendations. The commission's response of 1970 cited the serious pollution problems, which by now were becoming rather obvious, and substantiated the fact that they were causing transborder injury to health and property in contravention of the treaty of 1909. The major sources and their recipient bodies of water were identified and the commission recommended that

> the two governments agree on the adoption of water quality objectives, programs for reduction of phosphorus discharge, controls and/or compatible regulations on dredging, solid waste disposal, oily, hazardous or toxic materials, and shipping wastes.[23]

As it had done in an earlier reference, the IJC again recommended that it be given the authority for the coordination, evaluation, and verification of the remedial programs and estimated the cost in 1968 dollars as $1,373 million for the United States and $211 million for Canada.[24]

THE AGREEMENT OF 1972

Thus the groundwork was laid for the signing of the historic Great Lakes Water Quality Agreement of 1972 in Ottawa by President Richard Nixon and Prime Minister Pierre Trudeau on April 15 of that year. The agreement was a direct outgrowth of the 1964 IJC reference and the 1970 IJC report, and reflected a joint effort to protect all use of the lakes.[25]

Differences in Approaches to Pollution Control

The intense period of bilateral negotiation preceding the signing of the agreement uncovered a number of fundamental differences in approaches to pollution control between the United States and Canada. Three major areas of difference include: allocation of responsibility for pollution abatement; phosphate reduction in the lakes; and the effort to secure specific commitments on pollution abatement. On the first and most important issue of abatement responsibility, the United States and Canada differed. Canada argued that each country had the legal right to pollute up to 50 percent of a boundary water's assimilative capacity (i.e., half of the pollution the lakes were capable of assimilating). The United States, however, host country of most of the major polluters, recognized that such an allocation would require a massive U.S. pollution control program (and only a small Canadian program). The United States thus rejected the fifty-fifty proposal and held that both countries were obligated to control pollution[26] (with no proportion stated).

Second, phosphate reduction was in the forefront of Canadian proposals throughout the negotiation. U.S. negotiators were unenthusiastic, which partly reflected the Nixon administration's domestic policies. The United States, however, did put forth a weaker proposal for joint efforts to curb this pollutant, which Canada ultimately accepted.[27]

Third, when ascertaining the significance of the debate over specific commitments on pollution abatement, it is helpful to bear in mind that in 1971 only 5 percent of the sewered Great Lakes population of the United States was served by adequate treatment, whereas Canada's corresponding figure was 60 percent.[28] Hence, bilaterally, the two sides started from very different positions. Canadian efforts to obtain specific commitments on pollution abatement were rebuffed by U.S. officials concerned over the high costs. Canada ultimately accepted a U.S. commitment that municipal sewage treatment programs would, if not complete, at least be in the process of implementation by December 31, 1975.

Additionally, a number of lesser issues were considered in the negotiating process, including the possible nature, function, and funding of a proposed IJC staff unit as well as such technical matters as vessel waste regulation. The negotiation of a federal-provincial agreement between Ontario and Canada (a necessary preliminary to Canada's signing of the international agreement) and a variety of domestic policy issues served to lengthen the negotiating period to two years.

Main Features of the 1972 Agreement

Significant portions of the agreement may be found in the appendix. Of central importance were the provisions for the establishment of water quality objectives, a schedule for the reduction of phosphorus loadings, and a nondegradation philosophy. The agreement provided for the revision or addition of objectives over time as conditions warrant, thereby introducing an important mechanism by which the IJC, through recommendations to both countries, would be able to render the agreement more stringent. Additionally, the agreement's Article VI incorporated two references intended to address major areas of uncertainty related to Great Lakes water quality at that time. The first of these called for an identification of existing water quality problems and needed remedial measures for the Upper Great Lakes (Superior and Huron)—the first time these lakes themselves achieved agenda status in the bilateral Great Lakes quality debate. The second reference addressed the pervasive, yet poorly documented, pollution of the lakes from nonpoint sources, calling for an investigation of the nature, magnitude, and possible control measures for pollution from land use activities, including agriculture, forestry, and urban development.

As a result of the 1972 agreement, the International Joint Commission had its role enhanced since it is the principal institutional repository of the international jurisdiction inherent in the agreement's mandate. The IJC was given responsibility for collation, analysis, and dissemination of data relative to water quality objectives and programs, for giving advice and recommendations to government, and for coordinating joint activities, including research. The commission is required under the agreement to report annually on progress achieved, may verify independently the data and information submitted to it, and is permitted to publish and disseminate its findings. Independent verification is a unique feature for an international agreement, which some claim was undermined by agencies in both countries, due to mutual agreement to weaken the IJC's capability in this area.[29] While the consequence may therefore be no verification, the precedent is at least established and may be useful in the future.

Implementation of the 1972 Agreement

To carry out its mandate, the IJC established a Great Lakes Water Quality Board (with subcommittees) to serve as principal advisor to the commission on all matters of Great Lakes water quality; and a Great Lakes Research Advisory Board (to advise on the coordination of research and dissemination of information). The Pollution from Land Use Activities Reference Group (PLUARG) was also established to determine the extent, causes, prevention of, and practicable remedial measures for pollution from land use. It has been involved in assessing problems, programs, and research on the effects of land use activities on water quality in the lakes. It was also instructed to inventory land use and land use practices for trends and projections to 1980 and 2020; to study intensively a small number of representative watersheds to enable extrapolation of data to the entire basin and relate specific land uses and practices to contamination of water quality by Great Lakes streams; and to diagnose degree of impairment of water quality in the lakes resulting from land use activities.

The Upper Lakes Reference Group was formed to carry out on lakes Superior and Huron precisely the same work earlier IJC boards had carried out on lakes Erie and Ontario and on the St. Lawrence River over the years. This assignment represented the first IJC institutional presence on the upper lakes (aside from connecting channels) and their surrounding communities. Finally, the agreement authorized the commission to establish a Great Lakes Regional Office, which was established in Windsor, Ontario, in 1973. This office provides permanent professional and administrative support to the boards and reference groups. The role of the Windsor office is to compile and disseminate data and reports, sponsor workshops and conferences, perform liaison with media, organizations, and citizens, and provide secretarial support. It answers to the commission in Ottawa and Washington equally, provides a liaison between the aforementioned institutions given birth by the agreement, and has a mandate restricted to Great Lakes water quality. Hence, it is an IJC regional office only in the context of the agreement and its mandate.

In spite of the establishment of these various new institutions, the responsibility for implementation of pollution control programs remains entirely within the respective governments. The responsible agencies are Environment Canada, the Ontario Ministry of the Environment, and the U.S. Environmental Protection Agency. To a lesser extent, eight state environmental quality or water resource agencies are charged with incorporating the water quality objectives recommended by the IJC and approved by each government as part of the agreement into a set of legally enforceable regulations.

In spite of complex interagency agreements, ambitious pollution control programs, and the investment of billions of dollars, the implementation of the 1972 agreement can only be called a mixed success. Some would argue, however, that the momentum to construct treatment plants was established as a result, and that the appearance of new problems due to advances in detection technology should not cloud the success of this first agreement. By the mid-1970s some improvements were documented, especially in near-shore areas, but significant new problems emerged.

While municipal and industrial effluent was formerly the main concern, by 1976 this problem was beginning to be preempted by toxics and atmospheric deposition of pollutants (including acid rain). Nonpoint sources (agricultural and urban runoff, for example) were beginning to compete with point sources for focal attention. U.S. domestic politics also contributed to this mixed record, since the Nixon administration's impoundment of appropriated dollars seriously delayed sewage plant construction in Detroit and Cleveland, two of the principal U.S. point sources of effluent, and made the December, 1975, deadline unachievable. Complicated planning and administrative procedures associated with the basic U.S. federal water pollution legislation (P.L. 92-500) caused additional delay in compliance. Meanwhile, Canadian programs were moving along well and Canada was achieving a sewage service rate much higher than that of the United States.

U.S. diplomats note that the higher Canadian percentage of sewered population was predicated on a lower level of treatment (only primary), while the United States was moving more slowly, but in the direction of a higher level of treatment (including secondary). However, levels achieved through primary treatment with chemical addition are often equivalent to secondary treatment. In addition, almost all sanitary and storm sewers in Canada are separate. Therefore, a higher percentage of contaminated flow is treated than in the U.S., where a high percentage of flow often totally bypasses treatment. Hence, Canadian loadings of contaminent are lower on a per capita basis. In addition, Canada controls phosphate in detergent, while the United States relies on treatment in municipal sewage treatment plants. If the plants are inadequate, there is no control on phosphorus loadings. This debate was an undercurrent of the negotiations.

A further problem in implementation of the agreement concerned the proper role of the IJC Windsor Regional Office and its jurisdiction. Munton views this office as having an underlying structural problem attributable to the office's relationship to the board and the commission.[30] The basic disagreement arose over whether the regional office was to serve the Great Lakes Water Quality Board or whether it was to independently monitor the governments' performance and be responsible, and have free access, to the commissioners. This issue became a major point of contention during the later renegotiation of the agreement in 1977 and 1978.

Problems also ensued over the differing national approaches to the control of industrial effluents. In this area the early years of the agreement were devoted to working out individual compliance agreements with industry throughout the basin, whereas the emphasis shifted in 1975 and 1976 toward direct monitoring and enforcement.[31]

THE AGREEMENT OF 1978

As the fifth-year expiration date for the 1972 agreement approached, the new concerns of toxics, atmospheric deposition, and nonpoint sources came into their own and words like *DDT, mercury, PCB, mirex,* and *dioxin* became familiar to followers of the Great Lakes scene. Simultaneously, the Upper Lakes Reference Group was beginning to discover that as much as half of all contaminants in those lakes were being deposited from the atmosphere and that some of this pollution was made up of highly toxic heavy metals such as mercury and lead. Further, realization was increasing that nonindustrialized rural areas were contributing a greater pollution load to the lower lakes than had been previously suspected through agricultural runoff and other nonpoint sources. All of these factors set the stage for the 1977–78 renegotiation of the agreement.

The period preceding the renegotiation has been summarized by Munton as follows:

> The tone of later IJC reports, based on those of its board, gradually became more firm and more critical. In the 1975 version, the commissioners underscored the board's conclusion that progress so far had been ''generally slow, uneven and in certain cases, disappointing.'' They noted among other points that the phosphorus loading reductions agreed to in 1972 were not being met. They began to identify specific

geographic "problem areas" by name, and urged more forcefully the establishment of U.S. regulations on detergent phosphates. The emphases also began to shift from municipal pollution questions. Based on improved data, reports began to focus more on industrial sources and, in particular, on toxic substances, which the commissioners suggested "may well be the most serious and long-term problem governments face in ensuring future beneficial uses of the Great Lakes." The political focus shifted as well. While a few American cities continued to contribute the bulk of the municipal pollution, the two countries were a little more even with respect to sources of industrial pollution.[32]

The Great Lakes Water Quality Agreement of 1972 contained a provision that it would be in force for only five years and was then subject to renewal or termination. The two nations decided to renew under new provisions, but the negotiation was so hampered by procedural delays that it was not completed until well into 1978 and not signed until November 22 of that year.

U.S. diplomats were remarking by 1977 that the United States had expended $5 billion on the U.S. Great Lakes for sewers and treatment plants, far outstripping the Canadians on investment and public works on the cleanup efforts (easy to do, however, when one considers the much greater net U.S. pollution contribution to the lakes) and were noting that the controversial Nixon impoundment of sewer money turned out in retrospect to be desirable since that much money could not have been expended that fast effectively. They cited a poor record for Toronto, Thunder Bay, and other Canadian cities and maintained that in percentage reduction of point source pollution discharges the United States was well ahead. Simultaneously, Canadian diplomats were showing concern over the slowness of U.S. sewage construction (attributed partly to the Nixon impoundment) and were recognizing that the basic makeup of the sewerage had changed from municipal to industrial with increasing evidence of the importance of non-point sources (such as agricultural and storm water runoff).

Evolving Pollution Control Standards
By mid-1977, the United States was proposing a change in control philosophy to be applied to effluent standards,[33] while Canada was opting to maintain the general water quality standards philosophy. If the change desired by the United States took place, further development on the Ontario shoreline of Lake Erie would be prevented, since the United States was already using well over 50 percent of its would-be allotment in the lake (or over half of the lake's assimilative capacity) under such a plan. Further, the U.S.-backed effluent standards approach would require the same level of

pollution treatment on either side of the lake and in all areas. Canada feels this is unnecessary and that its lakeshore communities cannot afford it. (It was not in any event being applied uniformly on the U.S. side, so Canada questioned the fairness of the United States in asking Canada to subject itself to such stringency.) Finally, the U.S. proposal was based specifically on U.S. domestic law (P.L. 92-500), and Canada does not wish to subscribe automatically to U.S. laws.[34] Another difference in philosophy was the U.S. desire to include tributaries (again reflecting domestic law), whereas Canada opposed such inclusion, believing that what goes on in domestic tributaries is not international business as long as the lakes come out clean.

Further, the United States at that time leaned toward closing the regional office at Windsor (the EPA wanted it closed, some allege, to move its activities to EPA's Great Lakes office at Chicago), a position strongly opposed by Canada. Also, in Canada's view, the United States was moving too slowly on phosphorous and detergent control, thus damaging the lakes more than necessary. On this later issue, Canada went the route of limiting phosphorus in domestic laundry detergents, the quick way of achieving the goal of phosphorous reduction, while the United States preferred the slower method of building treatment plants to contain it (which is now the prevailing Canadian approach as well). The United States did not accept the IJC recommendation to ban phosphorous from the lakes (even though a few U.S. states and municipalities have tried it). (Many detergent manufacturers have responded to public pressure by eliminating phosphorous from their products.) Conversely, on the matter of commercial vessel waste discharge into the lakes, the United States wanted a complete ban, while Canada wanted to permit some discharge. The IJC recommended against any discharge, and Canada did not accept this recommendation.[35] Canada's diplomats also viewed the U.S. system as being more stringent and having more statutes, but the U.S. was being slower in implementation than Canada. They also suggested that, while many regulations appear to be more stringent, they have poorer overall effects environmentally.

Features of the 1978 Agreement

With the signing in November, 1978, of the second Great Lakes Water Quality Agreement (see appendix), Ottawa and Washington reaffirmed their determination jointly to restore and to enhance water quality in the Great Lakes system. Importantly, by 1978 the idea of taking a more comprehensive approach to issues (that is, an ecosystem approach) vis-à-vis end-of-the-pipe waste treatment had emerged. The contrast between the two approaches is growing steadily and could emerge as a major issue in the 1980s. Reflecting on this signing, the IJC observed in 1980:

The Agreement reflects increased knowledge about a) the presence and effects of toxic and hazardous substances; b) pollution from the air to the water; c) the extent of land drainage pollution of the Great Lakes; d) and a significant improvement in understanding the technical and scientific aspects of water quality since the 1972 Agreement was signed.[36]

During formal negotiation of the revised agreement, the aforementioned difference in diplomatic views played a prominent role. The strong U.S. desire for effluent standards such as those found in P.L. 92-500 was pitted against the Canadian objectives philosophy, with Canada asserting that U.S. laws cannot be applied extraterritorially, that such stringent controls were not needed in Canada, and that enforcement of stricter standards on Canadian industries was regarded as unlikely, given Ontario politics. The U.S. proposal was ultimately rejected.

The further U.S. desire to include tributaries, rejected by Canada on sovereignty grounds (and likely with sensitivity toward potential problems in federal-provincial relations), was accepted in modified form, and the IJC mandate was thus spatially expanded in the basin. In fact, only those portions of tributaries near the lakes are included, and the federal governments may not "poke around" upstream.

The role of Canadian provinces assumed a new status under the 1978 agreement, for on the Canadian side a federal-provincial board has been established, under a federal-provincial agreement, to assure close cooperation. Ottawa cannot work with the United States without the full partnership, operationally if not formally, of Ontario. The federal government can provide administrative and financial support to the province, but when it comes to operations, and also to professional expertise and research capability, the province is in control.[37]

The fate of the IJC's Great Lakes Regional Office at Windsor became one of the most sensitive of the bilateral differences. The U.S. proposal to discontinue it outright was so strongly opposed both by Canadian and U.S. environmentalists that it was withdrawn and replaced by a modification of its mandate, namely to report to the IJC boards on administrative and technical matters and to the commission itself only on public information services. Canadian Chairman Maxwell Cohen strongly rejected this change, and Munton points out

Cohen's most basic objection was that the commission's ability to review independently the governments' progress under the agreement depended on its direct link with the Regional Office.[38]

Canadian Commissioner Keith Henry also criticized the attempted modifications of the Windsor office as an emasculation of the IJC and stated on CBC nationwide radio that he was angered because, by extending the board's authority over the regional office, "the governments are taking away our inspectors,'' thus precluding the IJC's independent surveillance and verification on the lakes.[39] A compromise ultimately resulted from this (at times) bitter debate—simply to review the office and all its functions, deferring any hard decisions to a later date. (A retired IJC official of many years' experience predicts that the regional office will fail because it has two masters—the commission and the boards. Some view it as merely de facto now, with the staff mere secretaries to the boards and their professional expertise ignored.)

Three Perspectives on the Agreement

The Steel Industry. There was much interagency review and internal disagreement within both national delegations—disagreement which had to be settled internally prior to acceptance of a national negotiating position. However, most public attention at the various public meetings, hearings, and statement submissions focused on the positions of industry and environmentalists on various points. One of the more representative of the industrial groups was a Canadian organization, the Ontario Integrated Steel Industry, representing three major steel companies, all located on the lakes: Algoma Steel Corporation at Sault Ste. Marie, Ontario, and Dominion Foundries and Steel, Ltd. (Dofasco) and Steel Company of Canada, Ltd. (STELCO), both in Hamilton, Ontario. These companies combine to produce 80 percent of Canada's steel and employ forty thousand people, a substantial percentage of the Canadian work force. This industry spent over $160 million to reduce water pollution from 1960 to 1978.

The concerns of the steel industry related to the agreement were expressed in the document *Comments of the Ontario Integrated Steel Industry on Canada-United States Agreement on Water Quality in the Great Lakes,* issued on July 14, 1977. The document was submitted to both the federal and provincial governments in Canada and outlined what this industry perceived as strengths and weaknesses of the 1972 agreement. Concerning strengths, the industry commended progress toward improved water quality achieved under the 1972 agreement and endorsed the concepts of nondegradation of the lakes, the designation of certain local areas as problem areas, and of mixing zones (that is, the recognition that areas exist at or near pollutant outfalls where specific water quality objectives cannot be attained by reasonable and practical engineering measures).

With respect to weaknesses, however, the industry challenged four aspects of the 1972 provisions and proposed recommendations for the 1978 agreement. First, the industry expressed concern over the setting of specific water quality objectives for "the most sensitive and beneficial use."[40] Noting the potential social and economic dislocation associated with the attainment of perceived unrealistic objectives, they called for impact studies to determine the social, economic, and environmental implications of establishing specific objectives. Second, the industry questioned the use of water quality objectives as an enforcement tool, pointing to the need for compromises and exceptions in instances where

- there is an "overlap of mixing zones";
- the "assimilative capacities of water bodies are unknown"; and
- "upstream and downstream users and dischargers must meet similar objectives."[41]

Third, while endorsing nondegradation in the 1972 agreement, they opposed efforts to substitute the principle of water quality enhancement in the 1978 agreement. In the words of the industry,

> enhancement of water quality, although a desirable environmental goal itself, may well inhibit or prohibit the attainment of other social and economic goals of equal or greater importance if carried too far.[42]

Fourth, the industry called for greater flexibility in "locating mixing zones as well as allowing for concentration gradients inside such areas."[43] They cited the inability of state-of-the-art techniques to specify the size, shape, and exact location of mixing zones, as called for in the specific water quality objectives. (However, such specification inevitably appears to be a political decision that cannot be determined on technical grounds alone.) They asserted that unique and changing natural features preclude a precise definition of mixing zones as well as the pollutant concentration gradients within.

In a later document, entitled *The Ontario Integrated Steel Industry's Submission to the International Joint Commission Regarding Group II Objectives as proposed by the Water Quality Board* (September, 1978), the industry challenged the proposed specific water quality objectives for ammonia, copper, iron, nickel, hydrogen, sulfide, cyanide, chlorine, and water temperature (the latter was included since it is expensive to cool the water before returning it). In its conclusions, the industry claimed that the Water Quality Board had searched for data to justify the lowest-level objective with little regard for "acceptable" levels. Also, they noted that the private sector had been excluded from participation in setting objectives (another point of view is that they were invited but chose to refuse participation) and asserted, once again, that the social and economic consequences of the proposed objectives had been ignored. Finally, the industry recommended:

- The Commission must recognize that many industrial processes are such that lake water quality objectives cannot be met at the "end of the pipe" because of the limitations imposed by available funds and technology. The Commission must therefore ensure that the mixing zone concept is not eroded to the degree that the lake water objective becomes in fact the "end of the pipe" effluent requirement.

- The Water Quality Board must be directed to include with their submission to the Commission an assessment of the social, economic and technical implications of their proposed objectives. This assessment should be done in consultation with industry.
- Industry must be represented on the Water Quality Board through committee membership to ensure that there is input from the private sector in the initial stages of formulating objectives.
- The International Joint Commission should respond to written submissions from the public. This would create a dialogue and assure those participating that their submissions were being considered by the Commission.[44]

Environmentalists. In general, environmental groups were in favor of strengthening municipal and industrial pollution control programs through the renegotiation process. Perhaps the most prominent of these organizations was the nonprofit citizens group, Great Lakes Tomorrow (GLT). GLT was formed in 1975 by a group of sixty scientists, citizen activists, government officials, and foundation representatives as a three-year project (since become permanent) to increase public understanding and input related to the future of the Great Lakes Basin. The project was cosponsored by the Lake Michigan Federation, a nonprofit citizens organization, and The Institute of Ecology, an international organization of scientists and environmental leaders. Through its involvement, GLT attempted to provide input to decision making through forums such as the Great Lakes Basin Commission and IJC-sponsored public meetings.[45]

Project management for the GLT project included one centralized office in Chicago and five smaller task forces to address specific concerns on each of the Great Lakes. Although the orientation of GLT was primarily toward environmental approaches to problems, additional concerns such as navigation, economic development, population, and fisheries were also integrated. The principal tools utilized by GLT to achieve public understanding and input includes newsletters, special action alerts, informational materials, and notification of public hearings.[46]

The institutional focus of GLT was decidedly binational, as supported by members and activities in both the United States and Canada. Indeed, this binational emphasis is evident in the published goals of GLT:

- to secure a strong, irreversible, and concerted commitment from the governments of Canada and the United States to rehabilitate and restore those areas of the Great Lakes which have been degraded, and to maintain in non-degraded conditions the remainder;
- to promote the necessary institutional and organizational changes needed to develop a shared Great Lakes perspective and binational concern for the consequences of decisions affecting the Lakes;

- to build an anticipatory-planning capability covering water, shoreline, and associated land resources;
- to mobilize binational cooperative responses to situations requiring it and to open up the decision-making processes to public inspection and participation; and
- to develop a strong binational constituency "for the Lakes" and to secure greater public participation in decision-making.[47]

Of particular significance in these goals is the expressed commitment to the rehabilitation and restoration of areas of degraded water quality. This outlook directly conflicts with the principles set forth by the Ontario Integrated Steel Industry. In pursuit of these goals, GLT was active in recommending measures including, but not limited to, stronger phosphorus controls, mixing zone limitations, pH controls, a PCB ban, an end to winter navigation, increased public participation, and others related to oil and gas drilling, diversions for power generation, and navigation.

The Canada-U.S. University Seminar. As a unique institution with input into the Great Lakes water quality issue, the Canada-United States University Seminar drew together faculty and students for investigations under the supervision of Professors Leonard Dworsky at Cornell University and George Francis at the University of Waterloo. The seminar produced a series of nine reports between 1970 and 1978, with the broad objective of "improving the management of the Great Lakes of the United States and Canada."[48]

Through the IJC and the respective federal governments, the seminar encouraged public discussion and debate related to Great Lakes water quality issues. In particular, the work of the seminar focused largely upon means of improving the institutional framework for addressing the problems of water quality in the Great Lakes. Through a series of courses and meetings in 1977 and 1978, the seminar produced a set of main conclusions:

- Continued degradation of water quality in the Great Lakes is unacceptable.
- Both countries should adopt as a policy goal the determination to achieve the ecological rehabilitation and restoration of the Great Lakes in the manner endorsed by the binational citizens group "Great Lakes Tomorrow."
- Both countries need to agree on the essence of a science policy in support of research needed to achieve this policy goal.
- Land-management issues in the Great Lakes Basin have to be addressed if the widely acknowledged water problems are to be resolved.

- Attention has to be paid on a binational basis to planning and management of Great Lakes resources on a comprehensive, multiple-purpose basis in order to achieve optimum public benefits to both countries.
- The International Joint Commission through its Great Lakes Regional Office in Windsor, Ontario, needs to be strengthened and given the responsibility to initiate investigation into problems before they become political crises.

These recommendations were further accompanied by a more detailed agenda for Canadian-U.S. bilateral negotiation.[49]

The Final Document of 1978
As finally signed, the agreement of 1978 bore the following provisions providing for the development and implementation of programs to:

- control pollution from municipal and industrial sources (with programs to be in place by December 31, 1982 and 1983, respectively);
- largely eliminate the discharge of known toxic substances to the Great Lakes system and to establish early warning systems that will identify those substances which may become problems;
- identify airborne pollutant sources and relative source contributions entering the Lakes system;
- identify and control pollution from agriculture, forestry and other land use activities (i.e., nonpoint source pollution);
- improve surveillance and monitoring to enable more complete assessment of the effectiveness of remedial measures;
- decrease the amount of phosphorus entering the Lakes system (interim phosphorus loadings were also set);[50]
- meet additional new specific water quality objectives; and
- prepare an annual inventory of compliance with pollution control requirements.

Thus, the 1978 agreement places greater emphasis on toxics, on nonpoint pollution, and on a broader ecosystems approach to basin management than the 1972 agreement while carrying the latter forward and encompassing more of the basin (vis-à-vis the tributaries, the connecting channels, and the atmosphere overhead). In the view of diplomats, however, efforts toward cleanup are going very slowly.

A basic difference in the 1978 agreement was the decision not to include a termination date, as found in the earlier agreement. It will remain in force for five years, may be reviewed after five years, and will continue indefinitely until terminated by either party. Hence, what the two nations witnessed in 1977–78 (regarding debate over renewal) is not likely to occur again.

Evaluation

In the view of a senior Canadian official, the agreement will provide a framework for evaluation of the programs put into place by the different parties—it is not a prescription for management, and its utility will be in its evaluation of activities for government. That utility is still to be demonstrated. As one official noted, the easy and obvious work has been done, and the hard work remains. The real benefit of the agreement will not be seen for a decade. The agreement is not precise in nature and is not meant to be; it is a general call to a higher plane and a conceptual document only.[51]

Ontario is concerned that the IJC is not fighting hard enough for Canada's rights to the assimilative capacity of the lakes.[52] The province regrets the reluctance of the U.S. EPA to support points in the agreement that require specific identification of actual pollution sources (so many of which are U.S. sources). Is this an EPA recognition that there is a large difference between regulated limits and the achievement of them, as asserted by some Canadians?

An IJC attorney points out that the agreement is not a treaty, and therefore local governments do not have to take account of it or honor its provisions.[53] This fact will work against full and effective implementation of the agreement, in spite of the best efforts of the national governments. The whole issue of noninvolvement of local municipalities in the IJC framework is important. From now on, not much more can be done without their involvement, especially in the United States where "home rule" tradition is stronger.

Since the advent of the first agreement in 1972, the Great Lakes Water Quality Agreements have often been portrayed to the world as models of one successful method to resolve bilateral differences and jointly manage a shared common property resource. Interestingly, many middle management diplomats and bureaucrats who were involved in the negotiations of the early 1970s, especially in Canada, are now senior members of their agencies, are proud of the agreements and their own role in their coming to fruition, and are interested in carrying the concept forward. The concept of joint technical working groups, utilized so well in the drafting of the 1972 agreement, has now been adopted as a model for the negotiation of a bilateral air quality (and acid rain) agreement (see chap. 11).

The International Joint Commission is commonly viewed as having had a major and critical role in the agreement of 1978, for several reasons: in addition to the basic work of the commission in the late 1960s and early 1970s, which became such an important foundation for the negotiation of the first agreement, the commission was given the mandate to implement the first agreement and was placed in charge of the field coordination (at Windsor), and the commission has again been maintained as the focal point for the bilateral coordination of the second agreement. However, the commission was deliberately isolated from the give and take of the actual negotiation and was simply given the agreement and the responsibility of its mandate by the diplomats following the signing.

As an example of industry misunderstanding, the Canadian steel company STELCO learned to its surprise that diplomats wrote the final documents and, in STELCO's view, felt no obligation to observe the IJC or its (STELCO's) recommendations submitted thereto.[54] STELCO felt misled in believing that the Ontario Ministry of the Environment was the only agency with which it had to deal on water quality only to find, in effect, that this was not true, The IJC and, more importantly, the federal governments must be dealt with, indirectly as well. This exemplifies a classic problem and challenge for modern industry: that many different levels of government, agencies, statutes, and personalities—often in substantial contradiction with each other—must be dealt with. When a matter is international in nature, the diplomatic level is added to the challenge. And, of course, confusion and its resulting uncertainty represent a cost both to industry and society. A Canadian industry preoccupation during the Great Lakes negotiation was the export of U.S. environmental concerns and control philosophies to Canada, which they (and others) claim has no need for such restriction and cannot afford the cost. As STELCO's Hugh Eisler relates,

> [W]e feel there is some duplication between the efforts of the IJC and Canadian jurisdictions. Although in law the Province has the regulatory authority in most areas of the environment, we have found that IJC criteria very quickly becomes a part of the local Provincial standards. Therefore, criteria established by the Great Lakes Water Quality Agreement, which was signed without taking into account the input of those affected, and which in some cases lacks scientific justification, quickly became part of the Provincial regulations. As a result, a major project which was begun under another set of criteria was judged for approval purposes against these new regulations.[55]

Environmentalists retort that it is not realistic to expect the world to stand still during the long drawn-out period it takes to complete a major project.

CONCLUSION

The crucial substantive changes in the second agreement are:

- the agreement that municipal and industrial pollution abatement control programs would be completed and operative no later than

1982 and 1983, respectively; (1982 being Detroit's deadline for sewage treatment facilities to go on line);

• a new, more stringent set of phosphorus-loading reductions for each of the lakes, with actual division of loadings to be negotiated within eighteen months, a contentious matter linked to the debate over equality versus equity in use of the assimilative capacity, and still unsettled;

• elimination of the discharge of toxic chemicals[56] and the listing of 350 hazardous substances of concern; and

• a new surveillance program and revised water quality objectives, including standards for radioactivity.[57]

Given the great size and complexity of the Great Lakes system, it is premature to determine the ultimate effectiveness of this second agreement. However, with the joint failure to reach the May 22, 1980, deadline on phosphorus loadings, the two nations are not off to a good start.[58] Given the increased realization, especially in the United States, that society does not have the unlimited resources necessary to attain a zero risk environment (as implied in some U.S. statutes), ultimate cleanup, at least to the level desired by the IJC, may continue to be an elusive, and possibly unrealistic, goal.[59] The increased concerns over the health of the U.S. economy and the concomitant rebellion against overregulation by government in that country may lead the United States to adopt a new pollution control philosophy less purist in nature, less environmentalist, and quite possibly closer to the Canadian philosophy of regional water quality objectives. Such a development would undoubtedly bring the two nations closer together diplomatically and alleviate or eliminate one major area of bilateral debate, though it will not necessarily resolve the point of the debate.

In the Great Lakes experience, the United States and Canada have demonstrated that they do have the ability and will to agree jointly on direction and initiate a plan to proceed. They have yet to demonstrate a capability to carry through, however, especially when a long period of time and changing governments are involved. In spite of the two bilateral water quality agreements, the Great Lakes are in some ways as bad now as in 1972, but they would likely be a lot worse without the agreements. The challenge is not just to put laws on the books but to solve bilateral differences and ultimately to clean up the lakes. In these respects, the challenge has just begun.

A major question developing in the early 1980s concerning the future water quality of these international lakes was the impact of the substantial Reagan administration reductions in the staffing and operating budget of the U.S. Environmental Protection Agency, on whose back so much of the future of the lakes' ecosystem depends. The actual reductions and projected future reductions have led some to conclude the intent is to abolish the agency and all or most of the U.S. national pollution control effort with it. And, although it is the stated intent of this presidential administration to transfer as much power as possible to the states, it is very doubtful if the financially strapped Great Lakes states (and especially Michigan and Ohio, among the biggest contributors to lakes' pollution) will be able to accept or financially carry out present federal responsibilities in this area. From the diplomatic perspective, it is doubtful if the United States can fulfill its international obligations under the Great Lakes agreements either in letter or in spirit, with the implementation of these reductions. The question must thus be asked: do such reductions violate the agreements? If so, does Canada have any recourse? From the perspective of water quality, the bilateral future of these Great Lakes ecosystems does not look bright and another issue, fulfillment of international obligation (under the two agreements and perhaps also under the Boundary Waters Treaty), enters the lengthening list of Canadian-U.S. diplomatic disputes.

As a complex challenge involving immensely important economic interests and hence with important political ramifications, the Great Lakes issues treated in this chapter rank with acid precipitation and perhaps with the difficult hazardous wastes–toxic substances questions treated in other chapters. Hence, it should not be too surprising that diplomatic progress is exceedingly slow. Just as enduring the problem is costly, so too are mistaken remedies. One must not begrudge diplomats for slow progress, therefore, but one can hope that they, aided by other concerned interests from all sides, will carry on to a successful conclusion—acceptance on both sides as to what can and cannot be done, followed by maximum reduction of pollution. The importance of these questions is such that they are worthy of maximum attention and commitment. A double reward is in the offing, for solution of Great Lakes quality questions will contribute toward solution of hazardous wastes–toxic substances questions, and vice versa. The two are strongly intertwined.

It is now up to the IJC and, perhaps more importantly, the departments of State and External Affairs, which hold such a tight rein over the commission's activities, to see that this important and pioneering effort bears fruit, not only for the future of the lakes but also as a forward-looking model to be applied to other bilateral challenges, where appropriate.

The emergence of the serious environmental issue of toxic wastes in the late 1970s–early 1980s, particularly in the Great Lakes Basin, has nearly overshadowed the long-standing conventional pollution concerns in the lakes. Although technically part of the overall Great Lakes water quality issue, toxic substances disposal and management has now become such a serious international issue in its own right that a separate chapter addressing this subject is in order. It is to that subject, therefore, that we now must turn.

NOTES

1. The author is indebted to his assistant, Roy Stever, for research assistance on this chapter.

2. Leonard B. Dworsky, G. R. Francis, and C. F. Sweeney, "Management of the International Great Lakes," *Natural Resources Journal* 14, no. 1 (1974):105. Dworsky et al. have stated,

> the whole matter of regulation of the levels and outflows of the Great Lakes cannot be other than international. Artificial control within the lakes of the water supply which comes from both countries cannot be undertaken without affecting the water use interests on either side of the international boundary. Changes in outflows and levels will, of course, have similar international effects.

3. William P. Endle, "Lake Ontario Regulation," *Water Spectrum* 11, no. 2 (1979):23. Endle has assessed Lake Ontario regulation in the following manner:

> Obviously, no single plan can provide maximum benefits to riparian, shipping, and power interests at the same time. As stated before, an increase in benefits to one interest frequently means costs or losses to another. *It seems unlikely, then, that any regulation plan will ever be considered a popular solution. It is the author's opinion, however, that the most equitable distribution of benefits to all concerned may be achieved when the lake levels most closely approximate their long-term norm.* [Emphasis added.] This would also have the least adverse environmental impact on the lake. In other words, a plan which tends to reduce the occurrence of either extremely high or low levels and tries to maintain levels of long-term seasonal averages might be considered successful.

4. Ralph L. Pentland, "Boundary Waters Management," in *Canadian-American Natural Resource Papers 1975–1976*, ed. John E. Carroll (Durham, N.H.: University of New Hampshire, 1976), p. 15.

5. International Joint Commission, "New International Studies on Great Lakes Levels and Flows," *International Joint Commission News*, July 15, 1977, p. 17.

6. Pentland, "Boundary Waters Management."

7. Ibid.

8. Ibid., p. 16.

9. International Joint Commission (IJC), *70 Years of Accomplishment: Report of the International Joint Commission* (Ottawa and Washington: IJC, 1980), pp. 54–59.

10. Institutionally, the U.S. Army Corps of Engineers is a driving force on bilateral Great Lakes levels and consumption matters, through its prominent leadership role on all IJC boards in this area.

11. William R. Willoughby, "The International Joint Commission's Role in Maintaining Stable Water Levels," *Inland Seas* 28, no. 2 (Summer, 1972):114 (hereafter cited as Willoughby, "Stable Water Levels").

12. General A. G. L. McNaughton, House of Commons Standing Committee on External Affairs, *Minutes of Proceedings and Evidence,* No. 10, June 14, 1961, p. 245.

13. Willoughby, "Stable Water Levels," p. 117.

14. Ibid.

15. International Joint Commission, *Further Regulation of the Great Lakes,* as reported in the IJC Annual Report, December 1979, p. 34.

16. International Joint Commission, "New International Studies on Great Lakes Levels and Flows," *International Joint Commission News,* July 15, 1977, pp. 6–7.

17. Canada Department of External Affairs, "Government's Response to IJC Report on Further Regulation of the Great Lakes," Communique no. 12, February 21, 1977; U.S. Department of State, "Great Lakes Levels Discussions," press release no. 78, February 22, 1977.

18. Richard J. Wagner and D. J. O'Neil, "Canadian Penetration of the American Political Process: A Case Study," Occasional Paper of the University of Arizona Department of Political Science, n.d. (later revised and retitled "The Chicago Diversion Case: A Hypothesis").

19. Critical vetoes were made by President Eisenhower in 1954 and 1956, both stopping bills designed to increase diversions through Chicago. Wagner and O'Neil write, "it was clear that for Eisenhower the fourth reason given in the veto message (i.e., interests of other states) was the primary one, that other considerations, despite lip service to Canada and the vigor of her protests, were secondary to domestic interests. Without powerful allied American interests, Canada's case would have been lost" ("Canadian Penetration," p. 16). But, "while in 1958 Senator Dirksen laid the blame for the 1954 veto message in other than Canadian hands, he did believe that the 1956 veto message was largely due to the influence of Canada. Considering the importance of the St. Lawrence and the Columbia Rivers, both shared with Canada, his impression was essentially accurate" (ibid., p. 22). And, "until Canada consents to such action or the other Lake States also agree, the issue is dead. Canada collaborating with allies in the American political process had scored a victory" (ibid., p. 39).

20. Canada Department of External Affairs, Communique No. 12, p. 3.

21. Ibid., pp. 2–3.

22. For insight into the earlier history of lake level politics and particularly power and navigation questions on the St. Lawrence, the reader is referred to the works of Professor William Willoughby, including his "The St. Lawrence Seaway Understandings" (*International Journal,* Quarterly of the Canadian Institute of International Affairs, Autumn, 1955); "Early American Interest in Waterway Connections between the East and the West" (*Indiana Magazine of History* 52, no. 4 [December, 1956]); "Power Along the St. Lawrence" (*Current History* 34, no. 201 [May, 1958]); and "The St. Lawrence Seaway: A Study in Pressure Politics" (*Queen's Quarterly* 67, no. 1 [Spring, 1960]); and his book *The St. Lawrence Waterway: A Study in Politics and Diplomacy* (Madison: University of Wisconsin Press, 1961).

23. Charles D. Gunnerson and Kenneth A. Oakley, "Review Paper: Binational Abatement of Boundary Water Pollution in the North American Great Lakes," *Water Research* 8, no. 10 (1974):713.

24. Ibid.

25. Professor Frank Stone of Carleton University, a former diplomat involved in the original Great Lakes Water Quality Agreement negotiations, contends that the Canadian objectives were to get the Americans to spend more money on municipal pollution control and to force better enforcement of industrial effluent standards, although he believes a subsidiary purpose was to set up a new body of environmental rules (personal interview, September 24, 1980).

26. Don Munton, "Great Lakes Water Quality: A Study in Environmental Politics and Diplomacy," in *Resources and the Environment: Policy Perspectives for Canada,* ed. by O. P. Dwivedi (Toronto: McClelland and Stewart, 1980), p. 159.

27. Ibid., pp. 162–63.

28. Pentland, "Boundary Waters Management," p. 17.

29. George Francis, Professor of Environmental Studies, University of Waterloo, personal correspondence, June 5, 1981.

30. Pentland, "Boundary Waters Management," p. 165.

31. R. H. Clark, L. B. Dworsky, and G. R. Francis, *A Review of the 1972 Great Lakes Water Quality Agreement,* Canada-U.S. Interuniversity Seminar II, University of Waterloo and Cornell University, 1976, pp. 38–39.

32. Munton, "Great Lakes Water Quality," p. 167.

33. The U.S. approach was patterned after its own domestic law, the federal Water Pollution Control Act Amendments of 1972, which established emission standards at the point source of emission.

34. Depending on levels set and objectives, the U.S. advocacy of effluent standards could well lead to cleaner lakes but the cost and total value of such a system is what is debated, rather than the final condition of the lakes. In essence, Canada needs to protect the developmental potentials of its province, as long as it can avoid international harm in the process.

35. The vessel waste issue boiled down to a debate between flow-through treatment systems in ships (backed by Canada) versus holding tanks aboard ships (i.e., complete containment [backed by the United States]). Ontario requires holding tanks on pleasure craft, and Michigan believes all vessels plying her waters must have holding tanks (a position opposed by the EPA, which held that Michigan could prohibit discharge but could not mandate the tanks). Does this then mean that holding tanks must be aboard or simply that discharge is prohibited?

36. International Joint Commission (IJC), *Fact Sheet: 1978 Great Lakes Water Quality Agreement* (Ottawa and Washington: IJC, 1980).

37. Ron Shimizu, Great Lakes National Program Manager, Environment Canada, personal interview, Toronto, October 16, 1980. In recognition of this reality, Environment Canada moved its Great Lakes National Program Office from Ottawa to Toronto and is now budgeted at $10 million annually. The federal Great Lakes Program Manager of this effort, Ron Shimizu, has observed that the scientific community in the Great Lakes Basin is more influential than it realizes, an active community which government, in his view, hasn't fully utilized (although the IJC's Science Advisory Board is a start in this direction).

38. Munton, "Great Lakes Water Quality," p. 172.

39. Ibid., p. 178.

40. Ontario Integrated Steel Industry, "Comments of the Ontario Integrated Steel Industry on Canada-United States Agreement on Water Quality in the Great Lakes," Toronto, Ontario, July 14, 1977, pp. 2–3.

41. Ibid., p. 4.

42. Ibid., p. 5.

43. Ibid., p. 7.

44. Ontario Integrated Steel Industry, "The Ontario Integrated Steel Industry's Submission to the International Joint Commission Regarding Group II Objectives as Proposed by the Water Quality Board," September, 1978, pp. 11–12.

45. Great Lakes Tomorrow, *Improving Citizen Participation in Great Lakes Decisions: A Year End Report and Description* (Prepared by the Lake Michigan Federation) (Hiram, Ohio: Great Lakes Tomorrow, 1976), pp. 1–2.

46. Great Lakes Tomorrow, "Great Lakes Citizen," December, 1977, p. 2.

47. Ibid.

48. Leonard B. Dworsky and George Francis, *Toward the Future in the Great Lakes Basin: An Agenda for Negotiation, Canada-United States*, Canada-U.S. University Seminar, University of Waterloo and Cornell University, March, 1978, p. i.

49. Ibid., pp. 2–3.

50. An initial problem which arose under provision six was that the two governments failed to confirm phosphorus loadings into Lake Erie eighteen months after the agreement (May 22, 1980), as mandated in the agreement. According to the State Department, the governments requested help from the IJC to carry this out but the commission has delayed and the deadline was not adhered to.

51. Dr. Robert Slater, Director-General, Ontario Region, Environment Canada, personal interview, Toronto, October 16, 1980.

52. William Steggles, Advisor to the Ontario Minister of the Environment, personal interview, Toronto, October 16, 1980.

53. Sam Wex, Counsel, IJC, personal interview, Ottawa, October 22, 1980.

54. Hugh Eisler, Department of Environmental Affairs, STELCO, telephone interview, Hamilton, Ontario, October 14, 1980.

55. Eisler, personal correspondence, November 18, 1980.

56. Elimination of toxics cannot happen as the document is worded. The document only "bans" toxic discharges to the minimum level of measurement—a different concept from banning the discharge. As worded, an industry could know from its production chemistry that it must be discharging dioxin, for example. If it couldn't measure dioxin in the discharge, it would meet the letter of the agreement. It is questionable whether toxics can be controlled in this loose a fashion.

57. Munton, "Great Lakes Water Quality," p. 173.

58. It should be noted, however, that under the mandate of the second agreement the IJC has investigated and informed both peoples of the nature of the problem, important basic data for any future policy decisions. (These findings have been revealed in the following recent commission publications: *Water Quality in the Upper Great Lakes* [May, 1979]; *Pollution in the Great Lakes Basin from Land Use Activities* [March, 1980]; *A Perspective on the Problem of Hazardous Sub-*

stances in the Great Lakes Basin Ecosystem [November, 1980—Annual Report of the IJC Science Advisory Board]; and the 1980 Annual Report of the IJC Water Quality Board, entitled *1980 Report on Great Lakes Water Quality*.)

59. Another danger is that unrealistic objectives, although valid as a research tool, may soon lose their credibility as a device for achieving water quality improvements.

CHAPTER 7

Toxics at the Border

> *The bilateral toxics issue is imposing a serious strain on the Great Lakes Water Quality Agreement since it is predominantly a Great Lakes quality matter and falls under that agreement. The problem is highly visible, very sensitive in the political arena, and difficult and costly to solve, not unlike acid precipitation. A new type of diplomatic approach may be necessary to address the question effectively. As in the case of acid precipitation, the two nations appear to be at the threshold of their experience in this area, technically and diplomatically, and most of the story is yet to unfold.*

Just a few years ago the issue of toxic substances did not appear on any diplomatic agenda; now it is prominent.[1] Toxic substances are long-lived, nondegradable chemicals that accumulate through the food chains and are capable of inducing short-lived and long-term health effects including cancer and genetic change. It was not until 1978 that the issue of toxic or hazardous substance disposal was explicitly treated in its own right. It became officially a part of the revised Great Lakes Water Quality Agreement of that year. Since much of the border is composed of the industrialized Great Lakes–St. Lawrence system, it was inevitable that the problem of toxics would become an international issue soon after it became established as a source of concern domestically in the two countries.

The potentially severe impact of toxic substances on affected populations is so great that society is forced to pay attention to the subject and pressure public officials to move at a pace uncharacteristic of international diplomacy. Such pressure as has been applied in the bilateral toxics area may well lead to unconventional, even extreme, diplomatic behavior. (Such behavior, though not without risk, can aid, however, in the development of diplomatic innovation. Innovation is often a product of necessity.)

The issues raised by toxics questions are essentially new, rather than older ones in a new form, and they present a new challenge for the International Joint Commission and particularly for its Science Advisory Board (an appointed group of scientists, governmental and nongovernmental, advising the commission under the Great Lakes Water Quality Agreement of 1978). The latter has been especially active in identifying this issue and bringing it forward, a role it has also played with acid precipitation. The board's high level of activity was surprising in that it continued while the commission itself was in a position of enforced inactivity throughout 1981.

The Science Advisory Board thus prevented the 1978 Great Lakes Water Quality Agreement from failing completely in this area. Without needed leadership from the commission it would eventually cease to be effective. That leadership vacuum was partially remedied by appointments made in early 1982, though the Canadian Section remained without a permanent chairman as of mid-1982.

The Problem in the Great Lakes

The Great Lakes Basin and its drainage through the St. Lawrence constitute one of the world's most heavily industrialized regions and is a shared international basin in every sense. The basin contains a major concentration of chemical manufacturing plants, a host of industries that require toxic chemicals in their operation or create them as a by-product, and various chemical dump and disposal sites for both privately generated and governmental wastes. The stage is thus set geographically for a major bilateral confrontation, a confrontation assured when one realizes that one nation, the United States, contributes the major portion of the problem (though it is by no means alone).

Bilateral toxics issues can arise at any point along the border. One recent example was the large mercury spill in the Canadian section of the Columbia River and the furor it raised in Washington State. But for the most part it is a Great Lakes matter, involving essentially the same actors and many of the same interests discussed in chapter 6.

What differentiates this issue from that of Great Lakes water quality? First, toxic substances have much more serious health effects than have what might be called conventional pollutants such as municipal sewage or many other industrial effluents. They directly affect human health, both immediately and in future generations. In contrast, conventional less toxic pollutants, such as phosphorus, lead to eutrophication or oxygen demand (i.e., premature aging of lakes), are capable of some assimilation into the environment, and pose fewer health problems.

Second, public perception of the toxics issue and reaction to it has been much sharper than in the case of conventional pollutants. Third, there is also a limited air quality aspect to the toxics issue. Wastes incinerated to avoid geographically limited water quality problems can cause widely dispersed air quality problems. Airborne agricultural chemicals constitute a further air quality problem. Fourth, in addition to the seriousness of their impact and the difficulties of controlling their spread, toxics present an additional diplomatic challenge. They illustrate one of the main problems underlying bilateral environmental relations, namely the difficulty of working diplomatically with an evolving scientific data base.

Because disposal of toxic substances into the Great Lakes or their tributaries is an act of pollution, the original Great Lakes Water Quality

Agreement of 1972 prohibited it, implicitly if not explicitly. Of course, the Boundary Waters Treaty of 1909, to the extent it bars injury to health and property damage through pollution of boundary and transboundary waters, could be said to bar toxics dumping or runoff into the lakes as well. Thus, the IJC, the basic vehicle for implementation of both the 1909 treaty and the 1972 agreement, has played a central role in attempts to solve the problem.

Role of the IJC

In 1978, for the first time, the IJC was given an explicit mandate to monitor toxics in the Great Lakes system, perform research, and alert government to the extent and dangers of the problem. The commission embraced this task with enthusiasm, first under Canadian Chairman Maxwell Cohen, and then even more so under U.S. Chairman Robert Sugarman.

Under authority of the Great Lakes Water Quality Agreement of 1978, the IJC's Great Lakes Water Quality Board established a Toxic Substances Committee, which reported in November, 1980.[2] Its report, a necessary first step, identified the issue domestically in both countries, presented a framework for problem management, analyzed pertinent federal, state, and provincial legislation, and discussed data bases and information systems all preparatory to more substantial future work.

Simultaneously, the IJC Great Lakes Science Advisory Board, the body which was among the first to call attention to the bilateral acid rain threat to the Great Lakes, issued its study—*A Perspective on the Problem of Hazardous Substances in the Great Lakes Basin Ecosystem*. In its finding the board made the important point that not only had many new types of chemical-based products been developed in answer to consumer demands, causing part of the problem, but also that

analytical capabilities improved dramatically with the development of gas chromotography and mass spectroscopy. The positive identification of complex organic chemicals became easier, and detection limits were lowered making it possible to measure many chemicals at concentrations much lower than previously detectable in water and air.[3]

The board's study, combined with greater public willingness to support monitoring and control programs and concerns over health, significantly heightened public concern. Domestic and bilateral debates ensued.

The board's report, unlike that of the Toxic Substances Committee, also treated the human health effects, transport, and ultimate disposition, identified specific sources in the Great Lakes ecosystem and specific control alternatives, and presented a plan of attack recognizing that choosing a practical plan to deal with the most important substances is difficult, though essential.[4] It identified two especially critical concerns, toxicity to aquatic organisms and bioconcentration potential, and recommended focusing on the elimination of those substances creating problems in these areas.

The Science Advisory Board has urged the commission to declare its intent and commit its resources to support the board's approach. It views progress as slow, even undetectable, because the commission has not instructed the two governments regarding a hazardous substances strategy, something it recommends the commission should do now. The commission's response is not likely to come soon, however, given the number of vacancies on the commission (five out of six) through most of 1981, and particularly the new U.S. antiregulatory and antispending mood. Nevertheless, the Science Advisory Board did not refrain from making strong specific recommendations asking that the commission

- urge jurisdictions to institute programs to quantify the atmospheric loadings of hazardous substances to the Great Lakes;
- urge jurisdictions to recover hazardous substances for reuse and employ treatment technologies that destroy, rather than merely remove, contaminants from waste discharges;
- encourage dischargers to seek ways to reduce the use or loss of hazardous substances that may find their way into air or water effluents;
- urge jurisdictions to identify and inform populations in the [Great Lakes] Basin which may have higher than average exposure to hazardous substances as a result of their dietary habits or living conditions, and that the jurisdictions expand their efforts to identify any cause and effect human health relationships associated with the consumption of Great Lakes fish and wildlife;
- request appropriate agencies in Canada and the United States to review the human health toxicity information on those hazardous substances which form residues in Great Lakes fish and wildlife, and establish tolerance levels for those substances as they are identified;
- strongly urge governments to establish programs to develop routine fate and effects information needed for predictive hazard assessment;
- centralize an information system to collect, store, sort, and dispense data needed by the jurisdiction for control of hazardous substances;
- recommend research for developing methods to determine net benefits as a necessary consideration in future decision making in the Great Lakes Basin Ecosystem.[5]

These recommendations constitute a major effort for the IJC and represent a most comprehensive and a most specific bilateral response to the transborder toxic issue.[6]

The Niagara River Issue

The other major toxics issue in the bilateral environmental relationship has been that of the Niagara River, which separates New York and Ontario. SCA Chemical Waste Services, Inc. of Lewiston, New York, applied for and was granted a permit by the New York State Department of Environmental Conservation to dump up to one million U.S. gallons of treated chemical wastes per day in the Niagara River beginning in the spring of 1981. The United States contends there is no health danger, but the river is already considered highly polluted both by toxics and by more conventional pollutants, and the reaction in adjacent Ontario was immediate and strong. This attempt to set back the goals of the Great Lakes Water Quality Agreement of 1978 so soon after bilateral approval of that agreement was well publicized in the Toronto and local press, and the matter quickly became politicized in Canada. It became a subject of debate in the parliamentary elections of early 1980 when the prime minister made reference to it in a speech.

After the election, opposition parties in the House of Commons kept up the pressure leading to the sending of at least two diplomatic notes by External Affairs.[7] The second note referred to a federal-provincial report on Niagara River water quality, which concluded that a variety of chemicals in the river already exceeded acceptable levels and had therefore violated the Great Lakes Water Quality Agreement several times during the past year. These chemicals were linked to U.S. industries along the river. After repeated U.S. assurances that Canada's interests would be given full consideration, the furor died down. In early 1981, however, New York decided to revoke its permit to SCA. This action, coupled with the release in January, 1981, of the IJC's *Special Report Under the 1978 Great Lakes Water Quality Agreement on Pollution in the Niagara River* (released on the last day of the Carter administration—a final act of the soon-to-be-fired U.S. commissioners), again focused much media and public attention on the issue.

The IJC complimented New York State for revoking the permit, noted that its own Water Quality Board had consistently identified the Niagara River since 1973 as a problem area where specific objectives were not being achieved,[8] and concluded:

> The Commission is concerned about the impact of additional proposed discharges to the Niagara River at a time when water quality in the River does not at present meet or is close to exceeding the Agreement objectives.[9]

After chiding the governments for not fulfilling their obligations under the 1978 agreement, the commission asked that

- a comprehensive study of the Niagara River be undertaken;
- a comprehensive monitoring program be established for the river and the west end of Lake Ontario;
- the two governments prevent any additional discharge to the Niagara River that would violate the specific objectives of the 1978 agreement;
- the two governments be more responsive to the commission's questions; and
- local jurisdictions inform the commission ahead of time as to the extent that pending discharge permits would result in a change in the amount of toxic substances entering the Great Lakes ecosystem.[10]

Unfortunately, not all jurisdictions follow regulatory procedures which are designed to achieve the 1978 agreement's water quality objectives. U.S. jurisdictions are not bound to do so since the agreement is an executive one rather than a treaty ratified by the Senate. Cumulative effects of many different sources of contaminants over time, either within or between jurisdictions, are not considered in the granting of discharge permits.[11] The only method governments have for ensuring compliance with the agreement is to prevent further discharges until specific objectives are met. This would likely require reduction of existing discharges before permitting new ones.

With the revocation of the New York State permit and the sharp rebuke by the IJC regarding additional river dumping, the issue died down, but not before graphically demonstrating both the extreme condition to which the Niagara International River had been permitted to deteriorate and the gross failure of both governments to respond either to the IJC or to their obligations as parties to the 1978 agreement.

During late 1981 and 1982, the debate again became inflammatory. The Reagan administration's budget cuts, the politics of the U.S. EPA, (an agency fighting for its survival), New York's fears of its inability to cope with the issue, and the increasing awareness of the Canadian public through constant hearings, and political and diplomatic charges and countercharges, all well covered by the media, are ensuring further polarization of this issue. Alleged continuing seepage of toxics from the Love Canal and Hyde Park dump sites in this area are also keeping the Niagara River toxics pollution issue elevated politically and diplomatically. By spring of 1982, Canada still believed that New York toxics were reaching the river, while the U.S. EPA held firmly that there were no identifiable U.S. point sources. The issue was at an impasse.

A Canadian newspaper editorial recently referred to the Niagara

River as "tormented."[12] It noted the high population around the river and the importance of adjacent lands for fruit production and of the river itself for drinking water and crop irrigation, painting a vivid picture of man's fouling his own nest. The editorial reminded readers that both governments have all the necessary tools to do the job and urged top priority bilateral attention be given to the river and its immediate cleanup. It failed to mention the costs of that cleanup, however, just as it failed to assign costs to a do-nothing alternative. Society today demands to know those costs before action is taken.

Dioxin

Another international Great Lakes toxics issue is that of the late 1980 discovery by Canadian Wildlife Service biologists of extremely toxic dioxin in the eggs of herring gulls on islands in the lakes. This dangerous substance is now believed widespread in the lakes system. Dioxin is a by-product chemical produced in the manufacture of 2,4,5, Trichlorophenol (a pesticide) and is today only made by Dow Chemical Company in Midland, Michigan. It was formerly made elsewhere in the Great Lakes Basin by other companies and is incorporated in bottom sediments. There are likely leakages from near-shore storage dumps. Thus its sources are numerous but all seem to have one thing in common: they originate in the United States, for there are apparently no sources in Canada. Hence the ingredients exist for a bilateral dispute, especially since Canadian scientists discovered the problem and are performing most of the research on it.

This matter, too, received substantial media attention, reaching a peak at the end of 1980 and forcing the Canadian environment minister to threaten drastic action if a response was not forthcoming. Concluding that dioxin is present where millions of North Americans depend on the water for drinking, fishing, and recreation, the minister demanded that the United States eliminate all sources, including old dumps and current chemical production. If it was found impossible to prevent formation of dioxin during chemical manufacture, he claimed he would push for a total ban on dioxin production.[13] The Canadian concern is further exacerbated by an earlier U.S. EPA finding that no dioxin should be permitted in water for human consumption, since there is no safe level for the chemical. The environment minister admits, however, that it has not been detected in drinking water and thus does not present an immediate health hazard to Canadians.

Dioxin is one of those chemicals we know much more about today than in the past because of greatly improved detection methods. There is, therefore, some debate, especially in industry, as to whether it has been with us (without unduly harming us) over a very long period of time but passed unnoticed through lack of detection capability. Hence it is difficult to know if society is overreacting or if the more extreme concerns are legitimate. Because of the suspected health effects, Canada's health minister joined the environment and external affairs ministers in the fray, the three demanding immediate action from the United States.

The change of administration in Washington decreased the chance of a firm U.S. response, but Canada did secure a commitment from the United States at least to locate the dioxin source. The United States also agreed to determine if Dow Chemical Company's manufacturing process was the source of the substance. That effort is ongoing. In bringing up the possibility of legal action, however, the minister broke with precedent and may indeed have laid the groundwork for new diplomatic behavior. The issues of toxics and acid precipitation (see chapter 11) have become so sensitive that they may contribute to the creation of significant new precedents for international diplomacy.

In early 1981 the IJC issued its *Interim Report Under the Great Lakes Water Quality Agreement,* part of which dealt with toxics. The commission concluded that hazardous waste disposal sites were not the final answer and that government should encourage innovation in this area. It also requested an inventory of disposal sites in the region, studies performed on them, criteria for site selection, and examination of how future sites might be made safe.[14] It is clear that the IJC has been a major actor on the bilateral toxics front and desires to play a larger role in the future. It has always had both the mandate and the expertise, at least on the lakes, and may contribute to the solution of toxics problems elsewhere along the border as well.

Another facet of the toxics and hazardous substances issue is that of legal (and, occasionally, of illicit) movement of these materials across the international border. The issue causing most concern was the closure of the U.S. border to Canadian-source Polychlorinated Biphenyls (PCBs), forcing Canadian producers to develop domestic disposal methods rather than depend upon disposal sites in the United States.[15] With reciprocal arrangements having been agreed upon, the border has been reopened, and this aspect of the toxics question is no longer on the bilateral agenda.

Conclusion

Serious public (and, therefore, political) concern has developed over the health (and especially the carcinogenic) aspects of pollution, just as concerns over pollution aesthetics or property damage seem to be decreasing. Thus it may be that cleanup at all costs will be imposed on this one form of pollution. This would be a costly proposition but one which may become politically necessary. It is conceivable that government will bear some of the burden.

Industry has strongly applauded recent moves toward social and economic impact statements and the application of benefit-cost analysis. Such impact statements often lead to less ambitious cleanup objectives. Moves in this direction will likely succeed except in those instances of serious (or at least perceived serious) health threats, and this is the case, by definition, with virtually all toxics management. Thus, there will be no compromise with toxics. Toxic and hazardous substances will likely remain high and of continuing concern on the public environmental agenda.[16]

The bilateral toxics and acid rain issues have much in common, since they both represent serious U.S. threats against Canada as perceived by the Canadian people. Together they represent a changing scenario. The perception of more numerous but relatively small Canadian transborder impacts on the U.S. environment,[17] characteristic of the mid and later 1970s, may well evolve to a perception of less numerous but infinitely more complicated and allegedly more damaging U.S. transborder impacts on Canada in the 1980s. These issues, toxics in the Great Lakes and acid rain, are certainly ones where public perception (and the media attention that influences it) are of the highest importance to the present conduct and the future hope for Canadian-U.S. environmental relations.

To the west of the lakes and prairies, concerns over water scarcity and quality are replaced by concerns over aesthetics and wilderness preservation, as prairie grasslands merge into forested mountains, tiny turbid prairie creeks into clear raging torrents, and private crop and grazing land into national parks and forests. A wholly different type of bilateral environmental concern results therefrom, and the Cabin Creek–Flathead River issue in Montana and British Columbia and the emerging Alaska issues are indicative of that type. We now turn to these different issues.

NOTES

1. In November and December of 1980, Canadian newspapers were filled to overflowing with such headlines as "Knew About Hazards, U.S. Put Toxics Wastes into River, Report Says" (*Toronto Globe and Mail*), "Deadly Dioxin in Lake" (*Toronto Sunday Star*), "Dioxin Leak May Imperil Canadians: Roberts" (*Toronto Globe and Mail*), "America's Wasteland: Niagara Falls Contamination Found Reaching Across Lake Ontario" (*Toronto Globe and Mail*), "Canada Repeats Niagara Wastes Complaint to U.S." (*Ottawa Citizen*), "Roberts Considers Suing Polluters in U.S. Courts" (*Toronto Globe and Mail*), "Stop Dumping Into Niagara, IJC Suggests" (*Toronto Globe and Mail*). Similar radio and television broadcasts saturated the Canadian public at the end of the year and in the early months of 1981. U.S. media have also given much attention to toxics in the environment but, unlike the Canadian media, have generally ignored the bilateral aspects.

2. IJC Great Lakes Water Quality Board, *First Report of the Toxic Substances Committee* (Windsor, Ontario: IJC Great Lakes Regional Office, 1980).

3. IJC Great Lakes Science Advisory Board, *1980 Annual Report: A Perspective on the Problem of Hazardous Substances in the Great Lakes Basin Ecosystem* (Toronto, Ontario: IJC, 1980), p. 1. Both gas chromatography and mass spectroscopy have been in use for some years. It is the linking of the two that is significant.

4. Ibid., p. 47.

5. Ibid., pp. 51–55.

6. At the same time the IJC Joint Great Lakes Water Quality Board/ Science Advisory Board issued the annual report of their Committee on the Assessment of Human Health Effects of Great Lakes Water Quality and the Water Quality Board singly issued its *1980 Report on Great Lakes Water Quality*, both of which additionally addressed toxic substances in the Great Lakes Basin.

7. Diplomatic Notes were sent by External Affairs to the U.S. State Department in April, 1980, and again on November 28, 1980.

8. International Joint Commission (IJC), *Special Report Under the 1978 Great Lakes Water Quality Agreement on Pollution in the Niagara River* (Ottawa and Washington: IJC, 1981), p. 3.

9. Ibid., p. 4.

10. Ibid., pp. 8–9.

11. International Joint Commission, press release, Washington and Ottawa, February 11, 1981, p. 1.

12. "Act on Niagara Pollution," *Toronto Sunday Star* editorial, February 15, 1981.

13. Canadian Environment Minister John Roberts in early December, 1980, in statements reported widely in the Canadian press not only called for a production ban, but said that he was so frustrated by the lack of action of U.S. governments on dioxin and other toxics that "my officials are exploring the possibility of suing U.S. polluters in American courts" (Jock Ferguson and Michael Keating, "Roberts Considers Suing Polluters in U.S. Courts," *Toronto Globe and Mail,* December 4, 1980, p. 5. Also, "Canada Ponders Suit Against U.S. Polluters," *Ottawa Citizen,* December 5, 1980, p. 19). The threat of litigation is a significant move since it points out both Canada's deep-seated frustration over the issue and the inability of its diplomats to extract a satisfactory response from the U.S. government.

14. International Joint Commission, *Interim Report Under the Great Lakes Water Quality Agreement* (Ottawa and Washington: IJC, 1981), p. 8.

15. Based on interviews with U.S. diplomats.

16. Former U.S. IJC Chairman Robert Sugarman believes that the role of toxics in Canadian-U.S. relations will increase and that this is now the outstanding issue in Great Lakes water quality (personal interview, Washington, D.C., May 20, 1980).

17. Two possible sources of Great Lakes toxics which have received virtually no bilateral attention are radium from uranium mine sources on the Serpent River which drains into northern Lake Huron, and radioactive tritium from heavy water used by CANDU nuclear reactors bordering the lakes (Ralph Torrie, Friends of the Earth nuclear analyst, personal interview, Ottawa, October 6, 1980).

CHAPTER 8   Wilderness and Development

The Cabin Creek coal mine issue and its alleged threats to the quality of the Flathead River raises the broad question of whether one nation has a right to implement domestic legislation (National Wild and Scenic Rivers designation) which interferes with the right of an adjacent nation to achieve its own goals of economic development. It also raises the broader problem of incompatible environmental and developmental values and objectives at various places along the border and the changing nature of these values, even over a relatively short span of time. The Alaska–Yukon–British Columbia issues, largely potential at this time, are reminders of the length of this border and the occasionally high values at stake. They are also likely to pit development against wilderness protection. Furthermore, this area is more subject than most border regions to the whims of external world markets (for example, the increasing price of gold and the incentive this provides to mine and dredge with the inevitable ecological consequences).

Approximately four-fifths of all Canadians live within one hundred miles of the U.S. border, and nine-tenths live within two hundred miles. A much smaller percentage of the U.S. population lives this close to the Canadian border. The close proximity of so many Canadians to the United States is no accident, for here lies Canada's best and most productive lands and most agreeable climates. The strip of land paralleling the border might well be regarded as Canada's "development corridor," not only in the historic sense but also in terms of much future potential.

While it would be an oversimplification to imply that the corresponding border region in the United States is the opposite in every respect— untouched wilderness and likely to remain so—nevertheless much of it is undeveloped and held in high regard for its wilderness values and recreational potential. Herein lie the roots of a dilemma for, given this prevailing situation, the two countries are likely to adopt incompatible objectives, leading inevitably to friction.[1] The conflict over the proposed Atikokan power plant and its alleged impact on the Boundary Waters Canoe Area in northern Minnesota is one such example (see chap. 10). Other examples may be found farther west and north, including the proposed Cabin Creek coal mine in British Columbia, just a few miles from the Montana border and Glacier National Park, and a number of Canadian hydro dam and mining proposals in British Columbia and the Yukon which could affect

Alaskan wilderness. At present these are basically water resource issues, but air or possibly acid rain components could develop.

In understanding these "wilderness versus development" types of issues, no suggestion is being made here that either one of these two allocations of resources is superior to the other. Both are needed by society, but bilateral difficulties begin when one nation perceives a greater need for one while the adjacent nation longs for the other. While the two objectives can be pursued simultaneously, and even be complementary, they can also be mutually exclusive and thus incompatible. This is the current situation at Cabin Creek.

Cabin Creek

Ever since the existence of a mountain of metallurgical coking coal and thermal coal adjacent to British Columbia's Cabin Creek became known in the early 1970s, Montana residents, national environmental organizations, and lovers of nearby Glacier National Park have been uneasy. They fear this beautiful mountain wilderness is in dire danger of destruction. The so-called Cabin Creek or Flathead issue has yet to become a full-fledged bilateral diplomatic issue. However, the anxiety of environmentalists has been so evident and their reaction so well organized that U.S. diplomats, pressured directly by Montana officials and the state's congressional delegation, have continued to draw attention to this potential danger for the past four or five years. Since the decision to mine has not been made, Canada views the question as a nonissue, while the United States sees it as a constant threat.

Cabin Creek raises a major question. If one government sets special environmental conditions for a section of its own country, should another nation be deprived of its normal right to economic development in the adjacent section which would jeopardize the former's environmental conditions?

This issue, more than most others examined here, has also brought to the fore a major difference in national attitudes toward official secrecy versus "sunshine laws." Montanans are accustomed to much more openness in government and much less secrecy than are British Columbians. Thus, Montana environmentalists and government officials have been frustrated in trying to deal with British Columbia. The province has not really grasped Montana's concern and has felt somewhat threatened by the vigor of the state's reaction, not unlike Saskatchewan's reaction to Montana over the Poplar issue. Clearly, British Columbia felt directly threatened by Montana's new law which enables interests in the state to sue British Columbian interests for violations of the Montana environmental quality laws. U.S. diplomats have speculated that such state laws might be enforceable in light of recent precedents.[2]

Despite serious reservations in the United States, development of the extensive resource base in the area (on both sides of the border), which includes coal, timber, oil, and natural gas, is almost inevitable, and so is the likelihood of domestic disputes in both countries. These domestic issues will probably spill across the international border. At the same time, the ground is fertile for transborder environmentalist alliances.

The original proposal by Sage Creek Coal Company, owner of the mining rights to Cabin Creek, consists of the development of a large mine to extract high-grade metallurgical coking coal for export to Japan and other Asian countries. Such a development in the remote mountainous terrain of southeastern British Columbia requires the construction of a complete town site and transportation system. This costly infrastructure rivals the investment required for the mine itself, and its environmental impact would likely exceed that of the mine.

The quantity and grade of the coal, as well as its easily accessible near-surface location, make ultimate extraction certain, and U.S. diplomats have consistently operated under this assumption. Canadian diplomats, in contrast, have maintained that since no decision has been taken to mine, there is no issue. Two factors have led developers to postpone mining. One is the currently very soft world market (essentially in the steel industry) for coking coal and the consequent uneconomic price that the developer could expect in the early years of operation. The second is British Columbia's unwillingness or inability to subsidize infrastructural development in both the coal- and gas-rich northeast part of the province and at Cabin Creek simultaneously. The decision to focus first on the northeast has been ruled to preclude development of Cabin Creek for the time being. However, the world thermal coal market is improving and that may provide the rationale for the mine to go into operation.

With no improvement in sight for world demand in metallurgical coking coal, but with the value of lower-grade thermal coal climbing steadily upward, attention is now shifting in that direction and Cabin Creek is beginning to appear more as an energy issue.

At first sight, Cabin Creek appears to be a water quality issue, the problem being the pollution of Cabin Creek itself, the Flathead River, and ultimately Flathead Lake through acid mine runoff, and through general erosion and sediment deposition from town site and mine construction.[3] The greater underlying fear in Montana, however, appears to be the general impact of opening up this wilderness valley in both countries to a town site and the resultant increase in population, transport routes, and other aspects of economic activity. Furthermore, although the project is strictly in Canada, there is fear that new Canadian roads and traffic will increase pressures to build a U.S. highway north-south to the border, sounding the death knell (in the minds of some) to preservation of the wilderness in this area. Thus this is much more than a water quality issue.

The legal tolerance of the Flathead River to dilute pollutants was lowered when the United States designated it a National Wild and Scenic River. This designation, coupled with the fact that the Flathead forms the western boundary of Glacier National Park, gives the Cabin Creek question additional importance to many U.S. residents. As Canada discovered to its surprise in the case of Atikokan, posing a threat, however small, and whether real or perceived, to an area known to and loved by citizens throughout the land (in this case the Boundary Waters Canoe Area) can bring on a surprisingly vigorous and well-organized reaction. The result has been similar in the case of posing a threat, real or perceived, to the very famous Glacier National Park and to an equally famous adjacent river.

The Flathead Coalition, which formed quickly in Montana in response to this "threat," was able to respond effectively, with broadly based funding and other support, and found it could pressure the U.S. State Department directly through the Montana government and congressional delegation. Most of this effort was occurring at the state level at the same time that Montana was beleagured by Saskatchewan's Poplar power plant at the other end of the state. Hence, the Cabin Creek issue, known in some quarters as the Flathead or Sage Creek issue, probably received much more attention than it deserved, especially in view of its merely potential, rather than actual, status. In fact, U.S. domestic threats to Glacier National Park from energy resource development in northwestern Montana may well be greater than those from Canada, though this remains to be seen.

Montana's fear of deteriorating water quality in the Flathead River is not groundless. The absence of stream pollution is dependent upon the project's full adherence to British Columbia's environmental guidelines. However, these are only guidelines, not statutes carrying the force of law, and if past experience (anywhere) is a guide, they are not likely to be totally adhered to.

The question is twofold: how much pollution will actually occur, and how much should be tolerated within the spirit of the Boundary Waters Treaty and past international practice? Since Canada is the upstream state it has the right, under the U.S.-supported Harmon Doctrine, to do as it pleases with the flow of the river, and U.S. diplomats must start from this premise. Nonetheless Canada is bound by its position as a responsible neighbor to show concern over whatever damage it does cause.

Clearly, neither Canada nor British Columbia is bound by a U.S. decision to give the Flathead a special legally enforceable environmental status (National Wild and Scenic designation) since this is a domestic act that Canadians were not consulted on. Furthermore, under the precedents of international law, neither Canada nor British Columbia need consider the indirect regional impacts of their infrastructure on the wilderness in Montana. Not only is a sense of wilderness unquantifiable, but in this instance it is the personal values of Montanans which are at stake, toward whom Canada and British Columbia are not responsible. Moreover, in its Poplar water quality findings of 1981, the International Joint Commission determined that the Boundary Waters Treaty of 1909 does not preclude all downstream harm from pollution, only unreasonable amounts clearly causing property damage. Effluent into the Flathead River from the Cabin Creek coal mine and town site may fall into this category and the Poplar River precedent may be used by diplomats to reach a decision.

British Columbia's position has been that if Cabin Creek coal mining were to proceed, the province has the power to regulate the situation so that the Boundary Waters Treaty would not be violated. However, its reluctance to share information (and the company's desire to protect proprietary technical and economic information) has instilled strong doubts in various Montana circles as to the validity of this contention, as well as fears of environmental consequences.

In its defense, the province boasts of its Coal Guidelines[4] and the three-stage approval process that any applicant, including Rio Algom Ltd., must go through before they can mine. The British Columbia Environment and Land Use Committee (ELUC), the provincial government agency charged with administering the coal and other environmental guidelines, promised to see that the guidelines were honored, but ELUC was abolished in mid-1980. The guidelines remain in force, but their future is unclear, especially in that they are apparently being waived in the case of the large northeast British Columbia coal developments.

Three factors have been identified as relating to the continuation or termination of Cabin Creek as an issue: (1) the economic viability of the company, since the decision to proceed is theirs; (2) the firmness British Columbia shows in adhering to its own guidelines, which may well be determined by the strength of its desire to undertake this development and its ability to afford it; and (3) the strength and persistence of Montana and citizen environmentalists.[5]

A major concern of the company is the need to establish a "rule of law," clear and agreed upon guidelines and principles by which all concerned may proceed, perhaps with a treaty and mechanisms to back it up.[6] Such an organized approach is seen as far preferable from the company's perspective to the present ad hoc approach to disputes like that at Cabin Creek. The options available for resolution of the Flathead (Cabin Creek) issue are summarized in table 3.[7]

Bilateral and domestic pressure has already brought about some modifications of the mining company's plans, including a province-company agreement to find an alternative to the diversion of Howell Creek (a tributary of Cabin Creek), which responds to a specific concern of Montana resource managers. The removal of planned housing and service facilities to the nearby town of Fernie as an alternative to building a new town site at Cabin Creek, and the transmission of electricity to the mine

TABLE 3. Summary of Options for Resolution of the Flathead Issue

	Authority	Mechanism	Principal Actors	Comments
Scientific/technical information exchange	U.S./Canada Bilateral Information Exchange Meeting, Cranbrook, B.C., August 18–19, 1980	Direct and informal contact between Montana and British Columbia	State and provincial bureaucrats, resource managers, scientists	Favors cooperative approaches, local control, resolution based on resource concerns
Joint resource management	None for integrated programs; interagency agreements for local programs	Joint data collection; cooperative management plans for protection of fish and wildlife, ecological integrity	State and provincial resource managers, scientists	Practicable only on a local scale, to protect individual species or habitats
Bilateral negotiation	Boundary Waters Treaty of 1909, Article IV	Bilateral study, recommendations, and monitoring	U.S. Department of State, Canadian Department of External Affairs, state and provincial officials	Emphasizes institutional concerns over resource needs, potentially lengthy process with uncertain outcome
International Joint Commission (IJC) reference	Boundary Waters Treaty of 1909, Article IX	Scientific study, recommendations, and monitoring	IJC, advisory board	Favors resolution based on scientific investigation, recommendations and monitoring are externally set
Direct political pressure	Democratic process of government, open planning process	Public participation, journalism, communication of concerns	Montana interests	Provides input of local concerns, must be pursued in conjunction with other options for results
Legal recourse	Foreign Sovereign Immunities Act of 1976, international law	Legal suit in American courts; injunction enforced by Canadian courts	Plaintiffs in Montana, defendants in British Columbia	Constrained severely by lack of precedents, doubtful jurisdiction in Canadian courts

Source: Roy Stever, "The Flathead Issue in United States/Canada Environmental Relations: Institutional Responses in Montana" (Masters' Thesis, University of New Hampshire, 1981), p. 50.

from sources outside the area in lieu of building an onsite 40Mw coal-fired power plant to serve the mine's needs are further concessions to Montana pressure.[8] This evidence supports assurances by senior Canadian provincial and federal bureaucrats that Canada's environmental decision-making process is sensitive to Montana concerns.

However, Montanans have a further problem. This relates to the difficulties the state has encountered in seeking cooperation and coordination with the U.S. State Department. There was an initial lack of interest in the department in setting up a bilateral scientific information exchange, although this has now been established. Montana was further dissatisfied with the department's position requiring closed deliberations in bilateral negotiations in contravention of Article II of the state constitution, which requires open meetings and public access.[9] Both of these problems have been solved but, unfortunately, the Cabin Creek experience may well leave in Montana a residue of ill will toward the State Department. Given the long international border with three provinces which Montana shares, it behooves the State Department to work to improve its tarnished image in this important state. The formation of a joint bilateral monitoring group in 1981 was meant to defuse the issue, but Montanans were still upset, as evidence their testimony at public meetings in spring, 1982, where water pollution fears continued to surface.

Meanwhile, other events in the upper Flathead Valley may well defuse the Cabin Creek issue bilaterally more than could any environmental controls. The most important of these is the increasing pace of development in the U.S. portion of the drainage area. To the extent that the United States develops resource-exploitative activities such as oil and gas drilling, road building, and large-scale timber harvesting, on the U.S. side, it weakens its own credibility in demonstrating concern over Canadian resource exploitation north of the border and thus its negotiating position as a complainant.[10] Further, if any of the Cabin Creek coal finds its way to U.S. markets (for example, to a new coal-fired power plant in eastern Washington, as has been speculated), then U.S. diplomats must consider the welfare of U.S. customers as well as the protection of the Montana environment in any negotiations with Canada. Either of these developments might well cast a new light on the whole Cabin Creek–Flathead River issue.

Alaska–British Columbia and Alaska-Yukon Issues

There have been no actual transboundary environmental problems to date along the long and remote Alaska-Yukon and Alaska–British Columbia border, with the exception of the offshore Beaufort Sea issue discussed in chapter 4. However, the extent of the border and the area's great resource development potential (minerals, electricity production, logging, fishing, and various types of manufacturing and processing) virtually ensure that problems of one sort or another will develop and that the fast-developing state of Alaska will become a new actor on the transboundary environmental scene. The settlement of the Alaska land claims issue in late 1980, after a decade of intense national debate, concluded an issue which had dominated Alaskan politics and society throughout the 1970s, finally freeing the state to turn in other directions, including eastward to Canada. The coincidental rise in the value of energy, including hydro power, and in the price of gold have ensured that this region so well endowed with untapped large dam sites and deposits (even if small) of highly desired gold will be the site of new Canadian-U.S. bilateral problems.

HYDRO DEVELOPMENT

In the British Columbia–Alaska border region, both hydro- and gold-related problems are emerging. British Columbia Hydro (B.C. Hydro), a crown utility, is considering five separate large dam sites on the Stikine River (and its tributary, the Iskut), which rises in the province and flows through Alaska to the sea. One or two sites will ultimately be chosen, and Alaska fears that the changes in water flows resulting from the dam(s) might affect navigation or fisheries (though the impact could be positive as well as negative). A proposed National Wild and Scenic River designation for the U.S. portion of the Stikine could create further complications, although it appears that the recent federal wilderness designation for the river's shoreline will not enter into the debate.[11]

The Stikine issue is likely to emerge sooner than the Yukon Basin diversion issues, which are more distant in time. (A number of high-level meetings between Alaskan and British Columbian officials and heightened media attention in early 1982 belie this prediction.)[12] Three bilateral areas for potential issues have been identified in the Yukon Basin:

- the Mid-Yukon Project, a Northern Canada Power Commission proposal to build an up-to-1000 Mw dam at Carmack on the Yukon River, with major impact on U.S. salmon runs;
- an alternative diversion into the Taiya River from the Yukon headwaters;
- a diversion of the Yukon headwaters into the Taku River.

All three would alter streamflow affecting U.S. fisheries. Some of the hydroelectricity from these projects could be exported to Alaska and, ironically, could result in the development of aluminum processing and electroplating industries there, resulting in negative air quality impacts on Canada. This would create a reversal of the bilateral problem or perhaps result in neutralizing the issue bilaterally along with mutual economic benefits from the industries. Such an outcome would depend on whether export would occur (insuring a sharing of both costs and benefits), on the difference in timing of the two developments between the countries, and on the seriousness of environmental problems arising and their costs.

For most of the Alaska-Canada border, the United States is downstream and thus more vulnerable to water-related problems than is Canada, while the latter is downwind and thus more vulnerable to air pollution. This geographical fact might be useful for diplomats to remember, for it might provide the grounds for an important *quid pro quo*. Environment Canada has for years recognized the potential for future Yukon water quality problems in Alaska caused by mining (with heavy metals in river water) and planned industrialization of the Yukon Territory. It has thus established a data-gathering program at the international border to monitor and provide baseline data before real problems arise.[13] This effort in such remote country is expensive and signifies the importance assigned to future problems in the region. Some have argued that the Alaska-Canada interface would serve as an excellent laboratory to try out different early warning and notification systems now, before real problems arise—a chance for a fresh start.

MINING

With the recent dramatic increase in the price of gold, all three jurisdictions—Alaska, British Columbia, and the Yukon—are witnessing a renewed interest in small-scale (but very polluting) placer mining on a host of rivers and streams, many of which cross the border. This situation is likely to increase in intensity, and downstream Alaska will be most affected. Placer mining requires much dredging, causing great sediment movement which can affect the fishery (both in the stream and in the receiving estuary) and cause aesthetic, and sometimes limited navigation, problems.

The biggest placer mining issue to emerge thus far is the development of Scottie Gold Mines and other mines near Stewart, British Columbia, and Hyder, Alaska, with water quality impacts on the latter. Environment Canada has invested considerable effort in studying the environmental impact of these mines and has maintained liaison with U.S. and Alaskan agencies on the matter. The Canadian agency has also developed a series of detailed recommendations of a preventative nature in order to insure minimal environmental impacts as well as to avoid an international problem. A meeting of both governments was held in Juneau, Alaska, in late 1980 (following an earlier public meeting in Hyder, Alaska) to discuss the technical nature of the problem and alleviate U.S. fears of environmental damage to Alaska's Salmon and Leduc rivers from mine-tailings disposal. This meeting and the general attitude of Environment Canada appears to represent a positive effort on the part of the Canadian government to interact with all possible across-the-border interests. This effort is designed to head off potential problems and to insure that the principles of the informal early warning and consultation mechanism now in place between the two federal governments are fully implemented (see chap. 12).

Alaska also has water quality concerns from potential molybdenum mining activity on the Canadian side of the Taku River drainage, as well as from molybdenum mining in the Atlin, British Columbia, area. Canadian environmentalists have attempted to interest the governor of Alaska in actively opposing these developments on the grounds of radioactive waste contamination of the Yukon River. The latter is more of a domestic issue with indirect international effects, but is another example of transborder effort to achieve a domestic end.

Alaska-Canada environmental issues thus boil down to energy (hydro) versus fisheries (mainly salmon), metals mining and related industrialization versus water quality and fisheries, and gold placer mining versus sedimentation and pollution. Speaking of the important Yukon Basin, Maxwell Cohen has written:

> The optimum use of the Yukon system awaits some kind of joint image of the future potential of the basin in the interest of both countries; but already it is evident that there are very different perceptions, environmentally, upstream in Canada and downstream in Alaska.[14]

It is evident that both nations are at the dawn of a new era in relations in this region, and many of the precedents established now may have uncommonly great significance to future generations.

Conclusion

Cabin Creek is not the only 49th parallel border issue which sets up a clash between preservation of wilderness values and development of resources—there are others. It would be simplistic, however, to view such clashes in narrow terms of Canadian priority to economic development in a near-border development corridor versus U.S. priority to the preservation of wilderness values in a near-border primitive region. The level and increasing intensity of U.S. resource development in these same areas belies this premise. Equally simplistic is the sometimes espoused Canadian assumption that the United States regards Canadian-originated southward flowing rivers as prime candidates for Wild and Scenic designation, while treating U.S.-originated northward flowing rivers as public sewers. Indeed, it is not only simplistic but patently false (not to mention dishonest) to view environmental protection as somehow on a higher plane and more deserving of honor than the equally desirable goals and values of economic development. Both can and must take their place in the balance as both nations are called upon to practice one or the other at various times and places along the long and diverse border that at the same time links and divides the two countries.

NOTES

1. Accuracy demands acknowledgement that there are in fact quite a few proposed resource development projects in the undeveloped northern border regions of the United States which could well impact Canada, including Dickey-Lincoln Dam in Maine, a potential copper-nickel mine and peat mines in Minnesota, coal and coal-based synthetic fuels development in North Dakota and Montana, and a possible large pulp mill in northern Idaho. However, none of these has yet become a bilateral issue, perhaps because they are not yet close to development. The common perception, however, remains one of Canadian desire for development and U.S. desire for wilderness preservation, and it is in this context that current issues are developing. For further insight, see John E. Carroll, "Shadows on the Border," *Living Wilderness* 45, no. 156 (Spring, 1982):18–22.

2. 495 F. 2d 213 (6th Circuit, 1974), especially the Michigan case of *Michie* v. *Great Lakes Steel;* 401 U.S. 493 (1971), especially the Ohio case of *Ohio* v. *Wyandotte Chemical Corporation.* A secondary Montana concern over Cabin Creek relates to air quality, not from the mine itself but from electric power plants which might be built in this area to utilize part of the coal. This concern will likely be elevated by the continental acid rain debate and, regionally, by British Columbia Hydro's planned coal-fired Hat Creek electric power plant to the north, which Montana fears will impact some of its higher elevation conifer forests with acid rain deposition.

3. Although the sulfuric acid level would be relatively low given that this is low-sulfur coal.

4. British Columbia Environment and Land Use Committee, *Guidelines for Coal Development* (Victoria, British Columbia: British Columbia Environment and Land Use Committee, 1976).

5. Ed Nef, Assistant to Senator Max Baucus, U.S. Senate, personal interview, Washington, D.C., May 20, 1980. There is some Senate support for an Article IX reference of this issue to the IJC if Montana requested IJC involvement.

6. Bill Burge, Attorney for Lornex, Ltd. and Sage Creek Coal Co., Ltd. (subsidiaries of Rio Algom), personal interview, Vancouver, British Columbia, December 9, 1980.

7. In order to better prepare itself for what might well be inevitable, Montana obtained substantial federal support to conduct a comprehensive planning study in 1977, the Upper Flathead River Basin Study. Subsequent studies have been carried out since, aimed as much at U.S. domestic resource development in the North Fork of the Flathead, but serving a useful purpose and foundation of facts for the international problems as well. By way of further preparation, the private Flathead Coalition, in addition to performing typical lobbying functions, is performing legal research. Its focus is on the new U.S. Foreign Sovereign Immunities

Act of 1976, which holds forth some hope by expanding the potential for individual and private court action to check pollution from a foreign source (James A. Cumming, Attorney, Flathead Coalition, personal correspondence, Columbia Falls, Montana, September 18, 1977). The Flathead Coalition is a most unusual citizens organization in the nature of its approach and the forward-looking research it conducts.

Toward the end of the 1970s the Flathead Coalition went into dormancy when one of its major short-term goals—to establish a federally funded Flathead River Basin Environmental Impact Study (FRBEIS) with its own staff of professionals—came to be realized. Although not a citizens organization per se but a federally approved and funded study group composed of professionals sympathetic to the goals of the Flathead Coalition, FRBEIS was deferred to by the environmentalist coalition which began to see no reason for its own continuance. The means are available to activate the coalition once again should mining become imminent or should the study group fail to be funded at an adequate level. For the time being, however, its goals are being achieved without its active involvement. A new study has concluded the perception is widespread that the current level of scientific/technical information is sufficient, precluding direct political pressure. With a breakdown of this communication link, however, reinstatement of direct political pressure remains an option (Roy Stever, "The Flathead Issue in United States/Canada Environmental Relations: Institutional Responses in Montana" [M.S. thesis, University of New Hampshire, 1981], p. 53).

8. Ronald Cooper, statement prepared for Flathead River Basin Environmental Impact Study for submission to the Canada/U.S. Bilateral Meeting (Kalispell, Montana, February 15, 1980).

9. Stever, "The Flathead Issue," pp. 42–43.

10. There is a direct parallel here to the Beaufort Sea oil and gas drilling issue (see chap. 4).

11. The Canada-U.S. Environmental Council in its meeting of March 23, 1981, in Washington, D.C., passed a formal resolution opposing dams on the Canadian portion of the Stikine River, implying negative impact on the Stikine-LeConte Wilderness in southeast Alaska.

12. Professor Irving Fox, Head, Yukon Basin Project, personal interview, University of British Columbia Westwater Research Centre, Vancouver, British Columbia, December 10, 1980.

13. Based on interviews with Richard Millest and Robert Gale, Water Quality Branch, Inland Waters Directorate, Environment Canada, Ottawa, June 3, 1977, and Mac Clark, Regional Director, Inland Waters Directorate, Vancouver, British Columbia, August 2, 1977.

14. Maxwell Cohen, "Transboundary Environmental Attitudes and Policy—Some Canadian Perspectives" (Paper presented to the Harvard Center for International Affairs, Harvard University, Cambridge, Massachusetts, October 21, 1980).

CHAPTER 9

Prairie Water Issues

The Garrison issue introduces a new concern, that of biota transfer through diversion of one river system (Missouri) into another (Red River and Hudson Bay) with serious and negative foreseen ecological consequences. This issue may become best known for the exhaustive and perhaps definitive role of the IJC in the dispute and for the uncompromising statement of its findings, namely, that if the diversion and biota transfer occurs, the treaty will be violated.

The Poplar River issue involves the operation of a large coal-fired power plant close to the border, with all the benefits accruing to Saskatchewan while Montana bears many of the costs and receives no benefit. The issue recalls many traditional water apportionment questions and relies somewhat on past precedent. At the same time it establishes a new precedent by introducing the concept of equal sharing based on the flow of a system rather than of individual streams. It raises, but does not fully answer, the question of the extent to which future opportunity costs in Montana should be considered, given that they do not represent current damages. Treatment of the water quality aspect of the issue also establishes that some downstream pollution is acceptable and does not violate the Boundary Waters Treaty of 1909. Both the apportionment and the water quality aspects of the Poplar case provide classic illustrations of treaty interpretation and the functioning of the IJC under the treaty.

In that great semiarid region stretching westward from Lake of the Woods on the Minnesota-Ontario border to the foothills of the Rocky Mountains of Alberta and Montana lies a fertile expanse for transboundary environmental problems. Among the most important of these are the perennial conflict over the allocation of small but valued quantities of water in countless border streams. In the water-scarce prairie way of life, this water is extremely valuable. Another set of problems arises from deterioration in the quality of already marginal, highly alkaline water supporting populations living at the very edge of their tolerance for this necessary water. There is also the problem of "clean," but dusty, air that can carry pollutants in both directions across the border and contribute to more serious problems further east. Yet this vast region is also an area of economic opportunity, including that of increased food production through water diversion and irrigation, and of new energy development. Coal from large near-surface deposits is a major potential new energy source. These deposits can be strip-mined, causing transborder water problems, and converted into electricity on-site (mine-mouth generation). But this creates numerous local water and air quality problems and contributes to continental air quality problems.

The Garrison Diversion Unit in North Dakota, with its undesirable impacts on Manitoba, is one of the best known of all Canadian-U.S. transboundary environmental problems. It is also one of the few major environmental issues that is not energy-related. The Poplar Generating Station in Saskatchewan, with its associated coal mines and their attendant problems in Montana, is the other major bilateral environmental issue in the prairie region. A discussion of smaller but persistent regional water issues concludes this chapter.

The Garrison Diversion

The Garrison Diversion Unit (GDU) stands out as a giant among transborder issues. It has clear national connotations in Canada and regional connotations in the United States and is a classic example of a traditional type of transboundary environmental problem. No issue of its kind has received more public and media attention or has become more of a cause célèbre, especially north of the border. Garrison is a publicly funded, large-scale U.S. irrigation plan and river diversion that would permit large quantities of irrigation water with its share of pollution runoff and exotic fish and other biota to enter Canadian drainages. The project, initially designed and planned in the late 1940s, is one of the biggest irrigation projects on the U.S. northern plains. It has profound socioeconomic, political, and environmental ramifications, both positive and negative, in North Dakota, with some impact on neighboring states, and with allegedly profound negative impacts on much of Manitoba.

The Garrison Diversion issue has a number of distinguishing features relevant to this study. It has largely arisen because of the significant differences in awareness between the two nations and because of considerable disagreement over the validity of scientific data. It also has been subject to clear IJC findings which have not in practice been accepted by the United States. In these characteristics, Garrison is somewhat akin to another serious bilateral dispute, acid rain, to be described in chapter 11.

In the United States the GDU is purely a regional issue, largely restricted to the affected area, except for some national environmentalist concerns. In Canada, on the other hand, the issue is well known coast to coast, has been heavily publicized in national media, and has become a major symbol to many Canadians of U.S. insensitivity toward, and lack of concern for, its northern neighbor. For this reason more than any other, it deserves a place in any book on bilateral environmental relations. (For a detailed history of Garrison through 1979, the reader is referred to Carroll and Logan, *The Garrison Diversion Unit: A Case Study in Canadian-U.S. Environmental Relations*.)

Garrison presents a highly complex example of the simultaneous interaction of numerous actors and elements, domestic and international, within a major Canadian-U.S. environmental dispute. The political element of this issue operates on two separate levels. First, there is the U.S. and Canadian federal level, with U.S. federal activity being almost purely domestically oriented. Because of growing citizen involvement, there is a second political level within each country, composed of intense activity and interaction among the local, regional, and national citizens groups and their respective governments.

Scientific data has played an important role in the GDU controversy. Debate continues, especially in the difficult matter of biota transfer, about the validity of the data itself and about the nature of acceptable standards for transfer. (Canada's stand is clear and adamant on this. No transfer is to be tolerated.) Some argue that the transfers have occurred and the damage that will be done has already taken place. Arguments are also made that Missouri River biota could not survive in the Red River–Hudson Bay system, making transfer a moot point. Finally, some debate continues as to the relative effectiveness of fish screens and filters designed to prevent transfer. Such disagreement about scientific findings confuses both citizens and politicians and presents a serious challenge to both decision makers in general and diplomats in particular. There is no real excuse for such disagreement, however, for the IJC, which is highly capable and respected as a scientific research body since it employs the best of both governments' scientists, has issued its recommendations. They are not only clear but are the product of a totally bilateral body representing the interests of both nations. If new grounds for doubt have been uncovered or new data revealed, then the commission should be asked for further counsel.

A further element of special importance in this case study is the International Joint Commission. In contrast to its role in the Champlain-Richelieu issue, IJC findings in Garrison have been clear, strong, and highly contentious. Moreover, the commission completed its huge task rather speedily. However, some (including U.S. diplomats) argue that the commission's findings may ultimately come back to haunt it, as they have at Skagit. In the case of Garrison, the commission went as far as alleging treaty violation if the project were carried out as planned. The diplomatic community was so annoyed at this perceived overstepping of bounds that future references might well be written more tightly so as to restrict the commission's activity even more than in the past. Be that as it may, the IJC role in Garrison was a large one with undoubtedly historic consequences.

THE PROJECT

The Garrison Diversion Unit would transfer water across the continental divide into central and eastern North Dakota from the Missouri River.[1] (At this point the continental divide separates areas draining into Hudson Bay from those draining into the Gulf of Mexico.) The diversion would be accomplished by the Snake Creek Pumping Plant, located near the eastern end of Lake Sakakawea, the reservoir behind the Garrison Dam on the Missouri River (see map 5). The pumping plant would lift water into Lake Audubon, an impoundment (reservoir) adjacent to Lake Sakakawea. Water would flow by gravity from Lake Audubon through the McClusky Canal into the Lonetree Reservoir. Water in the McClusky Canal would pass through a screening device designed to prevent fish, fish eggs, and other aquatic organisms from entering the Lonetree Reservoir. A small region, the Lincoln Valley Area, would be irrigated directly from the lower end of the McClusky Canal. The Lonetree Reservoir would form behind the Lonetree Dam, which is being constructed across the Sheyenne River, and the James River dykes along the headwaters of the James River. This reservoir would be able to divert water by gravity into the basins of the James, Red, and Souris rivers, as well as into the Devils Lake Basin. The Snake Creek Pumping Plant, the McClusky Canal, and the Lonetree Reservoir constitute the GDU's distribution system. Although primarily de-

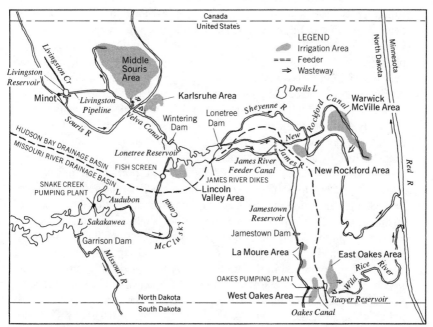

Note: not drawn to scale

Map 5.   Schematic Representation of the Garrison Diversion Unit. (Adapted from International Joint Commission, *Transboundary Implications of the Garrison Diversion Unit* [Ottawa and Washington, D.C.: International Joint Commission, 1977], p. 12.)

signed for irrigation, the GDU would also provide substantial quantities of water for urban-industrial purposes to Minot and thirteen other communities.

Garrison becomes an international issue because of the question of return flows (i.e., the flow of the diverted water to another river). The bulk of the return flows would enter Canada. Most of the remainder would pass south via the James River and reenter the Missouri River in South Dakota. It is the polluted nature of these return flows, resulting from agricultural pollution (fertilizers, agricultural chemicals, phosphates, nitrates), which may endanger the water supplies of a number of Manitoba communities. Biota transfer (i.e., the transfer of living plant and animal organisms, including diseases from one basin to another) allegedly threatens the substantial commercial and sports fishery of lakes Winnipeg and Manitoba, perhaps risking, directly and indirectly, the livelihood of a substantial number of Manitobans. Biota transfer has emerged as the main (and least tractable) bilateral issue of the Garrison case because of its threat to an established commercial fishery.

RECENT DEVELOPMENTS

With the loss through retirement in 1980 of Garrison's most influential supporter, Senator Milton Young, opponents assumed that the project was dead. The Senator did, however, get a further $9 million appropriation for Garrison as one of his last acts in office.[2] The Interior Department was thus presented with the dilemma of how to spend these appropriated dollars, as they were bound by law to do, and, at the same time, to abide by the Senate resolution to avoid spending money in such a way as to harm Canada. (The U.S. Senate, although appropriating funds for Garrison, has resolved that no funds shall be spent on any aspects of the project harming Canada.) And, there have been recent efforts by North Dakota to get started and fill the McClusky Canal, a key element in the diversion. It is impossible to spend more than a few million dollars on those portions of the project which only affect domestic waters, leaving the bulk of the appropriation to be spent on waters flowing into Canada. The bill requires that the money be spent on actual construction, not on further studies. The difficulty, of course, lies in the definition of harm to Canada, over which project proponents and Canadian opponents have consistently disagreed.

This dilemma led to threat of a suit by the Garrison Diversion Conservancy District, builders of the project, if a construction schedule was not announced and the funds spent as required by law.[3] Predictably, it also led to a Canadian Diplomatic Note (dated October 1, 1980) to the State Department asking how the funds were going to be spent and for renewed assurances that the funds would not be spent on any parts of the project affecting Canada.[4]

A new force arrived on the anti-Garrison scene in 1980, the Manitoba Indian Brotherhood. Little attention had previously been paid to this affected group, in spite of the fact that much of the fishing on lakes Winnipeg and Manitoba is either Indian subsistence fishing or commercial fishing carried on by Indians. An estimated seventeen thousand Indians in twenty-eight communities would be adversely affected, mostly around Lake Winnipeg. One-third of the lake's fishing and earned revenue is Indian, and the revenue from this fishery was $1.9 million in 1979.[5] The catch is 75 percent whitefish, walleye, and sauger, and these species allegedly will be decimated by 50 percent starting in 2033, as a result of biota transfer from the Missouri River system. With completion of the GDU this represents an estimated annual loss (present value of future losses) of some $48 million and an ultimate collapse of the fishery. These Indians also depend on fish as the basic element in their diet and earn additional income from guiding sports fishermen. Both means of livelihood may be threatened by a reduction in the fish population.[6] Representatives of the Manitoba Indian Brotherhood stress that this means the end of a way of life and of a culture and that it will inevitably lead to these peoples joining the welfare rolls to become a liability to society.[7]

With the election of a western-oriented Republican U.S. administration which is generally sympathetic to publicly subsidized western water projects, Garrison proponents received a boost. Election of a Republican administration in North Dakota and of a governor who had a longtime pro-Garrison involvement as attorney general was similarly perceived. However, the 1981 court victory of the National Audubon Society, which blocked further construction, together with the souring of North Dakota farmers over loss of their land (to provide acreage for new federal waterfowl refuges, to mitigate losses of such acreage to the project, as required by law, as well as for project construction), indicates that the matter will continue unresolved domestically, perhaps for many years.

The recent return of the more environmentally oriented and more nationalistic New Democratic Party (NDP) to power at the end of 1981, with the accession to the premiership of Howard Pawley, long an outspoken opponent of Garrison, undoubtedly means further polarization in Manitoba and, likely, Ottawa as well. Indeed, one of the first acts of the new NDP government was to lay plans for establishing a Manitoba government office in Washington, D.C., for the sole purpose of lobbying the U.S. Congress, the administration, environmentalists, and the U.S. media to promote the province's concerns over the project. Though this has not come to fruition, Manitoba's Garrison concerns are now directly represented at the Canadian Embassy in Washington. Several close appropriations votes in Congress and the Reagan administration's penchant for budget cutting lend encouragement to this diplomatically unconventional action. Consideration is also being given to establishing an office to achieve similar goals in Bismarck, the seat of North Dakota government. Thus, while perhaps unresolved, this bilateral issue will remain a visible one for a long time to come.

A DIPLOMATIC CHALLENGE

In spite of continued and strong opposition to the GDU project by Manitoba, there is some concern in the province that aspects of the issue have been overplayed and that the provincial and Canadian federal reaction has therefore been too strong. The transfer of disease and parasitic organisms in fish may not be as much of an issue as portrayed, and may be more hypothetical than real. It is further suggested that the undesired fish may have already been transferred but failed to take hold. The water pollution aspect and its impact on southern Manitoba communities can perhaps be answered by the drilling of new and deep groundwater wells, with the costs paid by the United States. However, Manitobans are convinced that no filter can be designed to prohibit biota transfer and that the only answer to this problem is institution of a closed loop on the Souris, thereby directing all return flows back into the James River system.

Perceived fears of the results of the GDU may have pushed Canada to take a stronger diplomatic stance than is appropriate. For example, Lonetree Dam could be built without an outlet to Canada, and yet much formal Canadian diplomatic opposition has revolved around construction of this dam, which may well be a U.S. internal matter.

Also, does Canada have a right to dictate to the United States how to use the U.S. portion of the Hudson Bay drainage? Is Manitoba too purist in being unwilling to accept any damage? In keeping with Canada's national behavior on other transboundary water quality issues, should Manitoba be willing to share its assimilative capacity? It has been suggested that this inconsistency between federal and provincial positions will cause future problems if Ottawa is pressured politically to support the provincial argument. In all other cases Canada opts for water quality objectives and the full use of the assimilative capacity. Might this inconsistency of Canada's nondegradation demand over Garrison return flows someday backfire? These are important questions that have essentially been lost in the fiery public debate over Garrison.

In viewing North Dakota's very strong and consistent support of the project and its opposition to any real compromise or modification, in the face of severe opposition, one naturally suspects that there may be reasons other than those publicized to account for the state's passion for this project. The Garrison Diversion Unit is normally thought of as an agricultural irrigation project which would provide supplementary benefits to municipal and potential industrial users. However, the real thrust behind the project may be conversion of the state's economy to a more diversified industrial base rather than enhancement of its agricultural base.

A basic rationale behind North Dakota's determination may well relate to its desire to ensure rights to Missouri River water and to a belief that the best way to guarantee these rights is to demonstrate a use for the water before other states upstream or downstream (Montana, South Dakota, Nebraska, for example) establish a claim. Diverting the water to Canada after irrigating its own land may be the means to accomplish this end. The state has established a right to 3.5 million acre-feet of Missouri River water, and some believe it will fight hard to protect that right. In other words, the drive behind Garrison may not be to secure the project itself or the ensuing irrigation advantages, but to acquire the water rights and perhaps ultimate freedom to use that water for industrialization and, possibly, energy-related developments.

If the end user loses the water right, then the supplier, in this case the Garrison Conservancy District, has no end use for the water. Under prior-appropriation-water law and tradition in the western United States, lack of a legitimate use leads to a forfeiture of the right to the water and permits upstream or downstream interests a chance to establish their own right and to appropriate reasonable quantities of water under that right.

There were a number of articles in the North Dakota press in 1980 that raised the fear of loss of Missouri water rights to North Dakotans, reflecting increased political interest and speechmaking. One such article quoted former governor Arthur Link that the state's claim to the Missouri River water goes when the project authorization goes.[8] State Engineer Vern Fahy is quoted as saying "First in time is first in right" and that many downstream states have suggested using the Missouri River to recharge their ground water aquifer.[9] He noted that as long as work continues the North Dakota claim stands.

Other state leaders could be similarly quoted. North Dakota politicians have often raised the spectre of downstream and eastern coal interests demanding water from North Dakota, and the National Audubon Society has been accused of being in league with coal and barge interests to keep the Missouri water going downstream. The state's congressional delegation has maintained an attitude that the project was to get and keep Missouri River water rights and that the project should be built in order to establish those rights. However, if it relies on the GDU project to establish this right, then North Dakota could only claim 871,000 acre-feet, not the 3.5 million authorized since the former is all the project can absorb.[10]

What this discussion of water rights suggests is that a solution to the bilateral dispute may well lie in some form of federal-state agreement guaranteeing North Dakota a firm right to all or most of this water until such time as it can use it, while at the same time deauthorizing the Garrison project or at least the great portion of it that affects Canada. With such a guarantee it is conceivable that North Dakota would drop the fight for Garrison. The water in question could be used to develop the state's extensive fossil fuel resources, including the substantial tight gas (or deep gas) formation that is reputed to underlie most of North Dakota. Recognizing a historical commitment to the state, Washington could not only guarantee the water by act of Congress but could also grant the state a federal appropriation to develop this natural gas and quite possibly a new synthetic fuels industry as well.

There are, therefore, ways to resolve this serious bilateral dispute before it causes more harm in Canadian-U.S. relations and more enmity and bitterness on both sides. The United States is coming perilously close to violating the Boundary Waters Treaty of 1909, and even the perception of treaty violation, if widespread in either country, will be enough to cause long-term harm and probable repercussions in other areas of the relationship. Striving to find a solution that will end this bitter controversy is a good investment in the future for both peoples.

Poplar River

The Poplar River issues of water apportionment, water quality, and air quality rank second only to the Garrison Diversion as the most serious bilateral environmental disputes on the prairies. These disputes are serious because:

- there are three separate problems originating from the same power plant, representing all three principal types of transboundary environmental conflicts;
- there have been two separate references to the IJC;
- the reaction of the state of Montana has been highly politicized;
- persistence and stubbornness have characterized Saskatchewan's response.

Poplar is a major prairie environmental issue, and it has left a legacy of lessons to both countries in its wake.

One major legacy will be the IJC's definition of pollution versus injury. In Poplar, the IJC decided that pollution that does not affect the overall interests of the downstream country does not constitute injury in the treaty sense even if some injury occurs. ("Overall interests" was not defined.) Injured downstream interests, however, must be compensated for their injury. This decision may well establish an important precedent affecting other cases for many years.

To follow the actions and reactions that have characterized this issue from its inception, one must understand the context in which this large one-to-four-unit, 300-to-1200 Mw, coal-fired power plant is being built. For most of its history, Saskatchewan has had a one-crop economy—wheat—that was dependent on volatile world markets. The province also has an unpredictable climate. With its single dependence on wheat, Saskatchewan was just as disadvantaged and powerless as the cotton-dependent U.S. South of the nineteenth century. When the Great Depression hit North

America in the 1930s, there were few places that were harder hit than Saskatchewan in terms not only of high unemployment and economic disruption but of widespread malnutrition, near heatless homes in the depths of some of the continent's worst winters, and of some of the greatest hopelessness to be found anywhere. It was in this environment that many of the province's present leaders were born or have childhood memories. In addition, Saskatchewan has been a "have-not" province within the Canadian confederation for most of its history.

Today, for the first time, Saskatchewan sees not only real prosperity on the horizon but also its first real opportunity to break the stranglehold of a one-crop economy. In addition to hopes for uranium, potash, oil, gas, farm machinery, and other manufacturing development, the province views its vast lignite coal deposits near the U.S. border and the inexpensive electric power this vast storehouse of coal can yield (for perhaps as long as five centuries) as a key to its future industrial development. Development of these resources offers a hope of reversing the tide of out-migration, providing residents with productive jobs, and attracting others to further diversify its economy. But this development could be threatened by strict environmental controls, for water and particularly for air (including sulfur scrubbers).

Saskatchewan's Grant Mitchell aptly brought this view of his province's role to life when he said:

> As a developing province, we view with concern approaches which would apply the best available technology in pollution control regardless of the cost to all new developments when no agreement has been reached to apply such standards to existing developments. It must be kept in mind that our new industrial and resource projects must compete with existing operations.[11]

Older developed regions in the East naturally opt to limit the application of controls to new developments so that the controls do not threaten already established facilities. It is equally natural for developing areas such as Saskatchewan to object to the unfair burden this places on their attempts to develop under the pressure of restrictive and expensive controls.

There is no question that the province's coal deposits are very large, near the surface, and thus capable of being strip-mined. Although deficient in heat (Btu) value, these deposits are sufficiently abundant to make up for the per-unit heat deficiency. They are also low in sulfur. The most economical way of developing this coal is mine-mouth conversion into electricity, with the electricity being transmitted to markets elsewhere. Although the province denies it, some of this electricity could conceivably be exported to the United States bringing positive cash flow to the province, at least until Saskatchewan industry develops to a level sufficient to absorb all the additional power. Finally, these lignite deposits insure that the province will be able to meet its own continually increasing demand.

It is in this context that the government-owned Saskatchewan Power Corporation (SPC) proposed in 1973 to build a large power plant consisting of four 300 Mw units. The plant was to be built on the site of a vast lignite deposit just a few miles north of Montana.[12] The Poplar Generating Station has been the object of two separate references to the IJC, the first on water apportionment, the second on water quality. It also has been a factor in the debate on air quality described in chapter 10. The apportionment reference was a traditional IJC-type exercise and will be described first.

WATER APPORTIONMENT
The three forks of the Poplar River—East, Middle, and West—are each considered as an international river. Therefore, any manipulation of the East Fork as required by the plant would have an impact across the international border. Such manipulation requires a Canadian federal license (as per the International Rivers Improvement Act of 1970). Such a license was granted by Environment Canada in May, 1975, to the Saskatchewan Power Corporation. Federal blessing of the project was given with the knowledge that there could be bilateral problems, but in the belief that none would occur. The federal environment minister did not reckon with Montana. Nor, apparently, did the U.S. State Department.

Any small stream in an arid or semiarid area, including the East Fork of the Poplar, is a highly valued asset because alternate sources of water are very expensive. This consideration, together with the fact that this particular prairie stream flows over a large vein of lignite coal, allows us to understand the importance of the issues at stake. The skeptic, who might otherwise consider all of the international debating over Poplar as excessive, is forced to reconsider.

As described in chapter 4, the United States has adhered to the Harmon Doctrine in international law. There are a few instances where the United States, as the downstream state, suffers a disadvantage under this doctrine's application. Poplar is one of them. The Boundary Waters Treaty of 1909 allows the upstream user (Canada) complete control over river flow on its side of the border, except that upstream interests (Canadian) are liable for any downstream damages the upstream uses may cause. Thus in the Poplar apportionment case, Saskatchewan had full rights to the water and was not obligated to share it with the United States. Neither the province nor Canada was obliged to ask for an IJC ruling or even to negotiate with the United States over the matter, except for carrying out its responsibility to compensate for downstream damages. Most parties in Saskatchewan were aware of this legal situation so certain U.S. objections inevitably angered Saskatchewan, given the province's right to proceed without negotiation.

Most Montana politicians, media, and citizens organizations did not seem to be aware of the inherent Canadian right and the lack of protection afforded them in water allocation in the Boundary Waters Treaty (resulting from earlier U.S. insistence, over Canadian objections, on protecting the undiminished sovereignty of the upstream state). While the furor in Montana over Poplar apportionment did not reach the levels it did with water and air quality, nevertheless some of the negative reaction that did arise assumed U.S. rights that did not exist.

There is sufficient water in the East Fork of the Poplar to support the needs of the first 300 Mw unit, and a second 300 Mw unit could probably be provided for in all but the driest years. There is a consensus, however, that the water supply is not sufficient for the third or fourth units and that, with present technology, new water would probably have to be diverted into the East Fork from other branches or from a wholly new source. Such a diversion could result in a net increase rather than a decrease of water crossing the border.

The Poplar River had never been apportioned at the border. The Montana Environmental Quality Council recognized that past practice involved an approximate equal division at the border;[13] while Saskatchewan's early board of inquiry report on the Poplar project mentions only that a riparian outlet in the water control structure would be installed for downstream releases and that SPC would make releases so as to approximate natural flow conditions.[14]

In early 1975, the IJC, under the continuing authority of its Souris–Red River Reference of 1948, announced a review of all impacts of the Poplar plant (and simultaneously all air impacts under its International Air Pollution Advisory Board's "watching brief" responsibility) for the purpose of bringing the growing problems to the attention of the two governments. The two governments endorsed the IJC apportionment activities, while deferring endorsement of air and water quality aspects. The IJC was thus launched (in a sense at its own request) on a study of the impacts of Poplar and appointed a task force to hold hearings in both the province and the state.

Saskatchewan emphasized at the IJC task force hearings that water control in general in this semiarid land was essential and that the province could beneficially use all the water of the Poplar River Basin for agriculture and energy. It rejected a traditional fifty-fifty division on the grounds that such a division does not reflect the potential for beneficial use on each side of the boundary, as required under the Helsinki Rules,[15] and that the province is thus entitled to more than a 50 percent share. Specifically, it opted for "permitting a great percentage of natural flow to be used on individual tributaries," and in fact up to 70 percent of the natural flow of such tributaries.[16]

The province also favored a reservation of existing water uses in Saskatchewan on the East Fork, including all water necessary for Unit No. 1, and maintained that this portion of the natural flow be subtracted before determining the U.S. share.[17] This obviously would net Saskatchewan a good deal more water than it might otherwise expect. In effect the province argued that the entire watershed rather than just one tributary stream should be considered for apportionment. In support of this view, the province again cited the Helsinki Rules, which

> define an international drainage basin as a geographical area extending over two or more States determined by the watershed limits of the system of waters flowing to a common terminus and consider that the drainage basin is *an indivisible hydrologic unit and must be considered in its entirety* [emphasis added] (Article IV, Helsinki Rules).[18]

Also cited was the IJC precedent in the Pembina River apportionment case, wherein the commission recommended that the basin's water be apportioned according to the average yield in each country (Canada: 60 percent; United States: 40 percent).[19] The foundation for the ultimate Poplar apportionment agreement, and perhaps for many to come elsewhere, was set in this request.

The provincial position called for a division of the total basin flow based on the average of five years of measurements rather than a single year to ensure greater accuracy. Saskatchewan officials emphasized to the commission the fact that prairie conditions require storage to trap the spring runoff so it can be used when most needed.[20] On highly flow-variable prairie streams a controlled flow available when needed is more valuable than a guaranteed percentage of the natural flow.[21]

Also to be considered is the matter of other streams in other basins. Should the fact that Saskatchewan gives 80 percent of Frenchman Creek to Montana and 90 percent of the Souris River to North Dakota (because it cannot utilize them at present) justify a trade-off on the Poplar, with Saskatchewan receiving more than 50 percent? The answer to this question revolves around whether the recipient area to be deprived (i.e., Montana) can make use of the flow that Saskatchewan would be retaining. In the Souris River case in the early 1940s, the IJC recognized the value of upstream storage, as well as the obligation of the upstream country to meet minimum existing needs, and this management regime survived intact for forty years. Considering this philosophy, Saskatchewan's final submission to the IJC requested either an equal division based on a five-year moving mean or, preferably, a seventy-thirty split favoring Saskatchewan based on annual flows. Montana ultimately accepted this thinking. From the beginning, Montana and the U.S. State Department wanted water quality integrated with quantity, but Canadian opposition insured separation of the two problems, at least for a few years. Hence the IJC received two separate references two years apart (1975 and 1977).

U.S. representatives on the IJC task force initially opted for a fifty-fifty split, but construction of the Poplar Plant made this unrealistic. Insistence on an equal division would have prevented reaching any agreement.[22] In any event, U.S. members ultimately recognized that a fifty-fifty division would mean less water for the United States in low-flow periods,[23] but they did insist on a guaranteed minimum flow in the East Fork and additional flows in the Middle and West Forks to compensate for East Fork losses. In the final analysis, the Harmon Doctrine was honored but at the expense of the United States. If only the water and air quality issues could have been disposed of so easily. Of course, this is not to say that all residents of nearby Scobey, Montana, with their concerns over loss of possible ground water recharge for their municipal supplies, are completely satisfied.[24]

The overall Poplar water apportionment issue has been complicated by the Fort Peck Tribal Council's raising the question of Indian water rights under the "Winters Doctrine."[25] The Fort Peck Reservation lies in the lower Poplar Basin, mostly south of the point where all three forks of the Poplar join, and just north of the Missouri River. The tribe claims that it needs the water for future irrigation, industrial, and domestic purposes, and maintains that satisfaction of this need has priority over any national or international obligation under the Boundary Waters Treaty or any other treaty. Diplomatic problems result from this demand. First, although Canada does not recognize U.S. Indian treaty rights, which it regards as a U.S. internal matter, any action it takes to assert its full treaty rights to Poplar water could force the U.S. government to intervene. Washington would thus have to either intercede with Canada on behalf of the tribe or find an equivalent water source for the reservation. Second, the tribe's prior rights exacerbate water pressures on the area's non-Indian population. The IJC in its apportionment deliberations can do little more than to be aware of the existence of this problem. A letter to the commission from Deputy Assistant Secretary of State Richard Vine recognized the validity of the Indian claim.[26]

One assumes such validation would not apply to Canada nor that Canada could in any way be held accountable under international law. However, some tribe members believe otherwise and would like the opportunity for direct negotiation with Canada to protect their interests.

The original IJC task force report was rejected by Saskatchewan in 1977 because it did not advocate either of the two principal goals of the province, namely the five-year averaging of the mean flow or the seventy-thirty split. This rejection caused problems in Montana, since people there erroneously considered the task force report as a negotiated settlement and became upset with Saskatchewan's negative reaction. This was a low point in state-province relations during the apportionment debate, but it did not last long, for the final IJC recommendations satisfied both sides.

On April 12, 1978, the IJC approved the task force's final recommendations, thus terminating its role in Poplar apportionment (although it continued to be involved in Poplar water quality until early 1981). The commission recommended equal division of the flow of the West Fork (with the flow of each tributary not being depleted by more than 60 percent of its natural flow). On the remaining streams and their tributaries, the total natural flow would be divided equally with Canada delivering to the United States a minimum of 60 percent of the natural flow of the Middle Fork, while on the critical East Fork, Canada would guarantee to the United States minimum sustained flows throughout the year varying between 1 and 3 feet per second, as well as volume releases on demand varying from 300 acre-feet to 1,000 acre-feet. The significance of these figures is that Saskatchewan may store a major portion of the natural flow of the East Fork, permitting it to operate one 300 Mw unit at all times and a second 300 Mw unit 97 percent of the time (that is, continuously, except when there are extended periods of extreme drought). Montana, in return, is getting more than its normally to be expected share of the Poplar Middle Fork as well as the guarantee of a continuous small base flow of the Poplar East Fork.

Saskatchewan had succeeded in achieving its minimum requirement: a flexible apportionment formula for the East Fork of the Poplar, while Montana was guaranteed an equal apportionment of the waters of the entire basin (the three tributaries combined). Thus reliance was placed on traditional equal division of the waters (but in terms of the whole basin) protecting both countries. The IJC apportionment focus was broadened to encompass an entire basin, maintaining the twenty-year precedent of the Pembina River decision and strengthening future commission flexibility in apportionment decisions; and the immediate needs of Saskatchewan were filled (although there is evidence that the province is not fully satisfied and that more problems may be on the horizon) while Montana was not harmed. The IJC, working in its traditional area of strength, had maintained and promoted good bilateral relations while simultaneously solving a difficult technical problem.

The commission also used this opportunity to warn both governments of their international obligation to give prior notice to each other and to engage in early consultation, an obligation that was not fulfilled in the case of Poplar. In view of this finding, the commission did not determine the station to be "an existing use" of water because the obligation of prior consultation was not fulfilled.[27] Thus Saskatchewan was chided for not engaging earlier in prior notice through prompt release of information and was penalized by losing the possibility of having Poplar Unit No. 1 excluded from the water demand calculations. Hence the province lost rights to that much more water.

The commission also warned governments of the difficulties that result when construction continues on a project during IJC review. A race ensues, continued construction biases the outcome, and the result is substantial political difficulties in correcting such problems.[28]

What about additional units at Poplar? The plant is being built to accommodate four 300 Mw units, whether or not all four will ever be built. Most people on both sides of the border agree there will be no more than two units within the next four years. After that, however, opinions range from "inevitable" to "no chance." The SPC and the province both spoke of four units before the international conflict emerged. The announcement that there would be a second unit, even though expected, raised a major furor, though this can be partly attributed to the fact that major air and water quality studies were still in progress. Interestingly, though, the IJC crafted its apportionment formula to allow for a second unit, even though one had not then been announced—a sign of reality and foresight on the commission's part.

Construction of the third and fourth units is threatened not only by fear of environmental effects or the likely Montana or U.S. reaction. While there is sufficient coal, solution of difficult water and possibly marketing problems must first be achieved. The apportionment formula clearly prohibits further units, as does the natural flow of the East Fork itself. Diversion of other waters, such as Wood River or Lake Diefenbaker, is physically possible. Since these are both part of all-Canadian drainages, diversion into the Poplar would constitute a transboundary diversion and a water export from Canada, something that is politically unacceptable at present. Semiarid eastern Montana might at some point view such an import diversion as desirable, however (as might the Fort Peck Indians), if it were not too polluted. It is always possible that Canada would relax its rigid antiexport views if so much power production were at stake. It would likely have to be convinced, though, that it could keep water export to a minimum.

Another option for SPC is a new technology called dry cooling, as yet unavailable, which would obviate the need for more water. The need for a larger market could be solved by rapid growth in demand within the province, coupled with electricity exports to the United States, either directly through the Basin Electric Cooperative's transmission line intertie in North Dakota or indirectly through Manitoba. (The adjoining provinces of Manitoba and Alberta do not appear to offer markets.) The temptation arising from hundreds of years of coal supplies lying just below the surface and the economies of mine-mouth conversion to electricity may well mean that a way will be found to overcome these problems and construct additional units, in spite of their effects on apportionment and air and water quality. Once the IJC apportionment recommendations are validated by a formal agreement, however, at least the first stage of this bilateral apportionment conflict will have concluded.

WATER QUALITY

The most vexing water aspect of the Poplar dispute has been that of quality, both in the context of the IJC reference and in diplomatic negotiation. It has been carried on simultaneously in both forums for a number of years (an unusual circumstance) and has been the cause of much acrimony and bitterness. Although hydrologically, economically, and legally, it is illogical to separate quantity and quality considerations relating to such a small stream, the two governments did just that. The early history of the Poplar water quality debate is a bilateral diplomatic one; its later history an IJC matter. In its final stages the debate returned to the diplomatic arena.[29] The artificial separation of water quality and quantity was apparently due to Canada's and Saskatchewan's concern over problems that might arise from integrating the two aspects. It was unsuccessfully opposed by Montana and the United States.

Prairie water quality under natural conditions is at best marginal (i.e., on the verge of unacceptability) for most purposes, being high in alkalinity and containing many trace metals and total dissolved solids. It does not take a large addition of pollutant (particularly agricultural runoff) or much evaporation in this arid environment to make it unacceptable. In addition the region's scarcity of water means there is little excess available to dilute pollutants and little surplus to turn to when a traditional supply is lost.[30]

Bilateral water quality problems did not come to be recognized at Poplar until well after apportionment questions were raised and even after air pollution discussions became commonplace. In early 1976, recognizing Montana's concerns over pollution, Canada suggested that a bilateral study group on water quality be established, but the United States rejected this in favor of a reference to the IJC. In July, 1976 a joint decision was made to refer the matter to the IJC. More than a year of disagreements ensued over the precise wording of that Article IX reference, and it was finally forwarded to the commission on August 2, 1977, with a request for the final IJC recommendations by December 1, 1978. This request was not fulfilled until January 20, 1981. Showing some foresight, the two governments agreed that all findings of the commission should be predicated on a 600 Mw power facility, a realistic assumption by the time of the mid-1977 reference. To carry out its work, the commission established its International Poplar River Water Quality Board.

Montana's concerns over the deterioration of the Poplar River center on impacts on livestock, irrigation, domestic use in Scobey and other downstream communities, and sportfishing, all of which the state believed were threatened by an increase in total dissolved solids (TDS) in an already marginal river. Lieutenant Governor (now Governor) Ted Schwinden testified that the effect from the new reservoir on these already high concentrations is Montana's immediate concern.[31] Worries over TDS were soon supplemented by new concerns over boron, a toxic heavy metal that occurs in mine dewatering. Concern over its effects began to focus on the consequences of using water containing a high level of TDS and boron for irrigation and its impact on crop growth. It was found that some types of crops are unable to develop and others suffer retardation in growth. Both federal governments soon grew to identify these two pollutants as the key Poplar water quality problems and, in response, the Canadian side endorsed construction of ash disposal lagoons to hold the contaminated water.[32]

Saskatchewan was very proud of this decision to build and line ash lagoons to minimize boron and TDS leakage as a response to international concerns and resented somewhat the fact that this decision did not dissipate anxiety on the subject in the United States.

When it started its work, the IJC board encountered serious difficulties in acquiring hard data, partly because the necessary data did not exist and partly because of alleged provincial reluctance to supply available data to the board. This difficulty delayed the board's investigation and became so serious the commission had to ask the U.S. Department of State (representing the complainant) to intervene with Canada to rectify the problem.

The International Poplar River Water Quality Board, after serious disagreement over technical issues including what constituted hard data and how it was to be interpreted, finally issued their report in July, 1979. This report provided the first hard (albeit controversial) data on water quality problems, including boron and TDS, and their likely effects with particular reference to irrigated crops. However, the board report made it clear that there was not only major disagreement among its members (including a suggestion of some division along national lines), but also that information was inadequate in many respects. Hence, there was a strong recommendation that a government-enforced monitoring system be established for all parameters. The board also recommended that future water quantity and quality considerations be examined simultaneously rather than separately.[33] The repeated differences of opinion among board members and the admitted lack of data meant that the board report could not be relied upon, which presented the commission with a dilemma. Some Montanans criticized the report for narrowness in restricting consideration to a small number of currently grown and irrigated crops (alfalfa and barley) and for failing to consider a greater number of potential crops. They also questioned its neglect of compensation arrangements for farmers who would suffer financial loss (though it is questionable as to whether the board had a mandate to deal with compensation).

The current Montana concern for loss of future agricultural options is thus brought into focus. However, Saskatchewan has noted that if downstream Poplar water use is for spring irrigation, the quality will be relatively poor, compared to use for summer or fall irrigation. Late season irrigation is a much more efficient use of water.

Meanwhile, the issue was further complicated by overpoliticization by both Montana and Saskatchewan politicians and by the commission's repeated delays in making its final recommendations. The U.S. State Department had on a number of occasions requested that Saskatchewan not approve construction of Unit No. 2 until the IJC issued its final water quality report, claiming that such approval at a time when there were outstanding questions would undermine the orderly processes of consultation embarked upon by both governments.[34] The SPC had informed the commission on a number of occasions that the energy needs of Saskatchewan dictated that the utility move forward with announcement of Unit No. 2. Finally, on November 22, 1979, Saskatchewan Energy Minister (and SPC board chairman) John Messer wrote that the expansion decision had been postponed long enough, SPC could wait no longer, and would proceed with Unit No. 2.[35] Montana was dismayed and turned to the IJC for action.

Thus a sharply disputed board report on water quality impacts, Montana dissatisfaction with the findings of that report, ignored State Department and Montana requests for a moratorium on Unit No. 2 construction, the SPC's dissatisfaction with the IJC and the drawn-out nature of the process had, by the end of the decade, brought the Poplar issue to a new low point diplomatically. The first year of the new decade, in contrast, was dominated by a positive event, and one of the few types of actions that all parties, albeit with some difficulty, could agree on: a joint monitoring arrangement.

Much of 1980 was marked by slow but steady movement toward a bilateral monitoring agreement and an institutional mechanism which may become the greatest positive legacy of the Poplar experience for both countries. The monitoring agreement did not come easily because of doubt as to what the IJC would recommend and when. There were also diplomatic disagreements for many months. The United States was concerned that the scope of the Saskatchewan monitoring effort was too narrow, being limited only to sites in the immediate vicinity of the plant and excluding background parameters. Canada was concerned that the entire effort on the Montana side was too elaborate, involving testing too far removed from the border and utilizing too many expensive, sophisticated models of questionable reliability. In February, 1980, Canada proposed a data exchange, something less than the bilateral monitoring agreement desired by the United States and Montana.[36]

Meetings in March, 1980, laid the groundwork, and the U.S. State Department issued a proposed draft agreement.[37] Shortly thereafter Montana objected to the exclusion of local government involvement, a point later remedied to meet the state's objections.[38] By April, Senator Baucus of Montana became critical of the State Department for delay, worrying that the monitoring program would not be set up before Poplar Unit No. 1 went on line in the summer. The senator agreed with the Canadian proposal and further recognized that Canada was not legally obligated to undertake monitoring but was doing so out of a recognition and respect for Montana's environmental values.[39]

Joint efforts continued through the summer and were finally rewarded on September 23, 1980, with the signing of a formal agreement. The IJC had not come through with its report by that date, and government was forced to preempt the commission and reach the agreement, since Unit No. 1 would go on line during the summer. The agreement as signed included the Canadian proposal for a new bilateral institution, the Poplar River Monitoring Committee, composed of a representative of each of the four principal governments involved (Montana, Saskatchewan, and the two federal governments), and a chief elected official from local government on both sides of the border as *ex officio* members to observe technical deliberations—a result of the U.S. compromise with Montana. The purpose of the agreement is to provide for formal exchange of data from the various monitoring programs at the boundary.[40] Dissemination and interpretation is also provided for, and the monitoring committee will be entrusted with ensuring the exchange of the data and ensuring the adequacy of the monitoring systems. The joint arrangement was made effective for five years, with amendment possible at any time and review at the end of the period. A curious aspect of this agreement was that it was reached before the IJC completed its report.

It was not until January 20, 1981, with Poplar Unit No. 1 in operation for half a year and with the joint monitoring agreement signed, that the International Joint Commission finally concluded its Poplar water quality reference of August 2, 1977. In justifying this long delay, the commission noted the difficulties encountered in fact-finding, the inadequacy of baseline data, and the incompleteness of information on existing uses of the water. The commission concluded that increased boron concentration would likely cause a small reduction in yields of irrigated barley and wheat but emphasized that only a small acreage is involved at present.[41] The IJC also concluded that future uses of this water in Montana would not change substantially, a conclusion that Montana finds difficult to accept since it forecloses its options. The commission also asked that downstream water quality objectives be designed and additional mitigation be provided for. It called for adequate assistance and compensation to be provided by a binational claims commission to those adversely affected. The report recommended specific interim boron and TDS objectives, and also rubber-stamped establishment of the bilateral monitoring committee already in existence. However, it advised a broadening of the role of the joint committee to encompass the recommendation of appropriate water quality objectives, a point beyond which the diplomatic negotiators chose to go. And, finally, in recognition of Montana concerns, the IJC recommended that U.S. federal agencies provide those Montana residents who might be adversely affected with technical advice and assistance to help them in adjusting to the new regime.

Much remains to be learned about boron. Future irrigation uses in Montana are unknown, thus preventing calculation of an opportunity cost. The total number of units at Poplar remains to be seen, as does the prospect of additional water diversion into the Canadian Poplar. Thus, the debate on water quality remains open. Further, the debate over the definition of a "significant" decrease in crop yields and whether such should be measured against the state as a whole (as favored by Canada) or against the impact on individual farmers (as favored by the United States) is as yet unsettled. It is clear, however, that Montana, the complainant, wants more than it has received. Many Montana officials and some federal officials in the state assume that ultimately all four Poplar units will be built, perhaps with dry cooling systems to overcome the problems of water supply. Therefore, further bilateral problems are in store. For this reason, the state wants the SPC to carry out a baseline study on water quality and its relation to crop yields, to continue studies while the plant is in operation, and to compensate farmers fully for losses incurred. The Poplar River water quality dispute may thus continue for some time, in spite of the efforts of the IJC and the diplomatic negotiators, as well as the continuing work of the joint monitoring group.

CONCLUSION

Poplar has been a trying experience for Canadian-U.S. environmental relations. It has combined three separate conflicts occurring simultaneously—water apportionment, water quality, and air quality. It was worsened by Montana's feelings of being "dumped on" by Canada environmentally at both ends of the state (the western problem being British Columbia's Cabin Creek coal mine, known also as the Flathead River issue described in chap. 8), and by Saskatchewan's failure to understand Montana's strong environmental sensitivities.[42]

Another basic problem in this as in other transboundary issues is the problem of balance. Balancing of concerns can be critical, and balances are often attainable within a federal jurisdiction. Hence, in the words of one Saskatchewan official, this is an important reason to maintain the involvement of the two federal governments, for balance often cannot be achieved by a single province or state in its transboundary relationships. More than one Canadian diplomat has shown sensitivity to the fact that Manitoba feels "dumped on" by the United States (Garrison, Roseau, Red, and lower Souris river issues), while Saskatchewan is "dumping on" the United States (Poplar and upper Souris river issues), believing that it would be a lot easier for the federal government if each province could experience both conditions. In other words, a narrower—provincial or state—jurisdiction may see all the benefits as being on one side of the border and most of the costs on the other. The government representing those people facing costs without benefits has little incentive to adopt a conciliatory position or to seek some middle ground.

For the moment, with international air quality negotiations underway, with the IJC's role in Poplar water terminated, and with the joint bilateral monitoring group in place, Poplar as a bilateral dispute is at least temporarily defused. However, it may take extraordinary care to maintain a state of harmonious relations.

Continuing Prairie Water Issues

There are other bilateral prairie water disputes not as well known as Garrison or Poplar which are recurrent items on the bilateral agenda. Collectively, these prairie water issues demonstrate continuous bilateral concerns in both water apportionment and water quality, in a vast region where quantity is scarce and quality is marginal. Unlike the more contentious aspects of Garrison and Poplar, these will not fade from the scene. Because of the high demand on this limited resource they will continue to be in need of joint bilateral management, but they present more of an ongoing managerial challenge than a diplomatic crisis. If insufficient attention is paid to them, however, they will in turn erupt into crisis. The waters in question are located in a region of vast future development, both in energy (coal, synthetic fuels, oil, and natural gas) and in irrigation agriculture. All of these developments typically lead to demands for diversion and other water manipulation.

During their periods of relative quiescence, these prairie water issues can serve as testing grounds in which new techniques of environmental diplomacy may be tried. Indeed, the Pembina River experience has already provided precedents which can be of use in international law and in determining the bilateral allocation of a scarce resource. The Red, Roseau, Souris, Milk, and St. Mary rivers could well do the same.

RED RIVER

The Red River rises in South Dakota and flows north, forming the boundary between North Dakota and Minnesota, and draining into Manitoba. It continues northward past Winnipeg and ultimately into Lake Winnipeg and Hudson Bay. It is a large river, especially for the prairies, and drains one of the most fertile agricultural regions of North America.

Like many prairie rivers, the Red is subject to chronic fluctuation in level and has been responsible for disastrous flooding, both upstream in the United States and downstream in Canada. A demand often arises in Canada following severe floods for flood control across the border in the United States to alleviate Canadian flooding. The disastrous southern Manitoba floods of the late 1970s saw a renewal of such demands and a call for the establishment of a "Joint International Authority" to carry out an international program of flood control.[43]

The unilateral action of Minnesota and North Dakota farmers in diking their riparian farmlands with sandbags to protect their properties has exacerbated Red River flooding in Manitoba.[44] Canadians voiced a diplomatic complaint, arguing that the United States should permit no diking, but U.S. authorities have been unsuccessful in deterring valley farmers. The irony is that improved protection upstream in the United States reduces the protection downstream in Canada.

International institutions were involved in this area from 1939 to 1948 under the terms of an active water apportionment reference on the Souris and Red rivers. IJC oversight activity has continued under an IJC International Souris–Red River Engineering Board of Control since that time, but most of its attention has been focused on the Souris and other points west of the Red River. Manitoba wants an international study of guidelines for flood control at the boundary and suggests this may ultimately go to the IJC as a reference.[45] North Dakota's state water engineer regards the matter of Red River flooding as the most serious topic in the basin and holds that it is basically a U.S. domestic issue. He also reports that North Dakota and Minnesota have reached agreement on dike criteria, so that the major problem will be enforcement.[46]

Pollution has been a much more important bilateral issue than water quantity on the Red River. Water pollution had become serious enough by 1964 to be the subject of an Article IX reference in that year. This reference created the International Red River Pollution Board.

The bulk of Red River pollution is from nonpoint sources (mainly agricultural runoff). Point-source polluters include food processing industries, mainly potato plants and sugar beet refineries. Downstream in Canada there is significant pollution from both agricultural runoff and municipal pollution from Winnipeg and other communities. Being downstream, the matter of Canada's fouling its own nest does not concern the United States, but it does affect the bilateral debate because it is prevalent enough to weaken Canada's negotiating position.

The quality aspect is viewed as being reasonably under control at present, since it is generally predictable, and some efforts are being made to reduce it over time. However, contentious incidents occur, such as the 1977 loss of dissolved oxygen in the Canadian portion of the river resulting from a molasses spill by a North Dakota sugar beet refinery. The question of sugar beet effluent relates to a concern over the dikes designed to contain it. The IJC Red River Pollution Board has urged better maintenance and inspection programs of all sugar beet processing plants in the valley, a practice which should reduce the risk of similar international incidents.

A further problem arises from the existence of differing philosophies and differing capabilities within the two U.S. states in dealing with the river. North Dakota is viewed as more lax than Minnesota on statutes and enforcement and spends much less money. Manitoba views North Dakota as likely to accept the idea of water quality objectives for the Red River, but the U.S. government has been cool to the idea. However, North Dakota sees no serious problem arising from water pollution.[47] Minnesota is satisfied with its own pollution controls and is quick to raise the question of severe Canadian pollution downstream near Winnipeg. The two federal governments are seen, however, as gradually moving toward the adoption of water quality objectives at the border (that is, toward the Canadian position), as recommended by the IJC. Ottawa endorses continued coordination on the Red but would not support the suggestion of joint river management.

ROSEAU RIVER

The Roseau River flows from Minnesota to Manitoba, emptying into the Red River at a point some miles north of the border. Though a small stream, it drains a better-watered agricultural environment on the prairie-forest interface. Its levels fluctuate and it causes flood damage in both countries.

Bilateral attention to the Roseau started with an early IJC reference in 1929 as a result of U.S. fears of damage from Canadian flood control activity downstream. This proved to be a false alarm, and the river was not heard of again until 1949 when the United States requested the IJC allow the diversion of a portion of it for a wildlife refuge, a diversion which was carried out with no difficulty. Then in the mid-1960s the U.S. Army Corps of Engineers became interested in the alleviation of flood damages through channel modifications on the U.S. side. The IJC's International Souris–Red River Engineering Board investigated and concluded that there would be adverse impacts in Canada; it recommended "remedial alternatives."[48] With a renewal of efforts by the Corps of Engineers to secure funds and authority for the flood control project in the 1970s, the IJC established its International Roseau River Engineering Board and directed it to develop coordinated plans for the basin, determine the transboundary effects of those plans, and identify required mitigation measures.[49] The commission found that channel modification, the principal method proposed for flood control upstream in the United States, would attain its purpose, but only at the cost of increased downstream flooding in Canada. Thus it concluded that the United States must bear the cost of mitigating works in Canada if it were to proceed with this flood protection technique in Minnesota. Channel modification should therefore go ahead simultaneously with U.S.-financed Canadian mitigation measures and, in deference to the more environmentally sensitive nonstructural flood control philosophies of the day, recommended that Minnesota and Manitoba implement appropriate land use controls to supplement structural flood protection.[50] The U.S. Fish and Wildlife Service and U.S. environmentalists applauded the suggestion of land use controls and attacked the largely structural approach of the Corps of Engineers, believing it would cause the loss of 237,000 acres of wetlands and waterfowl habitat.[51] This concern is still present.

With U.S. acceptance in principle to fund the mitigating works in Canada, Canadian diplomats, favoring the IJC recommendation, viewed the project in 1977 as going ahead with no negative bilateral repercussions. However, Manitoba showed increasing concern over water pollution which would result from the flood control works, both from increased sediment loads from channelization and from greater pesticides runoff from the increased farming intensity resulting from flood control. In its testimony to the IJC at the Roseau hearing in 1976, Manitoba contended that the transborder pollution clause of the Boundary Waters Treaty's Article IV was not being adhered to since it did not consider water quality deterioration or increased erosion in the Manitoba reach of the river.[52]

In spite of all this effort, there was no movement on Roseau in the latter part of the 1970s, since the Corps of Engineers did not obtain the mandate to proceed. Minnesota environmentalist opposition intensified, and Canada recognized that, as in the Garrison case, it had a natural alliance with U.S. environmentalists; both parties shared the belief that the best Roseau River flood control was to be found in the natural upstream marshes. Manitoba will not now approve any aspect of the project without an environmental impact statement.[53]

The total forseeable impact in Canada is perhaps not as great as implied in the concerns raised by Manitobans in the affected area. However, because the province has had such a bad experience with the United States over Garrison, Winnipeg politicians and officials must pay close attention to inequities resulting from across-the-border water manipulation, whether major or minor. Lesser issues like Roseau can have other than environmental repercussions if they demonstrate how or how not to work together or if they produce models that can be applied elsewhere.

As the 1980s began there were increasing signs that the Army Corps of Engineers would make a strong effort to get this project off the drawing boards and also, ominously, some sign that Manitoba was moving away from the concept of accepting monetary mitigation, as recommended by the IJC. Thus, the makings of more intense Roseau River transboundary environmental problems were on the horizon.

PEMBINA RIVER AND ASSOCIATED STREAMS

The Pembina River and associated streams have the dubious distinction of being the only transboundary environmental problem to become violent, having involved a few shooting incidents by irate farmers. However, the incidents could have as easily occurred domestically in either country under similar circumstances.

The Pembina River rises in western Manitoba, flows half its length in that province, then crosses the border into North Dakota and runs parallel to the border through exceedingly flat terrain to its juncture with the Red River just south of the border. Flooding in the North Dakota section is frequent, and because the river is at all points close to the border, that flooding often backflows onto Canadian as well as U.S. farmland.

The Pembina was the subject of a 1961 IJC reference with respect to apportioning irrigation water. The 1961 IJC study on this issue was most unusual in the approach it took. The commission suggested that the project go ahead on economic grounds, ignoring the international boundary. These grounds involved the payment of cash, an approach advocated by former IJC commissioner Anthony Scott (see chap. 3). Pembina provides a useful insight into the kinds of recommendations governments are willing to accept from a bilateral agency which might be unacceptable from other sources. However, the commission's recommendations have yet to be acted upon. At the time, the IJC found dams to be economically unjustified. More recently, Canadian farmers diked the river (in the form of a road) at the border to prevent floods on their side.[54] U.S. farmers blew up the dikes in retaliation, since they naturally worsen flooding on U.S. land. Canada would like the U.S. Army Corps of Engineers to build a flood control dam (the Pembilier Dam) to rectify the problem, but the Corps will not become involved unless Canada contributes financially (for which there is no precedent). Canada does appear willing to build improvement works in its adjacent Au Marais River Basin that would enhance the value of the Pembilier Dam on the Pembina, but it is not willing to invest in the U.S. Pembilier Dam itself. The United States is willing to proceed with drainage improvements. Manitoba argues that these may be visible with tangible domestic political benefits but do nothing to benefit Manitoba. The province would like the United States to build the dam unilaterally and then collaborate jointly with Canada on various channel and drainage improvements to the Pembina, Au Marais, and another stream, the South Branch Buffalo Creek. Should the United States ever conclude the Pembilier Dam is justified on domestic grounds or in the interests of bilateral relations or should Canada be able to assist in financing the dam, then the way will be open for possible joint efforts covering several basins. Without the dam, however, little movement will occur unless pressure builds from the farmers or an ugly incident ensues. Congress has been considering legislation to appropriate advanced planning funds for Pembilier Dam. The North Dakota state legislature supports this congressional action, thus perhaps signaling an end to this issue.

SOURIS RIVER

The Souris River presents a more complicated situation in the prairie region, since each nation has both upstream and downstream portions owing to the nature of the river's course. Although the Souris has become caught up in the Garrison debate, for many years it constituted a transboundary environmental issue in its own right.

The Souris rises in Saskatchewan, flows through North Dakota, then recrosses the border into Manitoba before draining into the Red River–Hudson Bay system. Hence, Saskatchewan is upstream, North Dakota is both upstream and downstream, and Manitoba is downstream.

As the subject of a 1939 water apportionment reference (at which time an IJC board of control was established) and of the 1948 reference on the Souris and Red rivers previously mentioned, this meandering little prairie river has had more of a bilateral history than most and is still being monitored by an IJC investigative board. This activity is in addition to the intense scrutiny the Souris has received by both the commission and diplomats since Garrison burst on the scene.

The Garrison impact on the Souris would result from the fact that the river would carry most of the Garrison return flows with their pollutants (mainly dissolved solids) into Canada, creating drinking water, among other, problems at Melita and Portage la Prairie, Manitoba. Apart from Garrison, however, the Souris remains a source of some present difficulty and may be at the center of further problems in the future.

In its 1948 reference the IJC recommended that the river be apportioned fifty-fifty at the Saskatchewan–North Dakota boundary and that the province always release to the state a minimum of twenty cubic feet per second (cfs) in the driest months (June to October). That figure was based on anticipated uses not in North Dakota but in Manitoba, the ultimate destination of the water. Manitoba believes that twenty cfs is not nearly enough and does not know why Canada agreed to it. The province's complaint centers around the fact that most of the water now goes to national wildlife refuges in North Dakota—to ducks instead of to Manitoba farms. Since upstream Saskatchewan could conceivably be entitled to much more than a fifty-fifty division (theoretically as much as it can store) according to the Harmon Doctrine's favoring upstream states, it is not happy with this obligation to downstream uses either, particularly when the trade-off is irrigation or power versus ducks. Saskatchewan is also dissatisfied that it is not permitted to name a representative on the IJC's International Souris River Board of Control. Ottawa monitors the river and keeps track of the flow, but when shortages develop, Saskatchewan claims, the federal government comes to the province and orders a release of more water. Because of a lack of direct representation, the province feels its interests have not been met. Further, North Dakota has proposed a joint Souris River management arrangement, but Saskatchewan cannot respond because of its lack of representation. The province would seek IJC supervision over any such arrangement, however, while it views North Dakota as wanting to work more directly outside the commission.[55] The secretary of the North Dakota State Water Commission reports that with few exceptions water quality is as good when it leaves the state as when it enters.

A currently minor Souris water quality issue is a potentially important one, since the province anticipates that North Dakota may well become more "activist" on water quality and blame Saskatchewan for deterioration. However, there is some question as to whether the state would be in a position to complain, given the pollution of its own portion of the river. Alternatively, some argue it is cleaner when it goes to Canada than when it comes into North Dakota. In North Dakota's view, flows have been adequate.

Another Souris issue is that surrounding the serious flooding experienced around Minot, North Dakota, and the proposal to construct Burlington Dam above the city to control that flooding. The dam would cause some flooding in Saskatchewan, mandating application to the IJC for an order of approval. A further bilateral problem is resultant flooding in Manitoba, since most operating plans for the dam lengthen the period of flooding in the downstream province. However, the state claims that no backflooding into Canada is involved under current design,[56] and no major disbenefit accrues to Canada. The key appears to be in the operating plan, but optimality for Minot and optimality for Manitoba may be mutually exclusive.

MILK AND ST. MARY RIVERS

The Alberta-Montana border has been one of the quietest of all border segments. Given Alberta's ambitious plans to develop a worldscale petrochemical industrial complex, Montana's goals in energy development, and the scarcity of water in this arid region, however, that present quietude cannot last much longer.

To date there have been only a few minor disputes relative to stream diversion and irrigation needs in this dry area. However, the St. Mary and Milk rivers were specifically cited in the 1909 Boundary Waters Treaty itself, which stated the two rivers are to be treated (and apportioned) as one for the purposes of irrigation and power. The treaty apportioned these rivers (which rise in Montana with its diversion of the St. Mary into the Milk, flow through Alberta, and then return to Montana, draining into the Missouri River system) and mandated that the Canadian Milk River may be used at the convenience of the United States to convey water via a diversion from the St. Mary. The original IJC board on the apportionment of this system has long since been disbanded, but the commission's 1921 order of approval did establish an International St. Mary–Milk River Board of Control which monitors the apportionment and reports back to it each year.

The diversion of the St. Mary into the Milk within Montana before the latter crosses the border raises a sore point with Canada since, in this exercise of the Harmon Doctrine, the diversion was made first and then the agreement was reached as to how much should be diverted. Prof. Arleigh Laycock, of the University of Alberta, believes there is no rationale in international law for this diversion.[57]

Today, Alberta perceives pressure in Montana to develop the head-waters of the St. Mary, and sees the St. Mary and Milk diversion as a trade-off for equal apportionment of the two combined.[58] There may well also be water quality concerns, in that Alberta is developing usage on the Milk and will likely build a dam to implement this usage and demonstrate through storage a claim to the river before upstream Montana institutes greater demand. The tributaries are also being utilized increasingly by the United States, and Alberta feels the United States is overdoing it.[59] Further, Canadian interest in the Milk is not confined to Alberta. Saskatchewan is interested in the Milk's eastern tributaries and wants input into Milk River apportionment to protect its interests, since greater Montana or Alberta demand will preclude future Saskatchewan claims to those tributaries.

With construction of a Milk River Dam and with increased demands on this area's water resources from industrialization on both sides, further Alberta-Montana disputes are inevitable and Montana, the only U.S. state bordering three Canadian provinces, will find itself spread even thinner in trying to manage and resolve disputes with all three provinces to its north and simultaneously dealing with Ottawa and Washington.

Conclusion

From the experience of Garrison, Poplar, and other prairie water issues, a number of conclusions can be drawn:

- the problem of agreeing on data (especially across international boundaries) is more difficult than most people realize. Thus, efforts at joint fact-finding, such as that represented by the IJC, are of real value and should be supported and broadened;
- water quality and quantity questions should be considered together wherever possible and especially on smaller water bodies;
- a federal role is necessary, even critical, to help states and provinces balance issues, so that solution may be achieved;

- environmental attitudes vary on the prairies, particularly from state to state;
- the IJC has played and is playing a crucial and largely successful role throughout the prairie border region;
- all parties to prairie disputes fear decisions that preempt their future, a problem not limited to this region but one which is especially applicable to border problems in this region, suggesting the need for newer more flexible compensation arrangements, perhaps freer transborder litigation opportunity, and the need for the continuing presence of an institution to monitor problems and adjudicate differences.

Water quantity and quality are not the only transboundary environmental issues plaguing the prairie region. With increasing interest in energy self-sufficiency in both nations and with an abundance of coal near the border, air quality is forming an additional component of prairie border problems and of problems to the west and east as well. It is to these international air quality problems that we now turn.

NOTES

1. International Garrison Diversion Study Board, *Report* (Ottawa and Washington: International Joint Commission, 1975). The material in this section was derived from pp. 6–16 of the report.

2. This appropriation was tacked on as an amendment to the Mount St. Helens Disaster Relief Act of 1980 (and added to another $3 million from earlier appropriations). Canada and Manitoba protested strongly and lobbied very hard, but the Senate as a final tribute to their North Dakota colleague passed it and President Carter reluctantly signed it.

3. "Garrison May Sue Interior Department," *Jamestown Sun,* Jamestown, North Dakota, November 12, 1980.

4. Canadian Embassy, Diplomatic Note No. 498, Washington, D.C., October 1, 1980, pp. 3–5.

5. Rieber-Kremers and Associates, Ltd., *The Impacts of the Garrison Diversion Unit on Manitoba Indian Communities: A Report for the Manitoba Indian Brotherhood* (Winnipeg, Manitoba: Rieber-Kremers and Associates, 1979), p. 1.

6. Ibid., pp. 2–3.

7. Colin Holbrow, Four Nations Confederacy, personal interview, Winnipeg, Manitoba, November 4, 1980.

8. "Link Denies Scarce Tactics on Water Rights," *Jamestown Sun,* April 4, 1980.

9. "Ogallala Plan said No Threat to Water Rights," *Jamestown Sun,* September 13, 1980.

10. U.S. Department of the Interior, Bureau of Reclamation, *Master Contract Between the United States and the Garrison Diversion Conservancy District for the Garrison Diversion Unit, Missouri River Basin Project*, Washington, D.C., 1965, p. 16. On the subject of default and loss of water rights, the master contract between the federal government and the local project managing entity, the Garrison Conservancy District Board, would seem to indicate possible water forfeiture.

11. Grant C. Mitchell, "The Transboundary Environment: A Provincial Perspective," in *Proceedings of the Canada-United States Natural Resources and Environmental Symposium*, ed. John E. and Diana C. Carroll (Durham, N.H.: University of New Hampshire, 1978), p. 14.

12. This was the corporation's second such endeavor, with the construction of the Boundary Dam Station and coal mines to the east having preceded Poplar. This plant causes minor air pollution in North Dakota which has never been questioned.

13. Ronald J. Schleyer, *The Transboundary Effect: Safeguarding the Poplar River in Montana* (Helena, Montana: Montana Environmental Quality Council, 1976), p. 28.

14. Province of Saskatchewan, *Report of the Board of Inquiry—Poplar River Power Project* (Regina, Saskatchewan: Province of Saskatchewan, 1975), p. 62.

15. The "Helsinki Rules" are rules of international procedure drawn up by the nongovernmental International Law Association in Helsinki, Finland, in 1966. They are also known as the Rules on the Uses of Waters of an International Drainage Basin.

16. Province of Saskatchewan, *A Brief Respecting Water Apportionment in the Poplar River Basin*, submitted to the IJC task force public hearing, Coronach, Saskatchewan, May 27, 1976, p. 24. Saskatchewan Power Corporation did become somewhat concerned about the project location, given cross-border reactions, but it is located fully in accordance with Canadian and provincial law as it existed at the time, if a bit less than sensitive to the diplomatic repercussions. For some time Saskatchewan worried about trade-offs which Ottawa diplomats might make which the province could not influence, and determined to regard any such trade-offs not in its interest as forced contributions to international goodwill, on which it would keep tabs. The province always felt that a departure from a fifty-fifty apportionment rule would benefit not only itself but Montana as well on other streams, and it determined early that if the IJC rejected its apportionment needs, then it would recommend that Ottawa reject the decision, recognizing both that any apportionment agreement is a departure from the philosophy of the treaty but a justifiable one, and that the Poplar project is premised on the treaty as written, under which Canada, being upstream, has a right to the water.

17. Province of Saskatchewan, *Water Apportionment in the Poplar River Basin*, p. 25.

18. Grant C. Mitchell, *A Submission to the International Joint Commission Respecting Water Apportionment in the Poplar River Basin* (Regina: Saskatchewan Department of the Environment, 1977), p. 11.

19. Ibid., p. 12.

20. Ibid., p. 14.

21. Ibid. On prairie streams 80 percent of the total downstream flow passes

in a few weeks and cannot be contained, justifying a more flexible apportionment formula than fifty-fifty.

22. Orrin G. Ferris, *Testimony of the State of Montana on the Apportionment of Waters of the Poplar River*, Public Hearings, Scobey, Montana, May 26, 1976, p. 2.

23. Ibid.

24. They lose the spring freshet, the value of which for ground water replenishment is debated.

25. This doctrine of Indian water rights, first expounded in the 1908 Supreme Court case of *Winters* v. *United States* (207 U.S. 564; 1908), stands for the proposition that Indians are entitled to sufficient water to satisfy the purpose for which they reserved their land. It has been repeatedly applied to reserve water for the irrigation of Indian reservations. Massie has noted that the Winters Doctrine contradicts and supersedes the western water law of prior appropriation since the water need not be employed for existing beneficial purposes but is in fact whatever amount is needed to fulfill present and future desires (see Michael A. Massie, "Guarantee and Controversy: The Winters Doctrine and Indian Water Rights" [Paper presented to the Western Social Science Association Annual Meeting, Albuquerque, New Mexico, April, 1980]).

26. Letter from Richard G. Vine, Deputy Assistant Secretary of State, to the International Joint Commission, June 3, 1977.

27. International Joint Commission, *Water Apportionment in the Poplar River Basin: An IJC Report to the Governments of Canada and the United States* (Ottawa and Washington: International Joint Commission, 1978), p. 71.

28. Ibid., pp. 71, 72.

29. This return of water quality to the diplomatic arena coincides with an IJC-oriented apportionment aspect and an air quality aspect which is totally diplomatic though not without IJC recognition.

30. These hydrological facts with their attendant economic realities have exacerbated the water quality aspects of the Garrison and Red River pollution issues as well as Poplar.

31. Testimony of Lieutenant Governor Ted Schwinden, State of Montana, to the IJC public hearings on Poplar River water quality, Scobey, Montana, November 2, 1977.

32. Canadian Embassy, "Poplar River Power Project," press release, August 7, 1979.

33. International Poplar River Water Quality Board, *International Poplar River Water Quality Report* (Ottawa and Washington: International Joint Commission, 1979), pp. xiii–xiv.

34. U.S. Department of State, Diplomatic Note, Washington, D.C., n.d.

35. John R. Messer, Minister and Chairman of the Board, Saskatchewan Power Corporation, Letter to *Toronto Globe and Mail*, November 22, 1979.

36. Canada Department of External Affairs, *Poplar River Basin: Proposed Canadian Data Contribution to International Exchange*, February, 1980.

37. U.S. Department of State, *Cooperative Monitoring Agreement for the Poplar River Basin*, June, 1980.

38. Montana's Lieutenant Governor Schwinden reacted strongly to the State Department's original denial of a local role, a denial justified on constitutional

grounds, and successfully sought the intervention of U.S. Ambassador Kenneth Curtis, resulting in an acceptable compromise.

39. Senator Max Baucus, press release, Washington, D.C., April 1, 1980.

40. Governments of Canada and the United States, *Poplar River Cooperative Monitoring Arrangement*, September 23, 1980.

41. International Joint Commission, *Water Quality in the Poplar River Basin* (Ottawa and Washington: International Joint Commission, 1981), p. xiii.

42. The differing environmental attitudes between North Dakota, with which the province had had positive environmental relations, and Montana were not recognized as they perhaps should have been. Saskatchewan now recognizes that the difference is such that North Dakota would likely accept sewage dilution as a legitimate use of a stream, whereas Montana clearly would not.

43. International Flood Control Coalition, *A Proposal to the Residents of the Red River Valley and Drainage Basin to Establish a Joint International Authority* (Montcalm, Manitoba: International Flood Control Coalition, 1980). A Red River Valley International Citizens Flood Control Coalition was established in 1979 with a membership of three to four thousand farmers in both countries. The coalition wants to achieve flood control by ending the drainage of U.S. wetlands and excessive road building and by promoting channelization. The coalition is strongly motivated by the human tragedy of Red River flooding, arguing the benefits of preventing flooding, even if at great cost.

44. Sandbag diking occurred first in 1975 on the North Dakota side of the river, causing increased flooding on the Minnesota side. To protect themselves, the vulnerable Minnesota farmers diked their shore resulting in dikes on both sides, a narrowed floodplain, and an inevitable increase in flood waters moving downstream toward Manitoba.

45. Tom Weber, Manitoba Assistant Deputy Minister for Water Resources, personal interview, Winnipeg, Manitoba, November 5, 1980.

46. Vernon Fahy, Secretary, North Dakota State Water Commission and State Water Engineer, personal correspondence, Bismarck, North Dakota, March 10, 1981.

47. Ibid.

48. International Joint Commission, *70 Years of Accomplishment* (Ottawa and Washington: International Joint Commission, 1980), p. 30.

49. International Joint Commission, *Coordinated Water Use and Control in the Roseau River Basin: An IJC Report to the Governments of Canada and the United States* (Ottawa and Washington: International Joint Commission, 1976), p. 1.

50. A similar recommendation was made to Quebec and Vermont in the Champlain-Richelieu case in 1981.

51. International Joint Commission, *Coordinated Water Use and Control*, p. 4.

52. Based on interviews with Dr. George Bowen, Assistant Deputy Minister for Environmental Management, and Tom Weber, Assistant Deputy Minister for Water Resources, Government of Manitoba, Winnipeg, Manitoba, June 27, 1977.

53. Manitoba Department of Mines, Resources, and Environmental Man-

agement, "Statement to the International Joint Commission in respect to Coordinated Water Use and Control in the Roseau River Basin," January 14, 1976, p. 2.

54. Some say these floods are exacerbated by the U.S. farmers channeling the water toward Canada.

55. Weber interview.

56. Fahy correspondence.

57. Professor Arleigh Laycock, personal interview, University of Alberta, July 11, 1977.

58. Dr. Thomas Sneddan, Research Secretariat, Alberta Department of the Environment, personal interview, Edmonton, Alberta, July, 1977.

59. Dennis Davis, Regional Director, Inland Waters Directorate, Environment Canada, personal interview, Regina, Saskatchewan, November 7, 1980.

CHAPTER 10

Air Quality at the Border

The experience of the early Trail Smelter case demonstrates the difficulties of reaching an equitable solution in a reasonable period of time, whether the interested parties have recourse to the IJC or to techniques of international arbitration. The Poplar and Atikokan air pollution disputes illustrate the vigor with which U.S. citizen environmentalist groups can react when air quality is at stake. The Atikokan power plant experience, in particular, demonstrates how nationally concerned those same environmentalists can become when a well-known, highly valued area such as the Boundary Waters Canoe Area is threatened (whether the threat is real or perceived). Cornwall Island shows that native peoples in both countries are also vulnerable to modifications of the transboundary environment. This issue also illustrates the problem of a government having to respond to an emerging foreign relations problem when as yet no domestic pollution law is being violated. This presents a dilemma for which there is as yet no answer. The long history of Detroit-Windsor air pollution perhaps best demonstrates the willingness of people to accept serious air pollution in their own home environment when job security is perceived to be at stake.

The preceding chapters have been largely concerned with water-related issues. It is now appropriate to consider the more challenging question of air quality at the border. All bilateral air quality problems are now subsumed under the current negotiation aimed at reaching a comprehensive bilateral air quality agreement. These negotiations are in turn dominated by acid rain concerns. It is the intent of this chapter to deal with the many local, "point-source," transborder air pollution incidents, while the following chapter treats the geographically broader and more complex subject of long-range transport of upper atmospheric pollutants, including acid rain. Much of the future course of both types of air pollution may well hinge on the outcome of the current negotiations.

Anxiety about transboundary air quality gives rise to new challenges not faced in the water quantity and quality issues, including:

- how to extend IJC coverage to air quality when this is not mentioned in the Boundary Waters Treaty;
- how to manage issues that overflow broad regions and have tangible socioeconomic impacts on large numbers of people over a wide area;

• how to fill information gaps (a much more difficult challenge than with water), and attack the withholding of information;
• how to clarify international obligations in this nontraditional area;
• how to handle conflicting standards and deal with "grandfather" clauses and other statutory exemptions; and
• how to avoid misjudging public opinion across large and diverse regions.

Ever since the Trail Smelter case over half a century ago, an increasing number of transboundary air quality problems have arisen at various points along the Canadian-U.S. border. In the 1940s and 1950s the Detroit-Windsor pollution problem surfaced; this problem expanded to the Sarnia–Port Huron area in the 1960s. There were also some air pollution concerns at International Falls, Minnesota–Fort Francis, Ontario, in the 1960s. More recent problems include the Poplar power plant in Saskatchewan-Montana, the Cornwall Island issue in Ontario–New York, the Atikokan question in Ontario-Minnesota, the minor air quality concerns of the Eastport issue in Maine–New Brunswick, and the much broader questions of Great Lakes water quality impacts from airborne pollutants and of acid rain and its bilateral ramifications; the latter two issues are the subject of a separate chapter.

The Boundary Waters Treaty of 1909 does not address transborder air pollution directly but does include it in an indirect manner. The treaty prohibits transborder air pollution which pollutes boundary waters, a not insignificant prohibition. By definition, then, the International Joint Commission and the Great Lakes Water Quality Agreements of 1972 and 1978 both have a role in, or pertain to, air as well as water quality.

Many air pollution incidents have not gone to the International Joint Commission because of governmental reluctance to involve the commission in broader bilateral questions where environmental impacts cannot be readily assessed or where the emissions at issue are not immediately visible (as in the case for acid rain). However, where the outcome is predictable and high credibility is called for, the two governments have not been reluctant to involve the commission. Examples are the Trail Smelter, Detroit-Windsor, and Sarnia–Port Huron. In 1966 the IJC was given its general standing reference to study any potential bilateral air pollution problems and to bring them to the attention of the two governments (incorporated in IJC Docket 85R, Air Pollution in Detroit–St. Clair River with General Observation Along Rest of Boundary, 1966). To perform this task, the IJC established its International Air Pollution Advisory Board.

On the surface it would appear that this reference represents a broad and substantial grant of authority, but in fact the commission has not had the funding to execute its responsibility effectively in this domain. What little alerting it has done has been all but ignored. It would seem that the commission's advice, when it is not specifically requested by government, falls on deaf ears.

The history of bilateral air pollution issues includes both those dealt with by the IJC under Article IX, and those more numerous and more significant issues which have been restricted to the domain of conventional diplomatic negotiation. The Trail Smelter case, to many the granddaddy of all Canadian-U.S. environmental problems and one of the few issues that has ended conclusively, teaches many lessons and is an appropriate opening case study.

Trail Smelter

There are those who argue that the early and lengthy Trail Smelter arbitration in British Columbia and Washington State is a classic example of how the IJC, the treaty, and the bilateral negotiating technique work, but others maintain that this same arbitration is a classic example of how the system does not work. The difference obviously depends on one's definition of success and one's opinion as to what would have happened in the absence of a treaty and commission. If the final outcome is what matters, the system was shown to work. But if some weight is given to the long period of time (and money) involved, the endless arguing across the border, the caution of precedent-concerned diplomats, and the suffering and uncertainty faced by the parties involved (the afflicted farmers and the smelter executives), it can be argued that the system did not work. It might be suggested that, if negotiation is the art of the possible, then the side one takes in this issue might well relate to one's view of what is possible.

During the 1920s, fumes from a zinc smelter on the border at Trail, British Columbia, were damaging fruit trees and crops in Washington. Article IX of the Boundary Waters Treaty was invoked, and the IJC received its Trail Smelter reference in 1928. Trail Smelter was the IJC's baptism of fire in the air quality area—the first time the commission had been used to solve an air problem. The commission concluded in 1931 that damage was occurring and recommended that the growers be compensated $350,000 for damages. At the same time it proposed an institutional approach by recommending the establishment of a scientific board to recommend necessary control measures to terminate "appreciable" damage in the United States.[1]

The state of Washington and local interest groups affected were unhappy with this solution, however, and prevailed upon the U.S. State Department to additionally assess accruing damage payments and ensure reduction of the SO_2 emissions until injury was eliminated.[2] After some delay, Canada accepted these terms. But increased fume emissions and damage in 1934, coupled with economic problems at the smelter, set back the negotiations.

The problem at Trail came to be linked to other current matters in bilateral relations, and the issue was transferred to the prime ministerial level in Canada. U.S. officials became convinced that unusual measures were necessary to end the conflict,[3] including a direct linkage with a trade agreement of great concern to Canada. The issue was therefore raised to these higher political levels, much give and take ensued, and it was not until the spring of 1935 that a bilateral agreement was reached. This paved the way for signing a joint economic agreement which had been linked to the pollution controversy.[4]

This did not end the matter, however, for in 1935 the two governments established an ad hoc arbitral tribunal, which did not make its final decisions on compensation until 1941.[5] The tribunal recommended that specific regulatory standards be established in addition to those recommended by the IJC. Both sets have since been implemented, and as a result the Trail Smelter is considered in some quarters to have the best air quality control in Canada.[6]

The Trail Smelter arbitration was an early example not only of transboundary environmental problems but also of cross-border collaboration of groups with a common interest. In the face of State Department opposition, U.S. companies, fearing similar pollution suits, contributed data and scientific personnel to the Canadian company's rebuttal of the claimants' case.[7] This presaged the modern-day phenomenon of transborder collaboration between citizen environmentalist organizations on the Garrison, Atikokan, and numerous other issues.

From the Trail Smelter issue, one of the few transboundary environmental conflicts which reached a clear conclusion, one learns that the wheels of international diplomacy move very slowly. Especially cognizant of this were the afflicted Washington farmers, who did not receive any compensation for more than fifteen years, in spite of the fact that their claims were recognized. Indeed, so much time had passed that the real estate executors rather than the farmers, who were by then deceased or retired, were the recipients of compensation. The case also points up the fact that, while once again the IJC conclusions were endorsed, the commission's role in the matter was an incomplete one. The real bargaining and significant decision making took place after it had disposed of the issue. Thus, although the Trail Smelter case can be presented as a success story in bilateral environmental relations, such an assertion must be qualified. The dispute might not even have been solved if the economic agreement had not been available to provide linkage.

Michigan-Ontario

The Detroit-Windsor and Sarnia–Port Huron transborder air pollution disputes are rarely mentioned today and are no longer active bilateral issues. The problems have not been solved, but there has been acceptance of deteriorated air quality and resignation on both sides of the border, particularly in the areas of most negative impact—Windsor, Ontario, and, to a lesser extent, Port Huron, Michigan. These deliberately linked issues do, however, have a rich history which has led to the development of some models of international cooperation (in the form of an early IJC joint air quality monitoring reference and in a model state-provincial agreement for mutual cooperation). This bilateral cooperation has been more an endeavor of mutual resignation than of mutual problem solving, but if a bilateral issue is "solved" through joint improvement of the air or by joint resignation to its deteriorated condition, it is in fact solved.

Since IJC findings indicate that Michigan-Ontario air quality has not improved and may even be worse, does this mean that the commission failed or that the IJC reference was useless? Or have the two federal governments lacked the will to promulgate and enforce necessary laws? Have the two source jurisdictions—Detroit (or Wayne County) or Sarnia (or Lambton County)—not done enough? Have the two recipient communities, Windsor and Port Huron, not complained loudly enough? A case can probably be made on any or all of these grounds, although the IJC is less to blame than other parties. Consumers who use autos and petrochemical products, electric energy, and many other goods all share some responsibility, as do those who are concerned for their jobs. The city of Detroit and Wayne County have many more immediate, and in the short run more serious, social and economic problems than that of air quality, including the recent depression of the automotive industry. Windsor shares in the latter, and naturally a concern for jobs at this critical time supersedes all else. Likewise, Canada's great dependence on Sarnia for its refining and petrochemical industries suggests that the solution to this Michigan-Ontario regional problem is in fact a national problem calling for national solutions. Detroit and Sarnia are nationally depended upon for the output of these industries and thus must be assisted by national efforts. In addition to the fact that the problems are too severe to be solved locally or regionally, there is a question of principle as to whether the federal governments should even expect local or regional solutions.

THE IJC'S ROLE

In 1949 the International Joint Commission received the second reference on air pollution in its history. This reference requested the commission

> to determine if the cities of Detroit and Windsor were being polluted by smoke, soot, fly ash, and other impurities, and to ascertain the extent to which vessels plying the Detroit River were responsible.[8]

The commission was limited to making recommendations regarding measures to reduce emissions from vessels. This limitation was criticized by the commissioners, who held that it diverted attention from the more serious on-land sources. They thus asked for a broader reference, but it was denied them.

The commission developed smoke emission objectives for vessels plying the Detroit River in the early 1950s and recommended in 1960 that the governments adopt these objectives, which they did. Six years of IJC monitoring ensued, followed by withdrawal of the commission from the issue by its own request in 1966, assuming the matter to be solved.

During this time, however, the IJC never modified its view that vessel smoke emissions were only a small part of the whole. Finally, in September, 1966, the two governments referred the general matter of Detroit-Windsor and Sarnia–Port Huron air pollution to the commission, asking if polluted air was in fact crossing the boundary in quantities sufficient to harm public health, safety, or welfare and if so to identify the sources, cite remedial measures, and indicate the probable total cost of implementing the measures.[9]

The Sarnia–Port Huron–St. Clair River question was added to balance U.S. pollution damage to a Canadian city with Canadian damage to a U.S. city. In the Detroit-Windsor case, the problem arises largely from metallurgical (that is, automotive) industries and power plants on the U.S. side, while the Sarnia–Port Huron case is largely a problem of odors from petrochemical plants, an oil refinery, and a power plant on the Canadian side.

The commission concluded in 1972 that pollution was crossing the boundary in sufficient quantities to cause harm, identified the sources, proposed preventive and remedial measures, and calculated their costs. Significantly, it concluded, above and beyond its terms of reference, that there was need for:

- binational contingency plans for reducing emissions during adverse meteorological conditions;
- uniform procedures for air quality monitoring, data exchange, and so on;
- uniform air quality standards on both sides of the border;
- individual point-source control in both countries;
- IJC responsibility for carrying out these tasks; and
- future research in a number of different areas, including fuel desulfurization and development of energy alternatives.[10]

These 1972 conclusions have proven to be significant forebears to more recent approaches and problems. Joint contingency planning, joint monitoring, and data exchange have since become integral parts of more recent binational air and water quality arrangements, although the two nations have yet to achieve uniform air quality standards.

The two governments accepted the 1972 report and conveyed an Article IX reference in mid-1975 which charged the IJC with oversight responsibility, as the commission had requested. The commission in turn established the International Michigan-Ontario Air Pollution Board in early 1976 as its permanent vehicle to carry out the terms of the reference and oversee the mandate to monitor permanently and report annually on air pollution control progress in the Michigan-Ontario region. The board was charged with advising the commission on trends and emissions for sulfur dioxide, total suspended particulates, and odors as well as on enforcement effectiveness, adequacy of surveillance, steps taken by local government, the adequacy of data exchange, and other matters.[11] The board has reported annually since 1976.

While the overall severity of air pollution in this region diminished considerably after 1972, virtually all of this reduction took place before the IJC began work under its 1975 reference. The reduction basically resulted from a major decrease in particulates as the effect of the U.S. and Ontario clean air legislation began to be realized. In its 1978 report, the commission noted that the ability to maintain particulate levels within the IJC objectives is elusive and is unlikely to be reached with existing control strategies.[12] The commission also identified carbon monoxide and nitrogen dioxide problems for the first time. The 1979 report noted little change in air quality and many violations of IJC objectives.[13] This report further recorded the new problem of the emission of toxic substances into the air from low temperature incineration. Overall, the commission found that "slight increases in pollutant levels may be occurring."[14] Thus the years since the IJC reference have been characterized as ones of maintaining the status quo or of falling a bit behind, with no evidence of solving or even of alleviating the overall problem. Each year has witnessed the addition of new air pollution concerns to the list, either as the result of the introduction into the area of a new problem or a recognition that a continuing problem is in fact more serious than once thought.

By the end of the 1970s the commission already feared that the deviations from the trend of improvement since 1972 signalled a critical period when the progress in air quality had slowed to a rate insufficient to meet its objectives or that air quality might already have begun to deteriorate once again.[15] In May, 1980, the commission informed the two governments that it believed it was obligated to broaden its mandate by reporting on substances not mentioned in the reference if it believed a problem was occurring.[16] This it justified by society's increased concern over toxic materials in the air, and it requested further guidance from government. Government has not, as yet, responded. The commission also noted that Michigan and Ontario had failed to meet their deadline for compliance with the objectives, and that progress with respect to suspended particulates and SO_2 had slowed considerably.[17]

What has been the value of the bilateral experience in this area? It has led to the development, through the IJC, of the hard data needed to understand the problem, has done so in a coordinated and internationally acceptable manner, and has focused much public attention on the nature of the issue. It has further provided new techniques and models for joint cooperation which have become even more important now that broader international air quality and water quality issues are arising in the 1980s. The commission has also benefited from the experience since it is now better prepared for other air quality tasks elsewhere along the border. And, in some ways most importantly, this reference to the IJC has caused government to request that the commission

> take note of air pollution problems in boundary areas other than those referred to in the reference which may come to its attention from any source. If at any time the Commission considers it appropriate to do so, the Commission is invited to draw such problems to the attention of both Governments.[18]

This, of course, gives the commission a standing reference, a "watching brief," to enter into any transborder air quality issue and, in an advisory fashion, to invite a reference, thus assuring it an important future role in the air quality area.

The tremendous socioeconomic problems in the Detroit-Windsor international metropolitan area will not go away quickly. Indeed, all indications are that they will worsen with the current decline of the U.S. and Canadian automotive industry and its numerous support industries in this area. It is possible that the air pollution problem will be at least partially solved by large-scale decline of the industry, although many other problems of greater magnitude will ensue. The concern for jobs and for economic stabilization in the region is as great and as legitimate as are environmental and health concerns attendant upon air pollution. Indeed, environmental problems will likely not be solved (unless by default) without first solving the basic economic questions.

It will be a monumental task for both federal governments, the province and the state, as well as the affected population. The State Department, External Affairs, and the IJC can contribute to this solution, perhaps first by recognizing that there is no comparison between Detroit-Windsor and Sarnia–Port Huron in terms of magnitude of the problem. It may be wise henceforth to avoid the temptation of linking lesser issues (for example, Sarnia–Port Huron, which is a Canadian domestic problem with lesser international ramifications) with major ones like Detroit-Windsor. Such linkage may make a neat balanced package and therefore be diplomatically desirable, but can be dishonest and misleading. Our two societies should by now have a mature enough relationship so as not to need this crutch of balance falsely equating damage from one with damage from the other. Presumably, the work of the commission to develop more knowledge about Michigan-Ontario air pollution will continue. A good record with sound, bilaterally agreed upon data is a necessary prerequisite not only for an ultimate solution in this region but elsewhere as well; most importantly, it must serve as an integral component of a future international air quality agreement. Such an agreement would likely subsume Detroit-Windsor and Sarnia–Port Huron and henceforth treat them as part of the overall continental problem, which they have in fact become.

Atikokan

What constitutes a transboundary environmental issue? As has been seen, a tangible negative impact across the border is not necessary for a transboundary environmental issue to develop. The mere perception of such a negative impact is sufficient. Thus the Atikokan issue, which is perceived by some U.S. residents to have negative impacts, at least on northern Minnesota, and perceived by many Canadians as not having transborder impacts, has nevertheless developed into an active issue which rose to a high place on the bilateral diplomatic agenda.

In September, 1973, Ontario Hydro, the crown corporation which provides virtually all of the province's electricity needs, announced that it was seeking a site in remote northwestern Ontario to construct a large coal-fired generating station. Such a station could provide substantial new power for the small west Ontario grid, perhaps constitute a base for new industrial development in a sparsely populated and economically underdeveloped region, and also take advantage of substantial prairie coal reserves and the rail lines linking the western coal-producing regions with western Ontario. After considering a number of sites, the utility settled on the small railroad and mining town of Atikokan and nearby Marmion Lake. By mid-1975 both the utility and the provincial government had approved Atikokan, welcome news in the depressed community of five thousand who feared the imminent closing of one or both of the area's iron ore mines owing to depletion of the resource. The plant site is twelve miles from the Quetico Provincial Wilderness Park and less than forty miles from the U.S. border and Boundary Waters Canoe Area (BWCA). These geographical facts were to become critical in the bilateral aspect of this dispute.

Ontario Hydro originally proposed an 800 Mw coal-fired power plant, consisting of four 200 Mw units, scheduled to go on line in 1983–84. It would have included a tall stack (650 feet) to disperse pollutants, electrostatic precipitators to control particulates, and a lake cooling system (Marmion Lake) to dissipate waste heat, but it would not have scrubbers to control SO_2 emissions.

The coal to be used would be relatively low-sulfur lignite coal. Such coal has a lower Btu (heat) value than much high-sulfur coal. Thus, greater quantities of this lignite coal must be burned to produce an equivalent amount of electricity. Since the precise source of the coal was not determined, it is not possible to predict the SO_2 emissions with precision, but they appeared to range from 125 to 225 tons per day; the project opponents stressed the higher end of the range, while proponents stressed the lower end.[19] A factor hampering assessment of the project was the long-held belief that much in the proposal was not fixed. One State Department spokesman referred to it as "dealing with a bowl of jello."[20]

Supporters of the project include Ontario Hydro, the Ontario government, and most of the people of Atikokan. Presumably the coal producers of the western prairies, the railroads that would transport the coal, and the advocates of new entrepreneurial activities in northwestern Ontario who could be guaranteed a new power source are also in favor of the project, although all three have been silent to date. Opponents of the project include a small number of environmentally concerned Atikokan area citizens, provincial and Canadian national environmental organizations, the Minnesota Pollution Control Agency (MPCA), which fears environmental damage in the northern part of the state, and U.S. regional and national environmental groups, many of which have struggled effectively over many years to protect the wilderness integrity of the BWCA from various threats.

U.S. REACTIONS

Minnesota's position has been to advocate use of SO_2 scrubbers on Atikokan. This is the policy for new plants within the state. Minnesota also makes clear that its concerns (and those of entrepreneurs) are more than environmental—it is concerned about the usage of air quality increments (i.e., pollution rights) as well, lest no room be left for future domestic polluting sources.[21]

Ontario Hydro and other Canadian proponents of the Atikokan project were surprised by the vigor with which U.S. environmentalists organized and fought against the project and the high level of success they seemed to achieve in a short time. An official of the Office of Canadian Affairs at the U.S. State Department reported that his office had received more mail on this subject than on any other Canadian-U.S. issue in memory.[22] It was not long before congressmen and senators reacted to the public pressure and in turn put pressure on the State Department to take a tough stand. The national media began to devote attention to the issue; in general, opposition in the United States coalesced more rapidly than did Canadian domestic opposition. The reason was environmental and little concerned with the impact on economic development.

The Boundary Waters Canoe Area within Superior National Forest is not only a designated wilderness, a reason in itself for high public attention at the national level, but is famous in its own right. It has been the subject of numerous books and articles over thirty or more years, read by two generations of U.S. citizens, and symbolizes the "great north woods" of romance and literature, encompassing endless lakes and forests, outstanding wilderness canoeing, and moose, bears, and waterfowl. Canadians have much of this environment on the Shield and so take it for granted. The United States has much less and its people tend to treasure and romanticize what they have. In addition, the BWCA had been in the spotlight for many years over logging, mining, and motor-boating initiatives, all of which were threatening the area's unique characteristics and were curbed by wilderness designation. With the advent of another "threat" to the BWCA, it was not difficult to rally many U.S. environmentalists and BWCA lovers to the cause.

There is irony to this outpouring of reaction to the Atikokan issue. More often than not, it is U.S. residents who do not appreciate the extent of the national Canadian concern over the Garrison project, the Skagit–High Ross Dam, or the coastal fisheries disputes and incorrectly and unwisely view these as regional issues of less consequence than they really are. With Atikokan, the situation is reversed and Canadians failed to realize the national awareness and concern in the United States toward the BWCA.

Strong protest and resultant pressure thus took Canada and the utility by surprise. As a consequence, Ontario Hydro and the provincial government grew defensive and reticent about discussing the project or releasing information. Allegations of staged information meetings, unwillingness to respond to inquiries, and general hostility to questioning have been made by parties in both countries, and the Ontario government's decision to exempt the project from its Environmental Assessment Act further angered opponents. The political commitment by the province to the Atikokan area as well as to Hydro was firm, however, and could not be readily withdrawn. Being convinced that the project would not have significant impact across the border and that it amounted to little compared to the U.S. emission of air pollutants on Canada, Hydro and Ontario soon grew resentful, perhaps even bitter, at this U.S. behavior. Ottawa, having to defend the province, soon grew to share in this attitude. Moreover, as the more recent concerns over acid rain came to the fore, the utility and the province soon perceived an added threat to their plans. The uncertain outcome of international air quality negotiations and the possibility that Ottawa would have to "clean up its own house" and curb domestic sources before taking a stronger stand against the United States created a problem.

Given imprecise knowledge on the source of coal and its sulfur content, environmental assessment of this project was a difficult task. From the beginning all official Canadian statements contended that the plant's emissions would not do significant harm across the border (though admitting that its across-the-border impacts could be measured). U.S. environmentalists believed that the harm would be significant, admitting that it would not take much of an increase in pollution deposition to upset the fragile ecology of this vulnerable lake and forest ecosystem. U.S. environmentalists further applauded the fact the BWCA was designated Class I, the highest air quality in the U.S. classification system, and that this would be violated by the plant. (However, the BWCA was Class II at the time the plant was proposed and was only upgraded in 1977 with the passage of the new Clean Air Amendments. It is generally acknowledged that the Atikokan plant probably would not violate the then existing Class II designation.) Canada, of course, does not recognize the U.S. air quality classification system as domestically relevant. The U.S. government, under great pressure from citizen environmentalists and the Minnesota government, questioned Canadian allegations of insignificant transborder effects, but did not fully subscribe to environmentalists' claims either. The result was the commissioning of U.S. EPA studies and ultimately a call for a reference to the IJC.

THE NEGOTIATIONS

The first Canadian–U.S. negotiating session on Atikokan took place in August, 1977. At that meeting Canada agreed to provide the United States with additional information for purposes of studying the environmental effects of the plant. Based on that information, EPA developed an SO_2 dispersion model confirming that the Atikokan Station would exceed SO_2 pollution limits under the significant deterioration provision and thus violate Class I, but only slightly. The State Department concluded the pollution threat was not sufficiently serious to ask Canada for a moratorium on plant construction or for the installation of scrubbers.[23]

Environmentalists attacked the SO_2 dispersion model used in the study as one which is designed to predict pollution dispersion only over a short distance. It is not a trajectory-type model which would be useful for predicting impacts for longer distances (over twenty-five kilometers whereas the BWCA is sixty kilometers from the plant).[24] These criticisms led to a demand for a new U.S. study which would include a point-source acid sulfate, SO_2, and mercury modeling effort and a more comprehensive regional model which would account for all sources of pollution in the region and determine specific terrestrial and aquatic impacts on BWCA. Steady pressure on selected congressmen and on the State Department resulted in the EPA National Water Quality Laboratory's conducting an expanded study to assist development of the U.S. position.

Environmentalists were less successful in urging the State Department to request a construction moratorium or investigation of alternatives. They were also ultimately unsuccessful in another goal, that of having the matter referred to the IJC, although they did succeed (with the help of Minnesota congressmen) in persuading the State Department to request such a reference, only to be turned down by Canada. The quid pro quo offered by environmentalists was referral to the IJC or installation of the best available scrubbers which would remove at least 90 percent of the SO_2 emissions. In response, the State Department compromised on this figure and asked for a 50 percent SO_2 reduction, which Canada again rejected. Negotiation then turned to the matter of referring the issue to the IJC without a moratorium on construction but with a program monitoring the effects. Canada agreed to consider this State Department proposal. This in itself was less than environmentalists desired. Fearing an IJC study would take several years, they wanted such an investigation coupled with a moratorium on scrubber installation.[25]

However, environmentalists did not even achieve the more modified goal of referral to the IJC without other conditions, for on March 20, 1978, the Canadian embassy delivered a strongly worded *aide memoire* to the State Department which concluded SO_2 concentrations crossing the border would not cause injury,[26] and on this ground denied the referral to the IJC.[27] The *aide memoire* also noted that the U.S. government had never disputed this conclusion. Further, Canada contended that, in view of the failure to demonstrate any potential injury—the traditional basis for considering transboundary pollution questions—there was no basis for considering scrubbers.[28]

Thus Canadian diplomats made it clear that they consider the U.S. request as one which asks Canada to comply with U.S. law, a request out of keeping with conventional international law which requires only avoidance of transboundary damage. On IJC involvement, Canada found that the commission should not be asked to recommend whether activities in one country should meet the legislative requirements of the other.[29]

In lieu of IJC involvement, Canada endorsed the Ontario government monitoring program to be carried out at the plants and assured the United States it would take necessary corrective measures if injury became obvious. Canada was also willing to discuss details of the monitoring system to make certain that U.S. concerns were incorporated into the program.

Just as Canada was surprised by the sudden U.S. request for a reference, so too was the United States surprised at Canada's categorical refusal to support one. The argument that Canada's firm belief that there would be no cross-border injury precluded its participation in a reference cannot be viewed as strong in view of the fact that the United States argued similarly when Canada requested a reference on Garrison, but went along with the reference out of concern for Canada's fears. Also, assuming Canada was on strong ground in its belief of no cross-border injury, would it not welcome IJC substantiation of this belief, substantiation which would presumably be more credible in U.S. quarters than Canada's own unilateral pronouncements? The real reason for the Canadian government's refusal to participate in this reference and to accept the diplomatic risk inherent in such a refusal may not be known for some years. However, plausible explanations include suspicions of linkage with the U.S. refusal in the early 1970s to join in a reference on the West Coast oil tankers issue, alleged suspicions or distrust of certain members of the commission, and fears of deeper federal-provincial relations problems with Ontario over IJC involvement.

RESORT TO A UNILATERAL EPA STUDY

Since the United States was not totally convinced of environmentalist allegations as to damage in Minnesota, it did not react to the Canadian decision. Rather, at the behest of environmentalists and Minnesota interests, it pursued previously announced plans for a more comprehensive EPA study while continuing to keep Canada aware of its concerns.

This comprehensive EPA study reached a number of conclusions as to air quality, terrestrial, and aquatic impacts. It found that U.S. Class I air regulations would be violated but very infrequently (two or three times a year).[30] The study also found that 70 percent of the sulfur emissions would leave the immediate plant site (thus raising acid rain concerns) and that there would be a small increase in particulate and gaseous pollutants in the BWCA. It further found that the expected 10 percent increase in SO_2 fallout must be viewed with considerable concern.[31] Thus acid rain concerns from Atikokan were further substantiated. The EPA also found that small additions of acidity to BWCA soils could affect the nutrient cycle outputs rapidly and irreversibly, especially when combined with natural releases of chemicals from weathering and when groundwater quality changes are expected to result. Also noted were possible effects on pine trees (retardation in growth) and lichens, on small herbivores, birds, and fish because of disruptions in the food chain, and on insects. But it pointed out that such possible effects are more likely only if combined with other regional pollution emissions.

In the aquatic area, EPA found that lakes are already acidifying and the process will accelerate from Atikokan loadings with the usual expected impacts on biota, enhanced by acid flushing at the time of the spring snowmelt. Increased toxicity from the release of aluminum and other trace elements from leaching of rocks and additions to mercury in fish tissues were also considered probable. Among possible aquatic impacts are ultimate loss of fish, reductions in productivity and diversity, and potential irreversibility of the process. The agency concluded that all of these are likely occurring now, but their rates would be accelerated with Atikokan's operation.

This long-awaited EPA report did not substantiate the more extreme fears of environmentalists, at least not in the short run, nor did it completely exonerate the Atikokan plant from negatively impacting the northern Minnesota environment. However, it came closer to justifying the Canadian position while telling a story of environmental decline already begun in this region, most probably due primarily to U.S. sources. Atikokan, it appears, would exacerbate but could not initiate that decline.

Ontario had responded to the first U.S. EPA study by noting that it showed even less impact on the BWCA than had the Ontario-sponsored Canadian study, further reinforcing Canada's position. Ontario concluded that the more comprehensive second study supports those findings of earlier U.S. and Canadian studies concerning the predicted air concentrations of sulfur dioxide of the Atikokan plant, but differs with the conclusions concerning the significance of further impacts. Ontario, however, departs from these findings in contending that the EPA sulfur dioxide and sulfate emissions predicted are unrealistically high; that EPA fails to subtract the loss of emissions from the mines that will close; that there is no substantive evidence to support adverse effects on forests; and that the station will not have a measurable effect by itself on the rate of acidification of lake ecosystems (though admitting it must of necessity represent an incremental addition).[32] Ontario and Canada have both emphasized in the Atikokan case that the plant is only a small incremental aspect of a much larger problem, that of long-range transport of air pollutants and acid rain on a continental basis.

THE END AND THE BEGINNING

Atikokan itself ended as a bilateral issue in 1979 for two reasons external to the project: declining demand for electricity in Ontario and the greater concern over long-range transport of air pollution and acid rain, of which Atikokan is only a tiny part of the whole.

Overoptimistic demand forecasting and significant excess generating capacity became by 1979 a hot political issue and even an embarrassment to both the utility and the provincial government. Elimination or modification of planned new power generating projects became necessary, and the Atikokan project was halved, from 800 Mw to 400 Mw, and set back in time, with the first 200 Mw unit to go on line in 1984 and the second in 1988. Some observers believe the second unit will never be built, as a result of a declining rate of growth in demand. At the level of only one unit there is broad agreement that the plant will not be a transboundary environmental problem. All the diplomatic debate of the mid and late 1970s was perhaps in vain, as economic forces outside the realm of diplomacy overtook and disposed of the problem.

However, while Atikokan as a single issue has declined, the much broader question of acid rain has taken its place; Atikokan remains a part of that broader issue. To some extent, it will be internalized within Canada as it comes under Ottawa's new hard line on domestic SO_2 emitters and Ontario Hydro may well be forced to do what the United States was unable to get it to do—put on scrubbers. Ontario's position is one of great reluctance, because of the expense involved, the imperfections of scrubber technology, and the ever-hopeful view that a more effective and cost-efficient technology will yet be commercialized.

Time appears short, however, if Ottawa is to bring Canada to a point of being able to credibly adopt a hard line against acid rain and SO_2 emissions from south of the border. Whether the disappearance of the market for Atikokan electricity or mandated scrubbers or other control technology will win out in the end remains to be seen. With the exception of the increased knowledge of SO_2 and acid rain effects that both societies gained from the studies and debate, the entire Atikokan experience may well have been a fruitless bilateral confrontation.

An answer to the question "What have we learned from Atikokan?" might be that

> the problem of acidic precipitation cannot be addressed by writing laws and implementing procedures to deal only with new developments—most importantly these efforts must be part of a far greater total commitment to correct existing sources. . . . We must not make it easy for existing sources to hide behind protective laws, with grandfather clauses, that delay achievement of new goals.[33]

This answer refers to the fact that Atikokan, a new plant, has been under heavy fire by a country which, in its own air pollution and acid rain policy, has exempted old plants, through a "grandfather clause," from the stricter requirements applicable to new plants. Thus the United States was asking Canada to apply strict rules to Atikokan as a new emitter but is itself exempting its many old polluters from much the same rules. This leads to one of Canada's and Ontario's principal concerns vis-à-vis the Atikokan issue. The United States is complaining to Canada about one plant in a large remote region while at the same time spewing much greater tonnage of the same kinds of pollutants from its urban-industrial complexes around the Great Lakes, the Ohio Valley, and elsewhere toward Canada. This fact has been a major contributor to the bitterness and defensiveness to be found in the Canadian reaction from the very beginning. The U.S. State Department is aware of the situation but has been unable to do much about it, being under pressure to respond to both environmental and economic pressures.

Cornwall Island

A brief look at the difficult transborder air pollution issue at Cornwall Island is instructive since it highlights the incomplete nature of present U.S. air quality statutes which, even when fully implemented, do not appear capable of solving this particular problem. An additional issue in this case, which no environmental legislation can address, relates to the confidence native peoples have or do not have in government to protect their interests. Indeed, uncertainty as to which government they are subject and according to what treaties has also revealed itself as a problem. These additional questions have further complicated the Cornwall issue.

A U.S. firm, the Reynolds Metals Company, operates a large aluminum smelter at Massena, New York, across the St. Lawrence River from Cornwall, Ontario. Near the two locations, occupying an island in the St. Lawrence River, is the St. Regis Indian band. The island is in the Canadian section of the river and therefore is Canadian territory, although the Indians do not identify any more with one country than with the other and have strong linkages to Indians on the New York side of the river. Most identify as members of their band and maintain treaty relations with Canada (and with the United States through brethren living on the New York St. Regis Reservation).

The Reynolds smelter began operating in 1958 and has emitted fluorides in both gaseous and particulate form since that time, according to the International Air Pollution Advisory Board.[34] By 1971 symptoms of fluorosis had appeared in Indian cattle, and the St. Regis Indians complained to the Ontario Environment Ministry. Their suspicions were verified by the provincial government, and a cash settlement between the Indian band and the Reynolds Company resulted, avoiding transboundary problems. A year earlier, in 1970, the company had complied with a New York State order to reduce fluoride emissions and thus was in full compliance with existing New York State statutes and regulations at the time of the 1971 St. Regis complaint. In general, fluoride emissions have fallen off significantly since that time (from 307 lbs./hr. in 1968 to 112 lbs./hr. in 1973 to 75 lbs./hr. in 1977).

In 1975 the affair reopened when Ontario's Environment Ministry received another complaint of fluorosis damage to cattle, but this time in an area of the island further from the smelter, raising a suspicion that fluoride contamination not only might be continuing but that it might be more widespread than originally thought. As a result of this suspicion, Ottawa became actively involved and Environment Canada began discussions with the U.S. EPA. The International Joint Commission also became involved in 1975 through the authority of its standing reference on air pollution at the border and expressed concern through its Air Pollution Advisory Board. The board subsequently held a meeting on the case and communicated its concerns to the commission. Beyond this, the IJC has not been asked by government to become involved.

INTENSIFICATION OF THE BILATERAL ISSUE

The appearance of more widespread contamination on the island and a general suspicion by the Indians that what harms cattle may harm human health raised this transborder issue to the higher plateau upon which it currently rests. This, coupled with the fact that the ambient levels of gaseous fluoride on the island exceed Ontario's air quality criteria for ambient air and would be even further in excess of Environment Canada's proposed federal objectives, have forced the issue into the bilateral arena. This is in spite of the fact that the levels do not exceed those of New York State. (The U.S. EPA has not yet issued ambient fluoride standards and may well develop a federal standard similar to New York's which would leave the company in compliance.)

The area of the island affected has diminished in both size and severity of impact through the 1970s, but the level found in forage in a limited area is high enough to make it unsuitable for animal feed, therefore constituting property damage. Also, there is chronic fluorosis in some of the island's cattle.[35]

The St. Regis Indian band argues that damage to its property is fully substantiated and suspects it is suffering health damage as well. Recognition of Indian claims of health damage were long in coming, but both governments have admitted that expensive sophisticated epidemiological studies are warranted. It has already been noted that the U.S. EPA lacks statutory authority to require the reduction of specific fluoride emissions. However, EPA has been monitoring the situation and has found the pollution control system used by the Reynolds Metals Company to be one of the best available.[36]

New York State fluoride standards are adequate to protect cattle and vegetation.[37] They are also the only legally enforceable standards for the area, and the IJC's International Air Pollution Advisory Board has concluded there will be no human health effects if they are met.[38] New York has further confirmed the accuracy of Reynolds data and finds that the company meets all the requirements of state law. New York has supported ongoing additional studies but would like to see them broadened beyond the immediate area of impact.

Reynolds Metals Company agrees with the data developed by the air and forage monitoring of Ontario and New York. The company's position has been that the pre-1975 damage claims to cattle were legitimate and were settled for cash; that the 1975 cattle claim was not related to Reynolds emissions and thus no claim was to be paid; and that human health effects of Reynolds emissions do not exist on Cornwall Island.

In its 1977 report, the IJC board found that Reynolds was the major

source of fluoride emissions impacting the island (and that Alcoa Aluminum Company was a minor source). It reported that there was no existing
human health problem, but that the island's inhabitants should be
monitored on a continual basis. It noted that fluoride levels in island forage
were in excess of New York and Ontario standards, but only over a small
area, and that the data on chronic fluorosis in island cattle was essentially
contradictory and more research was needed.[39]

Hence, the IJC found that while a U.S. company on U.S. territory is
the principal source of fluoride emissions affecting a Canadian island, it
also found that the company was operating completely within the law, was
not responsible for damage to human health (at the time of the 1977
studies), but was probably responsible for a small area of excessive fluoride levels in the island's forage and other vegetation. Given the general
agreement in the late 1970s between the various federal, provincial, and
state environmental agencies and the essentially positive report of the IJC
at that time, wherein lies the problem?

In early 1978 this issue was referred to as a "tragedy in the making," and it was written:

> St. Regis is well on the way to becoming the same kind of festering
> scab on federal politics that the mercury contamination of the
> English-Wabigoon River system has become in Ontario.[40]

The latter is a reference to the contamination of northern Ontario Indians
by toxic mercury poisoning of a river system by the pulp and paper industry. Similarities between the two issues include

> environmental contamination by industry, an incredible amount of
> political and bureaucratic dithering, and small and defenseless Indian
> communities.[41]

The Indians' cause was allegedly ignored by Canadian diplomats, causing
the band to become increasingly distrustful of government. In reaction, the
tribe raised the issue (with media help) into a cause célèbre, raised funds,
and focused their demands on two goals: immediately reduced emissions
by the Reynolds smelter and a full-scale epidemiological study, with sharply reduced emissions continuing until the study could be completed.
Ottawa finally agreed to fund an enlarged human health study, but it
would simply be an expanded version of work that had already been
done.[42]

The Cornwall Island issue thus established itself on the bilateral
environmental agenda. On February 15, 1978, Canada advised the U.S.
State Department that it believed there was evidence of harmful fluoride
pollution on the island and that the source was Reynolds Metals' plant.[43]

In a subsequent note to the State Department, the Canadian embassy provided substantiating documentation from academic and governmental scientific sources. While documentation was limited to effects on Canada, the embassy used the opportunity to remark that the Chief of Dental Services of the U.S. Bureau of Indian Affairs found a small percentage of island school children had dental fluorosis and that he was concerned about the air pollution from the aluminum plant.[44] The embassy confirmed that a health study was being conducted in Canada and, without waiting for that study's results, requested early emissions abatement to prevent further transboundary injury.[45] (Dr. Irving Selikoff, the noted epidemiologist, is in charge of this study, and has uncovered a new problem: the contamination of the island's residents by PCBs, possibly arising from transboundary—but as yet undetermined—sources.)

THE DIPLOMATIC DILEMMA

In a January, 1980, letter to the State Department, Reynolds Metals Company sharply disagreed with the findings of the primary documentation on environmental and health effects of aluminum smelting on Cornwall Island put forth by the Canadian government. The company also questioned the reports on the visit to the island of the U.S. Dental Services chief. Upon interviewing him, Reynolds learned that he had not studied the island children but had merely made a cursory examination of their teeth which turned up mild effects of excess fluoride in the environment and suggested that the Indians should investigate the source of the fluoride.[46] The company also made known its concern that the Indians had not permitted Reynolds to inspect for damage on the island.

Reynolds reiterated that they are in full compliance with all U.S. federal and state air and water pollution control laws; that the IJC board in 1977 found no health hazards; that they have made significant reductions in emissions over the decade; that their own employees have never experienced any health problems from the emissions; and that the St. Regis band has refused to cooperate with the company. It is the first of these points which creates problems for the State Department, for as long as the company is within the law, there is little that U.S. diplomats can do to get emissions reduced further. On the contrary, the department must protect the company's interests from across-the-border threats to their operation.

The dilemma which the Cornwall Island issue creates for diplomats is the problem of how to respond to charges of pollution damage from across the border when the alleged polluter is operating within all federal and local laws. It is to be hoped that the ongoing negotiations toward a Canadian-U.S. air quality agreement will address this problem, for there is little doubt that Cornwall Island and other such conventional transboundary air quality issues will ultimately come under the aegis of such an agreement.[47]

Poplar

The Poplar power plant in southern Saskatchewan, with its attendant impacts on Montana, has been described in chapter 9. It was noted there that this is the only transboundary environmental issue which has all three major environmental components: water apportionment, water quality, and air quality. While the water disputes were referred under Article IX to the International Joint Commission, there was substantial reluctance in both governments to refer the matter of air quality to the commission. Hence, the Poplar River air quality dispute has been handled by conventional diplomatic negotiations.

Poplar's transborder air quality impacts were first raised in 1975, following the earlier expressions of concern over water questions. Saskatchewan Power Corporation decided—for economic reasons and in the realization that local background air is not substantially polluted and that local human population is sparse—not to incorporate expensive scrubbers or other sulfur-control technologies. It did, however, install particulate-control technology (removing 99.4 percent of particulates plus gasses) and is building the plant in such a manner that sulfur control could be accomplished at a later date if desired.

The crux of this bilateral issue is attitudinal and legal. The more purist "no significant deterioration" attitude of the United States, regardless of the condition of the receiving air, comes up against what some would call the more realistic Canadian attitude which attaches greater weight to the existing ambient air quality and the ability of the receiving air to absorb (and dilute) additional pollutants. Current U.S. laws state that the best available technology must be used in this case (and would be if the plant were being built in Montana), regardless of need (that is, regardless of the condition of the local air). Canadian rules specify that the best practicable technology be employed, taking economics into account. In this case, since low-sulfur lignite coal will be used and Saskatchewan and Canadian SO_2 standards will not be exceeded, it is not legally necessary in Canada (and would be highly uneconomic) to install expensive SO_2 scrubbers.

In 1975 Montana state officials became worried that the emissions from all four units (if all four were built, for a total of 1,200 Mw) might violate the prevention-of-significant-deterioration (PSD) section of the federal Clean Air Act, since this region's air was quite clean to begin with. Thus Montana requested that the State Department ask the Canadian government to do an assessment. In response, a federal-provincial study concluded that the first two 300 Mw coal-fired units would not violate the PSD requirements for this area but that the second two units, if built, might pose a threat to the Montana airshed. The State Department position in early 1977 was that the first unit created no problem,[48] but that the second and subsequent units might well create one. Thus the bilateral difference was over the impact of the second unit.

At that time (before the passage of the 1977 amendments to the Clean Air Act), U.S. diplomats were also cognizant of potential violations of U.S. ambient air standards which might prevent future industry from developing in that area. (The 1977 amendments technically removed this threat, since foreign sources are legally excluded from the computations performed to determine violation. But in practice, pollution incurs health and property costs and thus political problems regardless of its source.) Specifically, Montana's Class II airshed designation was threatened. The U.S. EPA suggested lowering the Montana standards to Class III to allow for the extra pollution, but Montana opposed the suggestion. U.S. diplomats knew at that time that Canada would not impose further controls and warned Montana that Canada could not be forced to do so.

THE DEBATE CONTINUES

Through the period of late 1975 to early 1977, faced with repeated calls from Montana and from environmentalists for a construction moratorium until more information was available, U.S. negotiators experienced difficulty in obtaining information on the potential sulfur dioxide emissions from the plant. Part of the problem stemmed from the utility's inability to obtain appropriate coal for analysis, from Canadian tardiness in responding to requests for information, and from the physical difficulty of obtaining a realistic average sulfur content figure, given the variability in sulfur content within coal samples even from the same area. U.S. diplomats concluded early that the key factor in Poplar air quality is the sulfur content of the coal. Frustration mounted rapidly with suspicion that Saskatchewan was withholding data on its content. U.S. diplomats, caught between the Montana pressures and perceived Canadian reluctance to cooperate, were further frustrated by the persistence of the Saskatchewan Power Corporation, which never slowed down construction or showed any hint of hesitation in moving ahead with its project. Aware of this, Montana pressured Washington further.

In 1977 Governor Thomas Judge of Montana showed an interest in getting the IJC involved in the air quality aspects of the issue and communicated to U.S. IJC Chairman Henry Smith his desire for a formal framework for the discussion of the air quality issues.[49] In early 1978 he supported a State Department note requesting immediate consultations on certain aspects of Unit No. 1 construction premised on his belief that construction was not planned to avoid or minimize transboundary impacts.[50]

Further frustration resulted from contradictory Canadian information on the subject of particulate matter. Canada concluded in 1975 that there would be a transborder problem. In response, SPC agreed to improve particulate control on the first 300 Mw unit from 96 percent to 99.5 percent. Then, in 1977, Ottawa announced that total emissions would be substantially lower than previously thought, that particulate concentrations would be reduced significantly, and that NOx and SO_2 limits would not be exceeded even with a 1,200 Mw plant. Such news affected both negotiating positions, as it must with any dramatic change in the hard data upon which a position is based or strategy developed. It appeared that the utility's continued construction was further justified, making the case more difficult for U.S. diplomats.

The U.S. side did not have a monopoly on frustration, however, as Ottawa often found itself dealing with a tough, uncompromising, and perhaps recalcitrant provincial government at Regina. If left to Ottawa, such a plant would never have been located only four miles from the border, but the SPC's decision and Regina's approval was based solely on economics, namely, location of the coal deposits and ease of mine-mouth conversion of the coal to electricity. Difficulties in federal-provincial relations in Canada often bear upon bilateral environmental relations, and the Poplar issue is one such example.

Saskatchewan experienced its own frustration in a number of areas, including the unexpected virulence of Montana objections; the province's initial lack of understanding of the IJC and of diplomatic processes and access thereto; its sometimes poor relations with Ottawa on this and other issues; and the natural rivalry between a government-owned crown corporation, Saskatchewan Power Corporation, and a government agency, Environment Saskatchewan.[51]

By 1978 the United States had concluded from EPA analysis that

applicable Montana and EPA ambient air quality standards will not be violated at the *600 MW level* but a significant amount of the PSD Class II increment will be used up in Montana near the border, and applicable Montana and EPA emission standards will be violated for SO_2. [Therefore], justification for asking for SO_2 scrubbers on the Poplar plant would appear to rest primarily upon the emission standards argument. [Emphasis added][52]

Unit No. 1 was thus now viewed as a fait accompli, and some aspects of Unit No. 2 were being declared acceptable.

Montana's ambitious air quality data-gathering efforts will probably be more effective in defending the state against construction of Units No. 2, 3, and 4 than in accomplishing anything against Unit No. 1. This would be especially so if Saskatchewan should violate its own provincial standards and Montana should catch it in the act. There is clearly an attitude today in Saskatchewan, in the SPC, and in the Environment Department, that Montana is going too far in the direction of expensive and precise data gathering. It is possible, of course, that the state could develop a greater knowledge of Saskatchewan's environmental conditions in the immediate border region and one day find itself in the position of being able to inform the province that it has permitted violation of its own standards, should such violation occur. This would bring pressure on Ottawa and Washington to intervene (if one can assume that violation of provincial standards automatically implies violation of international commitments) and ultimately increase pressure on SPC to take remedial action: that is, to install scrubbers.

THE END OF THE DEBATE

The bilateral monitoring system described in chapter 9 applies to air as well as water, and, now that the IJC has concluded its work, the United States considers Poplar as one single issue while Canadian diplomats continue to view the air and water components as separate. Since all bilateral air quality issues have been subsumed in the ongoing international air quality and acid rain negotiations, the Canadian position of viewing air separately may be more realistic. Poplar air will ultimately be governed by the tenets of that agreement, and Saskatchewan Power Corporation and the province will have to abide by the final outcome of that negotiation, regardless of what happens in the water apportionment and quality aspects of this issue.

By mid-1978, the end of the Poplar air quality debate period and the beginning of serious bilateral consideration of a joint monitoring agreement (and prior to the formal announcement of Poplar Unit No. 2), the following developments had taken place:

- the SPC had agreed to install 99.5 percent particulate control on the first 300 Mw plant, to answer U.S. concerns;
- the SPC agreed to incorporate design flexibility into the plant (to accommodate scrubbers if ultimately necessary);
- the two nations formally agreed that the first unit would create no transboundary air pollution problems (though Montana did not fully concur);
- the two nations agreed that the first two units (600 Mw) would create no nitrogen oxide or fluoride problem;
- the United States contended that a 600 Mw plant would violate U.S. sulfur dioxide PSD regulations, while Canada contended U.S. standards would not be violated;
- the United States contended there might be a slight PSD violation for particulates at the 600 Mw level; and
- neither country would make predictions on a 900 Mw or 1,200 Mw unit (technically not proposed, but the plant is built to accommodate up to 1,200 Mw, a size which was once considered by the SPC and may be again).[53]

Both sides expressed the expectation that injurious transboundary environmental impacts would be prevented and that mitigation measures would be identified and applied. The United States reserved rights to seek compensation from impacts and Canada promised always to consider U.S. concerns.

UNANSWERED QUESTIONS

From a federal perspective, it would appear that the bilateral agreement on a comprehensive air quality monitoring network at Poplar has defused the issue. However, such a belief may be too sanguine. For example, Saskatchewan Power Corporation, which is paying the bill, has noted the corporation will rectify problems attributable to it which are detected by the monitoring network as long as the remedy is within its financial means.[54] If it is beyond its means, the issue could easily return to the bilateral arena. The SPC's manager of environmental affairs expresses a need for an international air quality agreement and fears that the bilateral monitoring committee may end up as a bilateral control committee. In terms of Montana's ambitious monitoring goals, Saskatchewan admits to a current challenge, namely, to contain the state to some modest degree of monitoring. The province also recognizes that, since the power plant will contribute to continental acid rain patterns, however little, the scrubberless plant may be threatened by the current acid rain negotiations. Accepting the philosophy of cleaning up its own house first, Ottawa may press Regina to require scrubbers lest the federal government be accused by the United States of double standards. Further, Saskatchewan's current ambient air quality objectives could conceivably be converted to U.S. point-source standards owing to Canada's hard-line bargaining position on acid rain and SO_2.

Montana is not convinced that the monitoring agreement will defuse the issue. Montanans still ask why, if scrubbers are required on Colstrip (a large eastern Montana coal-fired power plant), they are not on Poplar? A new Montana element in the drama is the recent application for Class I ambient air quality status for the Fort Peck Indian Reservation. Under the 1977 Clean Air Act Amendments, the tribe is entitled to such consideration, and it is now exercising that right, perhaps on account of its known dissatisfaction over the Poplar water apportionment decision. The likelihood of Poplar No. 2's violating U.S. standards is much greater under Class I than the present Class II, and the ultimate impact of this tribal application remains to be seen. Equity would seem to indicate, however, that such an ex post facto decision should have no effect on Canada and Saskatchewan.

A suggestion has been made during the past year that perhaps Montana would accept greater air pollution if it were guaranteed a greater flow of water from Canadian diversions from Lake Diefenbaker or elsewhere into the Poplar drainage. This might accomplish the dual task of providing sufficient water for Units 3 and 4 as well as provide an arid region of Montana with additional needed water. However, it is unlikely that this deal would be politically acceptable in Montana at present. The state is aware that the Poplar plants' lack of controls weakens Canada's arguments on the retrofitting of U.S. coal-fired power plants and may well incorporate this fact into the state's future Poplar strategy. Hence the U.S. EPA in Helena as well as the state government disagree with those who argue the issue has been settled with the monitoring agreement. They foresee the matter continuing indefinitely until scrubbers or other sulfur-control technology are installed. However, if a limit is placed on two 300 Mw units, then Montana's official concerns are answered, especially given the U.S. EPA Draft Environmental Impact Statement which found that there are no damaging air quality effects at the 600 Mw level. Montana appears satisfied with the monitoring agreement, as long as it can continue more comprehensive monitoring which could conceivably lead to "catching the Poplar plant in the act," as it were.

The Poplar issue raises a number of valid questions which await answers. Is it cost-effective to scrub low-sulfur lignite coal? Do the results of scrubbers justify an over 8 percent decrease in electricity output, a $50 million capital investment plus maintenance costs? Is the mining and moving of the limestone required by the scrubbers worthwhile? Is accelerated coal mining and additional strip-mined land sufficient to make up for the decreased output? Can a population base of one million sustain such cost?

As an individual issue on the bilateral environmental agenda, Poplar is likely to go the route of Cornwall Island, Atikokan, Detroit-Windsor, and acid rain, that is, to be governed by the outcome of the international agreement soon to be negotiated. Whether Poplar Units 3 and 4 are to be built or whether Saskatchewan or Montana will gain from that international agreement is yet to be seen.

Review and Conclusion

After 1970, with the passage of significant bodies of federal, state, and provincial clean air legislation and with greater awareness of air pollution costs to society, the number of bilateral air quality incidents began to increase. By mid-decade the late U.S. Senator Lee Metcalf of Montana and the Saskatchewan Legislative Assembly in Regina had begun calling for the negotiation of a bilateral air quality agreement along the lines of both the Boundary Waters Treaty and the Great Lakes Water Quality Agreements.

The immediate thrust for this call arose from the emerging concern over the air pollution impacts of Saskatchewan's Poplar River power plant on Montana, but the Atikokan issue a year or so later, coupled with the Cornwall Island problem and the attention drawn by the IJC's Great Lakes

Science Advisory Board to acid rain impacts on the Great Lakes all contributed to this thrust. All of these factors were operating before the cry for international acid rain control began to be heard across the land.

Not all of this thrust was motivated purely by environmental protection concerns. At least some of it stemmed from the desire of certain areas planning economic development to secure protection of their rights to develop and to insure an established set of rules by which they could operate and which would in turn reduce uncertainty for investors. International rules of the game, in air quality as in other areas, can be useful not only to provide protection for the environment but also to provide equal protection for developmental activities of one type or another. In this vein Saskatchewan asks why it should have to expend vast sums of money to avoid pollution and thereby be made less competitive with Ontario, while not enjoying those advantages that Ontario has had over the years to develop and pollute. Hence, Saskatchewan has fought for the applications of Best Practicable Technology (BPT) rather than Best Available Technology (BAT). In order to ensure the acceptance of this control philosophy internationally as well as domestically, it has opted for supporting the establishment of recognized rules of the game, for a guarantee of an ordered and orderly relationship, and for a protection of its right to develop. An international air quality agreement laying out such rules and thus guaranteeing order and providing the desired protection is clearly in the best interests of Saskatchewan.[55] Similar sentiments might well be expressed by Alberta, British Columbia, or even northwestern Ontario, or indeed by any heretofore underdeveloped border region of the United States which has been left blinking in the dust as rapid growth areas have expanded and diversified rapidly.

Although the Clean Air Act Amendments of 1977 exclude foreign source pollution from U.S. emission calculation levels and therefore these cannot technically be viewed as a threat to economic growth, nevertheless domestic pollution levels become that much higher with the addition of such foreign sources, with the health and environmental damage that implies. They therefore still pose threats to domestic growth, particularly in border states and provinces. Hence, there is good reason for alliances between environmental, health, and industrial groups. A political alliance of these often-divergent groups has succeeded in focusing public attention on transboundary air pollution and the need for international ground rules in this until recently largely ad hoc area. Sufficient interest developed in both countries[56] for the passage of a number of resolutions[57] on this subject in various legislative bodies. The U.S. Senate even attached a rider to the Foreign Relations Authorization Act of 1979 which mandated the State Department to begin diplomatic negotiations leading toward such an international agreement.[58] Then the public hue and cry over acid rain became so intense and the demand, particularly from Canada, for negotiations on this subject so outspoken that all attention focused on it. With the advent of formal negotiations in this area, the separate history of transboundary air pollution came to a close and the two issues were joined as one.[59]

It is to the overriding issue of long-range transport of air pollutants and acid rain that we must now turn.

NOTES

1. D. H. Dinwoodie, "The Politics of International Pollution Control: The Trail Smelter Case," *International Journal* 27, no. 2 (Spring 1972):227.

2. Ibid., p. 228.

3. Ibid., p. 232.

4. Ibid., p. 233. The linkage was to reciprocal tariff reductions in aid of the agricultural and forest industries.

5. Anthony L. Otten, "Current Institutions Used by the United States and Canada to Deal with Transboundary Air Pollution" (Thesis, Massachusetts Institute of Technology, 1978), p. 62.

6. Richard B. Bilder, *The Settlement of International Environmental Disputes* (Madison, Wisconsin: University of Wisconsin Press, 1976), p. 43.

7. Dinwoodie, "International Pollution Control," p. 234.

8. International Joint Commission (IJC), *Transboundary Air Pollution: Detroit and St. Clair River Areas* (Washington and Ottawa: IJC, 1972), p. 1.

9. Ibid., p. 3.

10. Ibid., pp. 57–58.

11. International Joint Commission, *First Annual Report on Ontario-Michigan Air Pollution* (Ottawa and Washington: IJC, 1976), pp. 13–14.

12. International Joint Commission, *Third Annual Report on Michigan-Ontario Air Pollution* (Ottawa and Washington: IJC, 1978), p. 2.

13. International Joint Commission, *Fourth Annual Report on Michigan-Ontario Air Pollution* (Ottawa and Washington: IJC, 1979), p. 1.

14. Ibid., p. 18.

15. International Joint Commission, "IJC Issues Air Quality Report for Michigan-Ontario Region" (Ottawa and Washington: IJC, February 28, 1978).

16. Ibid., p. 21.

17. International Joint Commission, "Michigan/Ontario Air Pollution Control Objectives Not Met by Target Date" (Ottawa and Washington: IJC, June 17, 1980).

18. John M. Leddy, Assistant Secretary of State, Letter to the International Joint Commission, September 23, 1966, as noted in IJC, *Transboundary Air Pollution*.

19. National Parks and Conservation Association, *Factsheet on Atikokan Power Plant* (Washington, D.C.: NPCA, 1978), p. 1.

20. U.S. Department of State, Office of Canadian Affairs, Environmental Officer, Interview, Washington, D.C., November 15, 1977.

21. The Clean Air Act Amendments of 1977 permit EPA and a state govern-

ment to subtract from their calculation of air quality increments any pollutants from a foreign source, but political realities reduce the value of this exemption, since pollution from domestic or foreign sources is just as damaging. (See P.L. 9595.) Minnesota has determined that

> the Atikokan power plant would appropriate between 25 and 75% of the Class II sulfur dioxide 24-hour increment. By implication, the Class I SO_2 24-hour increment would be violated.

(Aaron Katz, Minnesota Pollution Control Agency, Statement before the Ontario Royal Commission on Electric Power Planning, Toronto, Ontario, June 6, 1977, p. 3.) It fears that under U.S. clean air law future industrial growth would be severely limited in the region including proposed copper-nickel development which ironically may impact Ontario. It concludes that the only way these impacts can be reduced is through the use of SO_2 scrubbers. (Ibid., p. 4.)

22. U.S. Department of State, Office of Canadian Affairs, Environmental Officer, Interview, Washington, D.C., November 15, 1977.

23. National Parks and Conservation Association, *The Need for Scientific Study of the Atikokan Power Plant Plans* (Washington, D.C.: NPCA, 1978), p. 1.

24. Ibid., p. 2. Environmentalists are further critical of inattention to regional ambient air quality in Ontario and to the potential contribution of this plant to the continental acid rain loadings. Lack of attention to effects of air pollution on Boundary Waters Canoe Area ecosystems (beyond its simple occurrence) and to occurrence and impacts of transboundary water pollution were also cited as basic weaknesses of early EPA studies and resultant U.S. positions.

25. Joan Moody, "Fire and Rain at Atikokan," *Outdoor America,* May/June, 1978, p. 25.

26. Canadian Embassy, *Aide Memoire,* Washington, D.C., March 20, 1978, p. 1.

27. Technically, it is not necessary for both countries to agree to a reference before it can take effect, although many erroneously believe this to be so. However, in practice references have not gone forward without the agreement of both nations, and such a unilateral reference would undoubtedly result in serious difficulties, some of which might well test the flexibility and strength of the commission itself.

28. Canadian Embassy, *Aide Memoire,* p. 2.

29. Ibid., p. 3.

30. Gary E. Glass, *Impacts of Air Pollutants on Wilderness Areas of Northern Minnesota Draft* (Duluth, Minnesota: U.S. EPA Environment Research Laboratory, 1979), p. 7.

31. Ibid., p. 8.

32. Ontario Ministry of the Environment, *Review of Final Draft Copy of Impacts of Air Pollutants on Wilderness Areas of Northern Minnesota* (Toronto, Ontario: Ontario Ministry of the Environment, 1979), p. 8.

33. William Steggles, "Introductory Remarks to a Panel Session: Atikokan—Assessing a New Power Plant" (presented at the Action Seminar on Acid Precipitation, Toronto, Ontario, November 3, 1979).

34. International Air Pollution Advisory Board, *Transboundary Flow of Fluoride Air Pollution Affecting Cornwall Island* (Washington, D.C.: IJC, 1977), p. 2.

35. Ibid., p. 4.

36. Ibid., p. 5.

37. Ibid., p. 6.

38. Ibid.

39. Ibid., pp. 8–9.

40. Jim Robb, "St. Regis: Tragedy in the Making," *Ottawa Journal,* March 2, 1978.

41. Ibid.

42. *Canadian Press,* Ottawa, March 21, 1978.

43. Canadian Embassy, Note No. 585, Washington, D.C., November 30, 1979, p. 1.

44. Ibid., p. 3.

45. Ibid.

46. Richard E. Cole, Vice President and General Manager, Reynolds Aluminum, Letter to Sidney Friedland, U.S. Department of State, January 23, 1980, p. 1.

47. This issue appeared on the agenda of the Canada-U.S. Interparliamentary Group in May, 1980. "Both delegations (Canada and U.S.) agreed that this situation was a matter of concern and that further study is necessary." (Canada–United States Interparliamentary Group, Twenty-first Meeting, San Diego, California, May 23–27, 1980, *Report,* p. 22.)

48. The principal environmentalist organization in the affected area of northeastern Montana, the Three Corners Boundary Association, did not agree with the State Department's acceptance of Unit No. 1, arguing that this acceptance was premature, was based on an erroneous Canadian study (the Portelli Report), allowed insufficient public input, and was based on two debatable assumptions: that Unit No. 1 can pollute Montana air up to the maximum legal limit (Class II PSD) [which the association argues is bad strategy and unrealistic in that it allows no margin of safety, "pushing the people of northeastern Montana into a corner in which we do not want to find ourselves"]; and that the U.S. must not assume [as it does] that there will be no other pollutant source in the area, citing the greatly increased interest in potash mining and development in the region, presumably an activity which should not be precluded because the Saskatchewan power plant has used up all the air increment. (Letter from Dennis G. Nathe, Chairman, Three Corners Boundary Association and Montana State Senator, to Montana Lt. Gov. Bill Christiansen, September 24, 1976.) This organization has continued to place pressure upon the Montana government and the U.S. government to do all possible to deter the completion of all the units at Poplar.

49. Telegram from Governor Thomas L. Judge to Henry P. Smith, III, Chairman, U.S. Section, IJC, September 28, 1977.

50. Governor Thomas L. Judge, press release, March 14, 1978.

51. Provincial frustration was exacerbated by the shock it received from Montana's vigorous reaction. About twelve years earlier the SPC had built the Boundary Dam power plant and coal mines to the east at Estevan. This facility has air pollution impacts in North Dakota, but that state never complained. Saskatchewan has learned that Montana and North Dakota are very different states.

52. Letter from T. L. Thoem, EPA, Region VIII (Denver), to Karl Jonietz, U.S. State Department, January 15, 1978.

53. Based on the statement by the Canadian spokesman to the Poplar River–Nipawin Board of Enquiry, March 28, 1978; the text of a note of May 15, 1978, from the Canadian Embassy in Washington to the U.S. State Department; and the text of a note of March 14, 1978, from the U.S. State Department to the Canadian Embassy; and Annexes.

54. Fred Ursel, General Manager, SPC, personal interview, Regina, Saskatchewan, November 6, 1980.

55. Grant Mitchell, Secretary to the Cabinet, personal interview, Regina, Saskatchewan, November 7, 1980.

56. A Canadian member of parliament, Ralph Goodale of Saskatchewan, wrote in 1978 to U.S. Senator Paul Hatfield (appointed to fill the vacancy after Senator Metcalf's death),

I have recently become aware of an initiative taken by yourself and 18 other United States Senators to sponsor a resolution in Congress directing your Secretary of State . . . to begin negotiations with Canada toward an international treaty on trans-boundary air pollution problems. . . . I am certainly pleased to hear of your interest in this subject, and I hope you and your fellow Senators will enjoy considerable success in pursuing your resolution. I believe you will find considerable support in Canada for this idea as well [Letter of Ralph Goodale, M.P., to Sen. Paul Hatfield, U.S. Senate, June 8, 1978].

57. P.L. 95–426, 92 STAT. 963, Section 612, October 7, 1978.

58. The Montana legislature passed a joint resolution calling for Canadian-U.S. bilateral negotiation leading to both an air quality treaty and a new IJC-type institution to implement the treaty (HJR 0070).

59. The Joint Statement on Transboundary Air Quality by the two governments, dated July 26, 1979, and referring to previous informal bilateral talks held on December 15, 1978, and June 20, 1979, is transitional in nature between the "transborder air quality era" and the "acid rain era."

CHAPTER 11   Acid Precipitation

*Acid rain or precipitation is one part of a broader phenomenon known as
long-range transport of air pollutants (LRTAP). The pollutants transported
include toxic heavy metals, organics, ozone, and other substances, and
LRTAP occurs in both wet and dry form. The wet form, acid precipitation,
returns to earth causing acid deposition. Acid precipitation is defined as
naturally occurring moisture which has become acidified (i.e., has experienced a decrease in its pH[1] to lower than 5.0–5.6) by the addition of sulfur
and nitrogen oxides, SO_2 and NO_x, which have been emitted high into the
atmosphere, remaining there for a period of time, and traveling long distances before returning to earth.*

*The United States emits a significant quantity of both SO_2 and NO_x, a
significant portion of which crosses the border into Canada. Canada emits
some SO_2 and NO_x which enters the United States, but the proportion is
much lower. Canada is geologically much more vulnerable than the United
States, and harbors a vast commercially valuable forest resource, the base
of an important export industry which may be threatened by acid deposition.
Additionally, Canadians are much more aware of and concerned about acid
rain than are U.S. residents. Hence, asymmetry exists and a very serious
bilateral problem results.*

No other bilateral environmental issue has reached the level of intensity or
magnitude of impact on the Canadian-U.S. relationship as that of the long-
range transport of air pollutants, and especially, acid precipitation and
deposition. LRTAP is a major public policy question in Canada, and this
issue has significant diplomatic aspects as well. Within a few brief years it
has risen from a nonissue to among the most serious on the bilateral
diplomatic agenda. By most criteria, it has assumed the appearance of a top
priority bilateral conflict relative to other transboundary environmental
issues. It even supersedes the Great Lakes issues in complexity and conten-
tiousness, though the latter did take many more years to develop. Complicating the matter is the fact that, unlike the situation in the Great Lakes
or at Poplar or Garrison, the region that is the source of the problem is very
large, as is the area believed to be affected. Although westerners (from the
prairies to the Pacific) tend to regard the matter as an eastern problem,
there are signs in the West that the bases for a heightened debate are
developing there as well. The West is unlikely to escape the problem.

The stakes may be very high. If the impact of acidic precipitation is great, as many believe and as some of the evidence indicates, then the cost it inflicts will be high. Likewise, the economic costs to society of reducing the emissions thought to induce acid rain would unquestionably be high as well. Solution of the problem and reduction of emissions may jeopardize achievement of the national goals of energy self-sufficiency and reduced inflation presently pursued by both the United States and Canada.

Those who are hesitant to commit society to huge emission reduction expenditures are concerned about affordability. If decisions are made prematurely, or if scientific research on acid rain is done improperly, if the problem turns out to be a false alarm and all the control money has been spent, will it then be possible to get public support for real concerns in other areas? For example, airborne heavy metals from synfuels could turn out to be an even more serious problem, as could the long-range transport of oxidants, ozones, and organics.[2]

Given limited resources, there is a gamble either way. If there is an increasing problem which is irreversible, the do-nothing alternative may be dangerous. If, given limited resources, society makes the very expensive commitment to reduce emissions and then learns that the problem is not as significant as first supposed (or worse, that acidification is the result of processes other than LRTAP), the billions spent on controls would have been largely wasted and unavailable to respond to other needs.

Thus the economic stakes are much higher than those of other transborder environmental issues, and, given the great media attention and consequent public awareness of the issue, the political and diplomatic stakes have risen accordingly. This reminds us that if a bilateral border issue is widely perceived as such, then it is and must be treated as a real issue, regardless of the validity of the scientific debate or the qualifications of those debating.

Introduction to the Issue

The complex issue of acid rain—what it is, where it comes from, what it may or may not do—can only be summarized briefly in these pages as a basis for addressing the diplomatic aspects of the subject. There is controversy concerning the occurrence of this phenomenon. One group of scientists led by Gene Likens of Cornell University believes that precipitation in parts of the eastern United States and western Europe has changed from nearly neutral two centuries ago to a dilute solution of sulfuric and nitric acids today.[3] This assertion is not universally accepted by scientists, however, nor is his belief that the reason for the trend is a rise in sulfur and nitrogen oxides caused by fossil fuel burning. Likens has also found that nitric acid is a significant component of acid rain. Its contribution, now at

30 percent, is increasing in proportion,[4] which may have implications for regulatory control.

Sulfur dioxide (SO_2) is emitted from metal smelters and coal- and oil-fired power plants, while nitrogen oxide (NO_x) results from motor vehicles and high-temperature combustion. When SO_2 and NO_x mix in the upper atmosphere, join with already slightly acidic pure rain or snow, and return to earth, they can overwhelm the normal neutralizing capacity of water and soil and create an acidic condition.[5] A recent Environment Canada brochure on the subject notes that

> Long Range Transport of Airborne Pollutants refers to the way pollutants are carried by the winds over long distances, hundreds and even thousands of kilometers. These pollutants . . . are transformed into secondary products which then react chemically with the water vapor in the atmosphere and the result is acidic water vapor.[6]

Of the five classes of air pollutants—oxides of sulfur (SO_2) and nitrogen (NO_x), heavy metal particles, persistent organic chemicals, and photochemical oxidants—only SO_2 and NO_x cause significant acid precipitation. Three elements are necessary for these latter pollutants to become problems for society:

- an emission source, usually a collection of sources over a large industrial area, although single very large sources can in themselves create long-range transport problems;
- the right meteorological conditions which can carry the pollutants over long distances and which create the conditions for the transformations to take place; and
- a region of vulnerability, namely, one which lacks the ability to chemically buffer the acid deposition.

Western society has recognized and shown concern for the ramifications of the health and to a lesser extent the economic and ecological effects of local point-source air pollution and surrounding ambient air quality for many years. It is only very recently, however, that we have become aware that such local air pollution may actually be less significant than the total impact of long-distance dispersion of pollutants.

Harmful chemicals may move over hundreds of miles and descend on terrestrial and aquatic environments in either dry form (as solid particulates) or in wet form (encapsulated in a raindrop or snowflake). Of course, travel over great distances allows for the increased chance of crossing an international frontier, even in such large nations as those of North America. It is a paradox that the very same tall stacks which were seen by many industrial polluters as the answer to the elimination of local-area pollution and the ideal way to comply with air quality laws, and indeed were encouraged by government policy, have been guilty of dispersing pollutants far and wide at high altitudes, leading to inevitable acid precipitation at points far downwind. We are now learning pollution dilution over distance is a failure and, indeed, may lead to even more insidious problems.

Acid rain is inevitably an international issue, and an informal effort to agree on statements of international duties and procedures has been underway since 1978 in the form of continuing bilateral consultation and negotiation. There has also been an effort at joint fact-finding through the international scientific work of the United States–Canada Research Consultation Group on the Long-Range Transport of Air Pollutants (known also as the LRTAP Group). The LRTAP Group has produced two annual reports, in October, 1979, and in November, 1980, both of which merit close study since they were commissioned jointly by both governments. However, the validity of their methods and findings has been criticized (strongly in some industry quarters).

Industry criticism oscillates between two negative reactions: (1) the scientific data are not all in yet so it is too early to reach any significant conclusions; and (2) the LRTAP reports are out of date and based on questionable assumptions, and thus invalid. However, the LRTAP Group's methods and findings are a tangible example of successful international governmental cooperation to generate a common data base. The joint effort is similar to the traditional IJC approach, but, unlike the latter, it lacks industry support. Industry's concern over the findings of the joint bilateral research processes relates to speed of the work, reliance on questionable models and methodologies, lack of consultation with the industries likely to be regulated, and lack of peer review.[7]

EMISSIONS

Data developed by the LRTAP Group thus far indicate that: (1) sulfur dioxide causes about two-thirds of the acidity in acid precipitation and nitrogen oxides are responsible for about one-third; (2) Canadian-originated sulfur dioxide is principally a product of the metal smelting industry, while U.S.-originated sulfur dioxide comes mainly from thermal power generation; and (3) of the continental total of nitrogen oxide emissions, one-half comes from transportation, while the other half originates from power generation and other combustion sources.[8] The group noted that

> total SO_2 emissions for eastern North America will increase by 0–15% by the year 2000, mainly due to increased coal burning in electric utility plants and other industrial sources; while NO_x emissions will rise by 15–35%, again primarily due to increased fossil fuel combustion in industrial sources.[9]

The group observed that long distances from the point of emission and the passage of sufficient time (usually two to five days, according to some reports) are necessary to create an acidic precipitation condition. Prevailing winds are such that precipitation will cross the international border. They also pointed out that sufficient loading of the environment over a long period of continuous deposition is necessary for an acid rain effect to become apparent in the environment. This latter requirement implies that the United States and Canada have been acidifying each other's environments for some years but the magnitude of the impact is just now being realized. The LRTAP Group also notes that it is possible for both the United States and Canada to respect their own domestic existing air quality statutes and guidelines and still have a serious acid rain problem.

The damage arises not from air pollutant concentrations themselves but because of the loading of air-transported materials that are ultimately deposited on the natural ecosystem. Such atmospheric loading and deposition are not addressed in domestic legislation.

The LRTAP Group found that a large majority of the emission sources are in the midwestern and northeastern United States, with the highest density being from the upper Ohio Valley. A number of large power plants burning high-sulfur coal with little emission controls are located in this valley.[10]

Addressing Canadian sources, the LRTAP Group concluded:

Total Canadian sulfur emissions are about one-fifth those of United States sources, and are concentrated in the non-ferrous smelting sector which accounts for forty-five percent of total sulfur emissions. Power plants account for little more than ten percent. . . . Nearly half of Canadian emissions come from several small areas where non-ferrous smelters are located. One of these smelters, located in central Ontario, is the largest single sulfur dioxide emission source in North America, and is responsible for fully twenty percent of Canada's sulfur emissions.[11]

The total release of sulfur dioxide in North America into the atmosphere is currently 30.7 million tons per year, with about 25.7 million tons from U.S. sources and 5 million tons from Canadian sources. The total of nitrogen oxide emissions is 24.1 million tons, with the United States accounting for 22.2 million tons per year and Canada accounting for 1.9 million tons.[12]

The U.S. government does not necessarily endorse these precise Environment Canada figures, but it would probably agree with the orders of magnitude. These indicate that the United States is a far greater aggregate source, producing five times the level of sulfur dioxide as its northern neighbor, and ten times the quantity of nitrogen oxides. While some of this total is emitted in regions far removed from Canada, much is emitted from regions relatively close to the Canadian border which are both heavily populated and also heavily industrialized—the Midwest lake states and the Northeast. Almost all of Canada's No_x emissions are within 150–200 miles of the border, but many of its large SO_2 emission sources are well removed from the border.

Canadian SO_2 emissions from power generation are expected to increase somewhat in the near future, then decline. With expected growth in electricity generation from coal, in the near term, they may surpass nonferrous smelters as the most important Canadian source. But at present, while U.S. sulfur dioxide emissions are basically from electricity generation, Canada's are largely from metal smelters.

As regards nitrogen oxides, more than 40 percent of U.S. emissions come from transportation, while electric power generation accounts for 30 percent and other combustion sources account for 20 percent. Forecasts indicate significant increases from all of these U.S. sources in the future. Canadian nitrogen oxide emissions arise largely from the transportation sector (about 60 percent). Various combustion sources, including electric power generation, account for the rest, and future Canadian emissions are likewise expected to be much higher, partially as a result of expansion in electric power generation. A summary of current knowledge of U.S. and Canadian emissions is presented in table 4.

The source of much of the aforementioned data is the U.S.-Canada Research Consultation Group's report, a much criticized but nevertheless unique document to which both governments provided equal input and for which they both assume equal responsibility. While its findings do not necessarily receive the approval of these governments, they cannot be ignored. Their disavowal would constitute a vote of no confidence in the respective governments' own scientists. Such findings must, therefore, play a prominent role in any bilateral negotiation.

Although these findings rally the agreement of the participating scientists of both federal governments, they have not received the universal support of members of the two societies. For example, the American Smelting and Refining Company (ASARCO) considers the U.S.-Canada Research Consultation Group's report to be based on questionable assumptions and holds that it is a political rather than a scientific paper. The points that it believes erroneous include the conclusions that acid rain results purely from man-made emissions, that this rain has devastating environmental effects, and that stricter control of man-made sources of nitrates and sulfates will solve the problem. ASARCO cautions that the discussions of "clean" or natural rain is very misleading, in that "the pH of clean rain varies significantly, to at least a pH of 4.2"[13] (which is twenty-five times more acidic than pH 5.6, the frequently quoted pH for pure rain), and points out that the document ignores important natural emission causes, such as volcanic eruptions.

TABLE 4.   Canadian and United States Emissions

10^6 tons per year

	SO$_2$						NO$_X$					
	United States				Canadian 1975^a		United States				Canadian 1975^b	
	1950		1975^a				1950		1975^b			
	Short	Metric	Short	Metric	Short	Metric	Short	Metric	Short	Metric	Short	Metric
Utility combustion	5.4	4.9	18.6	16.7	0.7	0.6	1.2	1.1	6.8	6.1	0.2	0.2
Other combustion	13.5	12.2	4.4	4.0	1.1	1.0	4.1	3.7	6.0	5.5	0.5	0.4
Nonferrous smelters	3.4	3.1	2.8	2.5	2.4	2.2	neg.	neg.	neg.	neg.	neg.	neg.
Other industrial processes	1.2	1.0	1.9	1.7	1.2	1.1	0.3	0.3	0.7	0.7	0.1	0.1
Transportation	0.9	0.8	0.8	0.8	0.1	0.1	3.3	3.0	10.9	9.9	1.3	1.2
Total	24.5	22.0	28.5	25.7	5.5	5.0	9.0	8.1	24.4	22.2	2.1	1.9

Source: United States–Canada Research Consultation Group, *The LRTAP Problem in North America—A Preliminary Overview* (Ottawa and Washington: United States–Canada Research Consultation Group on the Long Range Transport of Air Pollutants, 1979), p. 5.

Note: neg. = negligible.

[a] Canadian sulfur dioxide emissions listed are based on 1978 data for major point sources and on 1974 (or later in some cases) data for other point sources and for area sources. United States point sources are at their 1977–78 emission rate while area sources are at their 1973–77 emission rate.

[b] Canadian nitrogen oxide emissions listed are based on 1974 data. United States point sources are at their 1977–78 emission rate while area sources are at their 1973–77 emission rate.

Atmospheric Transport of Pollutants

Latest research indicates that sulfur compounds remain in the atmosphere one to five days in eastern North America and other industrialized regions. The main factors determining this time span for atmospheric transport include the physical and chemical characteristics of the pollutants, the height at which they are emitted into the atmosphere (which relates to height of stacks), and the occurrence of precipitation. Other factors can also play a role.

Obviously, the movement or flux of pollutants transnationally depends not only on the amount of emissions in each country but also on the frequency and duration of cross-border winds in each direction. The dominant winds in eastern North America are the prevailing westerlies. These are modified by a stagnating summer high-pressure area and prolonged periods of brisk winds, and it has been found that the late-summer Maritime Tropical air masses have the greatest potential for the formation and transport of high concentrations of sulfate into the northeastern United States and eastern Canada. The LRTAP Group estimated transboundary flux (in millions of metric tons of sulfur per year) from Canada to the United States to be 0.5 to 0.7 (the figures differed for two different models),[14] and from the United States to Canada to be 2.0 tons (for both models). Both models assumed emissions from the North American region east of longitude 92°W.

ASARCO questions the relationship between length of time during which substances remain in the atmosphere and the height of their injection (suggesting that there is no causal relationship) and attacks the quality and usefulness of most of the data upon which this entire issue is based in North America, a widely shared view in some circles. For example, the Peabody Coal Company has expressed concern that

> conclusions about the cause, occurrence and extent of long range transport, and the resulting phenomenon of acid rain [are] accepted without adequate proof.

The company contends that

> the current state of scientific knowledge in this area is such that there is no acceptable way to predict that a reduction in regional, local or, for that matter, the total loading of sulfur dioxide emissions, will result in any appreciable change in the quality of precipitation in any other part of the country from which the emission was made.[15]

Peabody echoes ASARCO's concern that the transformation and deposition processes suffer from a lack of good scientific information and that "the situation for predicting the regional movement of pollutants is in a state of infancy.''[16] It states that *the source-receptor relationship cannot be defined and, thus, adequate control mechanisms cannot be developed,* in opposition to the fundamental conclusions of LRTAP and especially of the Canadian government.

Dr. Frank Frantisak, of Noranda Mines, Ltd., strongly contends that the data base is insufficient to justify a categorical statement that Canada's precipitation acidity has increased in recent years and rejects the assumption that precipitation acidity is mainly caused by SO_2, indicating that NO_x plays as important a role as sulfur oxides and is dominant in some areas and seasons of the year. Frantisak also rejects the position that policy decisions will have to be made now before all the data are in, arguing from his experience in northern Quebec that receptor lakes with low buffering capacity are not unlike equivalent unaffected lakes. Contrary to what one might suspect from these views, however, Frantisak has said,

> I firmly believe that acid precipitation is an environmental problem which could be of enormous proportions and could have far reaching consequences. However, I also believe that a mistaken acid rain policy could be as damaging as acid rain itself.[17]

He worries that ''we are being offered the solution before we know the causes.''[18]

DEPOSITION AND ITS EFFECTS

Deposition occurs following transborder flux, for what is in flux must come down. Deposition rates vary greatly from one region to another, as does the vulnerability of receptor environments. The LRTAP Group found that

> the total sulfur deposition in southern Ontario and southern Quebec calculated to be from sources in the U.S. was 50,000 and 68,000 tons S [sulfur] respectively for January and August 1977. This amount of deposition associated with U.S. sources compares with 110,000 and 100,000 tons S deposited within Ontario and Quebec originating within Canada. The total deposition in U.S. regions calculated to be from sources in southern Ontario and southern Quebec was 38,000 and 21,000 tons S for January and August.[19]

Volumes have now been written on the ecological effects of acid precipitation on aquatic and terrestrial environments and, to a lesser extent and more recently, on human health. Suffice it to say here that it is generally accepted that acid rain has negative effects on aquatic ecosystems, while its impact on commercial agriculture, commercially productive forests, and potable drinking water supplies is less certain. Much of Canada's continental land mass—the Canadian Shield—is inadequately buffered (by naturally occurring geological and chemical conditions which neutralize acids) and thus highly vulnerable to damage from acid precipitation. Canadian concern centers on acid precipitation's impact on lakes and their fish populations. Concern is also developing over possible impacts to its significant softwood forests—a serious economic consideration. Proportionately speaking, the U.S. land mass has much more buffering capacity and consequently is significantly less vulnerable, having far fewer lakes, forests, or agricultural lands open to the devastation which acid precipitation allegedly can cause. However, in some areas the United States is as vulnerable as Canada (for example, the Adirondacks, the New England mountains, and the upper Midwest).

Large portions of eastern Canada and the northeastern United States are composed of resistant granite and siliceous bedrock from which glaciation has removed the more recent calcareous deposits that once formed an overburden. Indeed, much of the original surface of Canada (which was a well-buffered surface vis-à-vis acid rain) now rests in the United States, having been removed from Canada by glaciation. (Since the United States is the source of most North American acid rain, it has been suggested that it is obvious whom God loves and that is the nation south of the 49th parallel! The image is one of moving Canada's needed protection to the United States, and then having the United States bombard a now very vulnerable Canada with acid rain!)

The LRTAP Group's study specifically cited:

• the loss of salmon populations in Quebec and the Maritime provinces' streams;
• the loss of 40 to 75 percent of the acid neutralizing capacity of many of the lakes in the Haliburton-Muskoka area of south central Ontario in a decade or less;
• the acidification of the waters of Quetico Provincial Park in western Ontario along with adjacent Boundary Waters Canoe Area in northern Minnesota;
• the general vulnerability of upper Michigan, northern Wisconsin, and northern Minnesota; and
• the advanced acidification of the lakes of the Adirondack Mountains of northern New York with a concomitant loss in the regional sports fishery, representing an annual economic loss from tourism of over one million dollars.

While data on the vulnerability of eastern North American terrestrial ecosystems is less specific, the group found that agricultural crops, particularly those on more acidic podzolic soils of northern and eastern regions of the continent are affected. The point is also made that while effects on forests may be cumulative and rather elusive, to await long enough to obtain say, a clearly demonstrated effect of some 15–20 percent loss in forest productivity could mean that a stage of site degradation has been reached that would be impossible to reverse. Several features of the regional situation suggests a threat exists now.[20]

Without question a great deal of northeastern North America is receiving large amounts of sulfate deposition, and much of the most productive forest lies within the area most affected by acid precipitation. Hard data recording effects on forests should be available soon. The LRTAP Group emphasized its concerns over irreversibility with the statement:

Continued pollutant loading of the region at present levels will result in continued degradation with extensive *irreversible* acidification occurring during the next ten years. Increased pollutant loading of the region will accelerate the rate of degradation. [Emphasis added][21]

In Canadian-U.S. diplomatic circles, such strong statements must be given some credence, given that they are the findings of the governments' own scientists. This lends greater significance to their role in, and impact on, the relationship.

The forest products industry is in a unique situation with respect to acid rain. On the one hand, the manufacturing and processing arm of this industry is an emitter of SO_2 and NO_x, the precursors of acid rain, and hence mandated control measures at the source will constitute a significant cost to the industry. On the other hand, the industry's raw material base is the potentially acid-vulnerable forests. Even a slight reduction in forest productivity from acid deposition would have a major adverse impact on the industry. The U.S. pulp and paper industry has concluded that the current state of knowledge does not allow definitive conclusions regarding the effects of acid deposition on forest productivity.[22] It concludes that effects on seedling germination, establishment, and growth and on forest biomass production can be beneficial, adverse, or not detectable. The industry also believes that direct effects on mature trees and forests (i.e., impacts on the foliage) are possible in a number of ways, as are indirect effects resulting from acid deposition on forest soil. To determine the nature and extent of these effects, it recommends an intensive five-year research program to provide sound bases for appropriate policy.[23] At present no concrete information exists on this matter, in the industry's opinion. It has found, for example, that available evidence does not justify general conclusions on the effect of acidic deposition on tree seedling establishment in those areas believed sensitive. They have also found that tree ring investigations do not enable the drawing of any definitive conclusions about the effects of acidic deposition on forest productivity.[24]

As regards the effects of deposition, ASARCO and others in industry contend that

the natural acidity of "clean" rainfall (pH 5.6–4.2) is certainly sufficient to greatly reduce the buffering capacity of lakes and significantly change the pH of bodies of water with naturally poor buffering capacity.[25]

They also view natural removal of materials from the soil by precipitation, and agricultural runoff, as contributing causes. These critics reject the data on terrestrial impacts, which are considerably weaker than those on aquatic impacts, and conclude that "any far-reaching programs to legislate more stringent pollution controls on this basis are very premature and scientifically unwarranted."[26]

Some of the principal matters of concern to industry, and especially to those subject to emissions reduction regulation, include:

- the spatial variability in precipitation composition, which is much greater for shower (convective) types of events than it is for large-scale storm precipitation events;
- the fact that precipitation composition can be significantly influenced by local emission sources;
- the absence of adequate long-term data to demonstrate conclusively any temporal trends or changes in spatial distribution patterns over the past thirty years or so (usable historical records are virtually nonexistent); and
- the large degree of uncertainty in quantitative estimates of natural source emissions, which is a major obstacle to understanding the role that natural sources play in acidic deposition.[27]

Citizen environmentalist groups appear largely to agree with the LRTAP Group's findings and have not commented on them in as much detail as have industrial spokesmen. The environmentalists reflect a desire for much quicker remedial action and a strong program of controls.

Yet many people, including officials of the Reagan administration, some U.S. federal and state bureaucrats, and the business community, believe there are insufficient scientific data to justify precipitous and costly action. Others, including many Canadian government officials and citizen environmentalists, while agreeing that much more needs to be known, fear potential irreversibility and therefore believe such "precipitous and costly action" is justified now. Much, though by no means all, of this disagreement divides along national lines, creating the foreign relations problem faced by diplomats. The next section will explore the diplomatic history of this issue before possible solutions are examined in the final section.

Acid Precipitation as a Problem in Canadian-
U.S. Relations

The roles of Canada and the United States vis-à-vis each other and the rest of the world have changed over the years. Such changed roles have called for a change in behavior. The people of Canada have long had to accept significant, though often subtle, U.S. influence over their foreign relations in order to maintain their materially high standard of living. Given Canada's overwhelming economic dependency on the United States since World War II, Ottawa has had to accept many U.S. positions on world issues or at least avoid differing substantially from Washington on various issues. Thus Canada has not often enjoyed the luxury of either openly opposing the United States, or of dealing from a strong hand in its direct relations with Washington.

The people of the United States are now learning what such enforced compromise in foreign affairs is really like. The influence exercised by the Organization of Petroleum Exporting Countries (OPEC) over U.S. society, with inflation and dollar devaluation effects of high-priced oil imports and the likely influence certain OPEC nations have had and will have on U.S. foreign policy, is a significant example in modern U.S. history of what dependence on foreign energy can do to the self-confidence of people in the United States. They may very well suffer further in ways other than economic, ways not yet obvious, as a result of their dependence on OPEC oil and seeming inability to free themselves from it.

This situation is, quite naturally, persuading the people of the United States to look more seriously at domestic energy sources heretofore untapped or whose potential has not been fully realized. This quest for a higher degree of energy independence has naturally resulted in proposals to increase the use of such domestic energy resources as coal. Reluctance to ''bite the bullet'' and practice energy conservation in a manner which would significantly reduce total energy consumption forces U.S. society to accept the potentially serious disadvantages and costs of large-scale rapid coal or nuclear development if continued vulnerability to OPEC is to be avoided.[28]

Large-scale development of coal-fired generation of electricity (through both old plant conversion from oil and new plant construction) will force the United States to accept certain costs in terms of risks to life and health from expanded coal mining. In addition, there will be ecological, economic, and social costs of coal strip-mining, and the inevitable deterioration in domestic air quality which will result if expensive scrubbers are not mandated to clean the effluent. All these costs are generally well known to those in the industry and in the environmental community, if not among the general public.

What is not widely known, however, is the foreign policy costs of such an increase in the use of coal, particularly with respect to Canadian-

U.S. relations. These costs, while exceedingly difficult to quantify, could conceivably be felt in many unrelated areas of the bilateral relationship and be magnified in certain vulnerable sectors of the economy and in different geographical locations.

Canadian government officials are well aware of Washington's stated commitment to expand significantly U.S. dependence on coal for electricity and are greatly concerned from an air quality and acid precipitation perspective. U.S. deposition of acid rain and snow on Canada because of the former's high energy demand may impose great environmental and economic costs on the latter, if new control technology does not come on line soon. (Canada's per capita consumption of energy is even higher than that of the United States, weakening its ability to criticize the United States. However, this consumption does not impose equivalent high environmental and economic costs across the border.) The Canadian government cannot accept the imposition of such costs lightly.

There appears little doubt that Canada is firmly committed to pressuring the United States to significantly reduce the amount of acid precipitation crossing the border, which necessitates reduction at point of origin. An impressive list of unconventional diplomatic initiatives (described in this chapter's section on the present diplomatic situation) gives evidence of this commitment. Given this determination, the failure to achieve such reduction will result in bitter feelings and a souring of relations, if nothing more concrete at the moment.[29]

An understanding of the diplomatic history of acid rain requires a knowledge of the roles and positions of the International Joint Commission, the province of Ontario (a key actor), and that of the two federal governments. Bilateral concerns were first raised by the IJC.

THE INTERNATIONAL JOINT COMMISSION

Maxwell Cohen, former Canadian chairman of the International Joint Commission, stated in a recent interview that

> while the scientific data on Canada-United States acid rain may be relatively new, the legal principles of air pollution, as they touch on the responsibility of states damaging each other, are today quite deeply rooted in international law. Indeed, the juridical parents of these modern doctrines are Canada and the United States [a reference to the 1930s Trail Smelter arbitration].[30]

The International Joint Commission warned of the dangers of acid precipitation for close to a decade and was among the first to do so. The IJC was, in fact, for some time the only bilateral actor in the acid rain arena. As Canadian chairman during the mid-1970s, Professor Cohen was especially vocal about the issue, and on a number of occasions warned of diplomatic and environmental ramifications if action were not taken. He early realized how difficult it would be to accomplish tangible results but indicated the effort to jointly achieve results itself had value.

The IJC has at least a dual role with respect to acid precipitation. The commission has a standing reference from the two governments to advise them of general air quality problems, including long-range transport of atmospheric pollutants, and thus acid precipitation, causing transboundary environmental deterioration. It also has a specific responsibility, under the mandate of the Great Lakes Water Quality Agreements of 1972 and 1978, to monitor and work toward the correction of water quality problems in the lakes, regardless of the source. When acid precipitation arises in one country and impacts on the international waters of the Great Lakes, the agreements are violated. It is then the IJC's specific responsibility to identify such violation and recommend remedial action.

The Great Lakes Water Quality Agreements called for the establishment under IJC auspices of a Great Lakes Science Advisory Board. The board's findings on long-range transport of air pollutants, though not binding on the commission, are as near as one can come to an IJC pronouncement on the specifics of the issue.

The board recognizes that the extraordinary nature of the problem dictates extraordinary responses. It documented specific effects on the Great Lakes Basin from airborne pollutants and noted that the basin exports pollutants to areas to the east. It called for specific data needed to understand the problem and supported the U.S. EPA's National Atmospheric Deposition Program to achieve this goal. The board noted the possible ameliorative effects of liming smaller lakes, stating,

> while it will not be possible to lime the vast areas of the Great Lakes drainage which are affected by acid precipitation, liming may be a means of maintaining natural biota in selected sanctuary areas.[31]

The board did recognize however, that the only effective means of reducing the problem is by greatly reducing emissions of sulfur and nitrogen oxides and concluded that "swift, decisive and wide-spread action is required."[32] It noted that the "tall stack policy" to encourage pollutant dispersement has been the only pollution abatement policy to date in many areas and that this policy has contributed to the acid rain problem rather than remedying it. The approach to air quality standards has been national, while the problem is international. Thus, ambient air quality standards can be attained by use of tall stacks which contribute to the long-range dispersion without solving problems beyond the immediate vicinity of the source.

The IJC Great Lakes Science Advisory Board recommended establishing an integrated acid precipitation program for the Great Lakes. The program would include detailed inventories of sources and effects within the basin, a survey of transboundary movements, a widespread public education program, a thorough economic analysis, and studies of the comparative human and ecological risks of various energy alternatives. Further, and most importantly:

> Every effort should be made to overcome the problems of "piece-meal" legislation which are sure to confound the control of emissions, many of which are in well-buffered areas where the immediate effects of acid precipitation will be negligible.

The Science Advisory Board formally recommended that

> the International Joint Commission immediately implement, as spec-ified in Article VII (6) of the 1978 Great Lakes Water Quality Agreement, liaison among institutions established under the 1909 Boundary Waters Treaty, appropriate U.S. and Canadian agencies, and international organizations which address concerns relevant to the Great Lakes Basin Ecosystem. . . . Further, the Board recommends that the Parties to the Great Lakes Water Quality Agreement be encouraged to formulate a reference within the context of an *eco-system approach* on the causes, effects and measures for the control of long-range transport of airborne pollutants with special attention to acid rain. [Emphasis added][33]

This board can only advise the commission, and its jurisdiction is limited to the Great Lakes Basin and to the mandates of the Great Lakes Water Quality Agreements of 1972 and 1978.

The IJC does have a separate International Air Pollution Advisory Board (IAPAB), also advisory but not as geographically restricted, which is obligated to advise the commission on the emergence or potential emergence of problems and suggest appropriate courses of action. Address-ing the general subject of transboundary air pollution, but applicable to acid precipitation, the IAPAB stated:

> The two countries have specific legislation aimed at controlling inter-national air pollution. To become operative, the U.S. legislation requires reciprocal arrangements in the foreign country affected whereas the Canadian legislation is dependent for its effectiveness on the existence of an international obligation. Because of this interde-pendence, neither section quoted can be used unilaterally for the purpose stated. Moreover, it is apparent from a review of existing transboundary air pollution problems that clear principles governing the obligation of one country to another in a given situation are

lacking. Accordingly, the Board recommends that this matter be further examined with a view to identifying appropriate legal mechanisms and principles. [The reciprocity arrangement has since been activated.][34]

Thus both IJC boards have recommended a strong IJC presence in air quality and acid precipitation, but neither government has been willing to give the commission the responsibility or funding required to play that role. Thus, though potentially a major actor, the IJC is largely a spectator in the emerging acid rain debate. There are some who believe, however, that it will ultimately be given the monitoring work once the diplomatic actors feel confident enough to turn over to the commission this very traditional role in which it has distinguished itself in the past.

ONTARIO'S ROLE

The province of Ontario is the principal Canadian source of emissions resulting in acid precipitation and one of the principal areas suffering the effects of both U.S.- and Canadian-originated pollution. It receives the acid-bearing winds from the U.S. Great Lakes and Midwest, and contributes acid to the winds blowing toward the northeastern United States and eastern Canada. The continent's (if not the globe's) greatest single source of SO_2 emissions are the International Nickel Company (INCO) and Falconbridge nickel smelters located at Sudbury in north central Ontario. Finally, a high proportion of Ontario's land, much of which is on the granitic Canadian Shield, is highly vulnerable to acid with only the extreme south having much natural buffering capacity. Ontario's numerous lakes magnify its problems from acid precipitation.

The province's specific concerns crystallize around the loss of a cottage-resort sportfishing industry in very popular lake districts, especially Muskoka Lakes and Haliburton Highlands north of Toronto—with a loss of jobs and economic development—but perhaps more importantly a significant loss of recreational opportunity for affluent middle-class Torontonians and other urbanites who place a great value on these recreational environments.

Under these circumstances, Ontario naturally is going to object to acid pollutants from the United States raining down upon its lakes, forests, and farmlands. But it is Ontarian business interests, economic development, and jobs that are at stake if Ottawa promotes or the United States demands a cessation, via agreement or treaty, of transboundary air pollution, for Ontario pollution cannot avoid crossing the boundary. Hence, Ontario fears the impact of air pollution restrictions on the already troubled International Nickel Company, one of the province's major employers with world headquarters in Toronto; on the metal smelting industry developing in the north;

on heavy industry in the Great Lakes Basin; and on Ontario Hydro's plans to build large power plants to generate electricity from coal with tall stacks which will presumably induce acid precipitation.

As has been described in earlier chapters, authority in Canada in the environmental area, including air pollution, is a shared jurisdiction, though in practice the provinces have lead authority. When one combines this great power of the provinces with the fact that Ontario, a superpower among them, is the most significant source of acid rain in Canada, it can easily be concluded that Ontario's position will be crucial to the development of a Canadian national position vis-à-vis the United States.

Ontario has been involved with acid rain in one way or another longer than any other area of Canada, and its inhabitants are far more aware of the matter as a public issue than are most other Canadians (although this gap is closing). They are far more exposed to it through mass media, and substantially more polarized, especially toward the United States. Former Ontario Environment Minister Harry Parrott remarked,

> We in Ontario are playing a major role in the war against acid rain. We have recognized the need for action on this phenomenom long before the full implications were recognized, long before the term "acid rain" was coined.[35]

Minister Parrott stated that Ontario has not shirked its duties in the area of research or public education on this subject, recognizing that this is not only a federal responsibility. At the same time, he continued,

> I do not minimize the absolute necessity for an international accord which permits the United States and its state partners, and Canada and its provincial partners, to work towards common goals, the most important of which is to minimize to the greatest extent possible the pollutant emissions at source and their harmful impact on the environment.[36]

The minister also noted that most of the new U.S. fossil fuel–fired generating plants will be built in the northeastern states, and believes this is the area which now generates most of the acid rain blown into Ontario.[37] The LRTAP Group and the U.S. State Department dispute this latter belief.

The Environment Ministry also believes that Ontario and the Canadian federal government are applying as much diplomatic pressure as they can on the United States under the circumstances. Noting in 1979 that the IJC has concluded that as much as a decade will pass before pollution control facilities can catch up with the problem, the environment minister remarked "the IJC's somewhat pessimistic assessment of the prospects for an international accord will not, and cannot, inhibit Ontario one bit in pressing for essential action."[38] The State Department disagreed with this statement in noting the assessment was not a position adopted by the IJC, but rather only an advisory report from one of its boards.

During the early stages of the bilateral debate (1978–80) Ontario rejected the idea that it must clean up its act domestically before it could legitimately complain about the behavior of other jurisdictions. But Ottawa as well as the United States early rejected Ontario's position on this matter. If the attitude prevails that one nation should not curb its emissions until the other does so, then nothing will ultimately be accomplished. More recently, the province has taken clear steps to indicate it accepts responsibility to put its own house in order.

Ontario believes the costs associated with a 50 percent annual reduction in SO_2 emissions in the northeastern United States are $5 to $7 billion, and the cost of a similar annual reduction in eastern Canada is estimated at $350 million.[39] Some argue that even such expenditures will not solve the problem, and indeed, recent economic conditions in both countries will result in political pressures to "go easy" in this area. Ontario recognizes these facts but claims it will push ahead positively and vigorously anyway.

Minister Parrott remarked that

the United States and Canada must reach an accord on what is to be done, but such an accord must be negotiated on the basis of strength from both sides. We in Ontario aim for an international agreement soon. But this agreement must be flexible enough to take advantage of new knowledge resulting from research as it evolves. This is especially true in the area of abatement technology. . . . We are dealing here not with any magic, short-term solutions. The problems are immense; many elements—including the huge emissions from the U.S.—are beyond our control.[40]

By virtue, therefore, of the provincial environment minister's frequent public speeches around the province, (and the Parrott record has been continued by Minister Keith Norton) along with media interviews to a market which cannot seem to get enough on this subject, more and more Ontarians are becoming sharply polarized (as well as educated) on this issue. The question remains, however, as to what effect this polarity will have on the ultimate settlement of the issue in Canadian-U.S. relations.[41]

In the past two years, Ontario has made a tangible commitment to operating in consultation with a strong federal stance. To wit, it promulgated in 1980 a widely publicized provincial control order limiting INCO emissions from 3,600 tons of SO_2 per day (representing 50 percent control) to 2,500 tons per day. A further reduction to 1,950 tons per day is required under the order in less than two years. Ray Robinson, former federal assistant deputy minister of Environment Canada, predicted that emissions will fall below 1,000 tons per day in the next half dozen years (by 1986).[42]

The province has also instigated pressure on its own (government-owned) crown utility, Ontario Hydro, whose coal-fired generators and tall stacks constitute the second most important emission source in the province. Under the pressure of "getting one's own house in order," in January, 1981, the crown utility announced it would spend $500 million to trim emissions leading to acid rain by more than 40 percent over the next ten years[43] (a goal toned down in 1982). The major part of this program involves the design and construction of two SO_2 scrubbers to be installed either at Nanticoke or Lambton generating stations, with an estimated cost of up to $283 million, and going into service about 1987. Further expenditures would be made with the installation of NO_x burners at large coal-fired stations, with further funds to be spent on low-sulfur coal and the purchase of hydroelectric energy from neighboring Manitoba which has surplus hydro. (Ontario could also increase its commitment to energy conservation and renewable sources.) Ontario's position is approaching Ottawa's position—a necessity if Canada is to present a strong and united front at the negotiations toward a formal international agreement. (It has been suggested, however, that Ontario may have surrendered valuable bargaining chips usable in coming negotiations.)

THE CANADIAN FEDERAL POSITION

While the Trudeau government began to recognize the acid rain problem in 1978–early 1979, it was not until Joe Clark and the Progressive Conservatives formed the government of Canada in May of 1979, and the new prime minister appointed an environmentally involved Vancouver attorney, John Fraser, as federal minister of the environment, that Ottawa really began to move. Mr. Fraser was well aware of the acid rain issue and maintained deep interest in this and other aspects of environmental quality. He soon began joint efforts with equally interested and highly motivated senior officials of his agency who had long been involved in the acid precipitation issue, as well as Canadian-U.S. environmental relations. The ministerial and bureaucratic levels combined to form a dynamic team and, as mentioned earlier, succeeded in elevating the international aspects of the acid precipitation issue to a surprisingly high level.

In October, 1979, Environment Minister Fraser presented his government's basic position on acid precipitation. He referred to the issue as an environmental time bomb and said it was a much bigger matter than anyone had previously envisaged. He announced a series of meetings with senior U.S., Ontario, and other government officials, and noted that a federal-provincial committee on control strategies for long-range transport of pollutants had been set up to focus on acid rain. (In the author's view, this is a most important step, and an absolute necessity if progress is to be achieved, given the realities of federal-provincial relations.) The committee has representation from all provinces and endorses federal negotiations with the United States. It is developing a work plan for analyzing specific abatement technologies and their respective cost estimates. On a basic point in federal-provincial relations, the minister stated:

> It is our present intention to rely on the provinces to put in place the needed control requirements. I am counting on my provincial colleagues and I am confident that they will do what must be done. However, members of the House have asked what will the federal government do should the provinces fail to act. The federal government has the constitutional authority to act for the good of the country in respect to controlling the interprovincial and international flow of pollutants. *We would not shrink from this responsibility.* Indeed, it is very much my view that the federal government ought to have at least residual authority set out in legislation to deal, as needed, with interprovincial and international pollution problems which is something we currently do not have, in both the air and water fields.[44]

This is a critical statement, for it means the federal government may be willing to buck the power of the provinces and assert its responsibility in interprovincial and international affairs. However, it is by no means clear whether Ottawa would in fact carry out this "threat," as it could be perceived by the provinces.

The minister also emphasized that political consensus is absolutely necessary. Scientific verification is not enough if an issue such as this is to move and gain government response. And he contended that consensus does exist in Canada. But he also recognized that "there are dangers at either end of the spectrum—environmental dangers in the case of understatement, economic dangers in the case of exaggeration."[45] Albeit given this recognition, the Canada Department of the Environment and its minister in the Clark government, John Fraser, clearly had a strong commitment in this area and a position which was much less equivocal than Ontario's at that time. (Ontario has since moved closer to this unequivocal federal position.)

By the time the Liberals under Pierre Trudeau returned to power in early 1980 and John Roberts succeeded John Fraser as environment minister, the acid rain debate had reached a stage where no change in direction was politically feasible, whether or not a change was desired by the new government. Key officials of Environment Canada continued in power, consultations with the United States continued as if there had been no change in government, and Ottawa continued to work with Ontario and other provinces to shore up a united front. Minister Roberts, however, found himself much less heavily involved than his predecessor, given his major responsibilities in the important matter of constitutional repatriation (as well as his dual portfolio in science and technology), and thus more responsibility accrued to the line agencies, particularly the Environmental Protection Service and, secondarily, the Atmospheric Environment Service, the latter being mainly responsible for direction and coordination of the major research effort ($41 million) announced by the minister early in his term. (Interestingly, it was a veteran senior negotiator on the Great Lakes Water Quality Agreement a decade ago who was given responsibility for the development of any new agreement negotiation involving air quality and acid rain. Since the direction of these negotiations and the working groups model which resulted from them are not unlike the similar bilateral negotiation ten years ago on the Great Lakes Water Quality Agreement, this assignment may have been more than coincidental.)

The general Canadian feeling is that Canada is the afflicted nation. Its stance has been a proactive one, in contrast to the reactive stance of the United States. The clearest evidence of this proactive stance has been the Ottawa decision to reduce SO_2 emissions by 50 percent by 1990, if the United States will agree to do likewise, creating for the United States a substantial challenge, as well as a heavy economic burden on Canadian society.

THE UNITED STATES POSITION

The U.S. congressional resolution requiring the State Department to enter into negotiations with Canada toward an air quality agreement was described in chapter 10. This was oriented toward two specific transborder air pollution complaints (Poplar and Atikokan) and had nothing whatsoever to do with acid rain. However, this resolution subsequently played right into Canada's hands as it formed the necessary foundation for the 1981 negotiations. Recognizing by late 1979 that there was both a domestic and international problem, President Carter endorsed the creation of a U.S. government interdepartmental acid rain group to hold further discussions with Canada while also approving $10 million in funds for a major new research program (at the same time Congress was unsuccessfully pushing for a higher expenditure). By early 1980 (if not earlier), however, circumstances changed dramatically as energy concerns clashed head-on with environmental concerns.

In early 1980, the U.S. Department of Energy (DOE) proposed, presumably with White House approval but without EPA approval, to convert more than sixty oil- and gas-fired utilities to coal, which would, according to DOE,[46] boost air pollution over the Northeast by 25 percent and sharply increase acid rain in Canada. This proposal came just after Canadian officials were reassured by the United States that coal conversion under President Carter's energy program would not increase air pollution. The reaction from Canada was swift and adamant as Canadian embassy officials descended within twenty-four hours of the release of the story on EPA, DOE, and the State Department. The proposal was withdrawn (and never enacted by Congress), and a few days later the *Washington Post* editorial cartoonist drew a cartoon suggesting that the United States' response to the Canadian rescue of diplomats in Tehran was gratuitously to give them more acid rain. An unfortunate aspect of this incident was that EPA administrator Douglas Costle just weeks before had assured the Canadians that SO_2 emissions in the Northeast and Midwest were decreasing and would continue to decrease. EPA Air Program administrator David Hawkins was later forced to admit that the figures on coal conversion were not available to Costle until after his meeting with the Canadians, a point of no little diplomatic embarrassment to the State Department. Hence, the national effort toward heavier coal dependence once again exacerbated the acid rain issue, and this time with direct international ramifications. The immediacy and strength of the Canadian response is further indication of the seriousness of Canada in maintaining a hard line on this issue.

In a strongly worded diplomatic note sent in response to the Carter administration's coal conversion announcement, Canada charged that impacts on its territory were ignored and specifically requested assurances that

- no program will be adopted which adds to the current damaging transboundary pollution levels or seeks to maintain the status quo through a policy of ''offsets'';
- any program proposed will be implemented in a way which would maximize the opportunity to reduce emissions of pollutants causing acid rain in both countries;
- the United States Government will study the transboundary implications of any coal conversion proposal, especially as these affect the prospects for reduction of transboundary air pollution; and
- the United States Government will provide to Canada the results of such studies . . . so that the Canadian Government might come to its own determination as to its transboundary implications.[47]

Further, when Environment Minister Roberts delivered his address on acid rain to the Air Pollution Control Association annual meeting in Montreal in June, 1980, he referred indirectly to the coal conversion program when he said,

the United States is not only predicting significant increases in SO_2 and enormous increases in NO_x *but is pursuing deliberate action to make sure it happens.* [Emphasis added][48]

Ohio politics has further clouded the U.S. acid rain picture. In addition to being the single greatest emitter of acid rain–type pollutants from its great concentration of high-sulfur coal–burning power plants in both the Ohio Valley and along Lake Erie, Ohio is a major producer of high-sulfur coal. Hence, employment in the mines and in coal transport as well as utility electric rates are at stake in any shift away from high-sulfur coal or mandated emission controls which might make its combustion less economically competitive. Ohio is a politically important state in any presidential election, a fact not lost sight of by the Carter administration when it granted pollution waivers to Ohio utilities shortly before the 1980 presidential election. Politics dictates that Ohio must be especially wary of any pressure to reduce emissions through regulation.

In the meantime, the U.S. Senate was busy promoting the Acid Precipitation Act, an ambitious effort to mandate the expenditure of significant funds in monitoring and research, calling for a master plan of attack in one year, and a ten-year research program (which U.S. industry wants shortened to five years). This statute was to be used as the basis for international agreements in this area. Initially given very little chance of passage, its backers, Senators Daniel Patrick Moynihan of New York and Max Baucus of Montana, masterminded its attachment as an amendment to President Carter's synfuels legislation, an action believed to significantly increase its chances of passage. The new amendment also mandated the establishment of a rather unique federal interagency task force composed of the heads of appropriate federal agencies and regional representatives from each quarter of the United States. Canada and Mexico would be able to appoint nonvoting but permanent members of the task force, a procedure without precedent. In the words of Senator Baucus,

> This is an acknowledgement that acid precipitation will require international as well as domestic cooperation. It gives our neighbors an opportunity to share directly any information they deem necessary and beneficial to our common interests and our common concerns. . . . Essentially, Mr. President, the modification which the sponsor of the amendment has so graciously agreed to provides that the Governments of Mexico and Canada, at their option, may appoint a person to the commission, the task force, without any decision-making authority—as advisors only—due to the international implications of the development of synthetic fuel plants, not only in America but certainly in southern Canada and Mexico as well.[49]

The Department of State was not happy with this unusual proposal for reasons of precedent and succeeded in getting it deleted from the amendment. It would have failed in any event, for the Synfuels Act ultimately failed to pass. The ambitious nature and magnitude of this overall proposal by Senators Moynihan and Baucus would have placed the United States on a more equal footing with Canada in terms of demonstrating major concerns over acid rain and would have changed its stance from one of reaction to action. The United States did not, however, choose to take this step, perhaps owing to the urgency of energy questions, and returned to a reactive state in the debate. With the advent of a new administration in Washington at the beginning of 1981, and with Republican control of the Senate for the first time in almost three decades, a whole new set of actors is now in control and it remains to be seen how they will ultimately handle this issue.

The Present Diplomatic Situation

Following the release of the first LRTAP Group report in 1979 a diplomatic breakthrough had occurred. The Department of State suggested that work groups along the lines of those established preparatory to the enactment of the two Great Lakes Water Quality Agreements be established to prepare the two nations for an international agreement on acid precipitation. Initially, although Canada supported the work group concept, its diplomats also wanted (and were under significant political pressure to extract) an immediate agreement or commitment from the United States that it would reduce emissions crossing the border. When it became obvious that this could not be obtained but that a Memorandum of Intent (weaker, perhaps, but at least tangible) was in the realm of possibility and that such a memorandum could establish scientific-technical groups along the lines of the earlier (and successful) Great Lakes groups, as well as include commitments to use all existing authorities, Canada agreed to give it consideration. Long and difficult talks occurred during the late winter and the spring of 1980, and the two governments reached agreement on the formal establishment of technical-level working groups to lay the groundwork for an agreement and the simultaneous establishment of a more senior coordinating committee to which the working groups would report and which would evolve into formal negotiating teams within a year.

THE MEMORANDUM OF INTENT

On August 5, 1980, the long-awaited Memorandum of Intent was signed in Washington. The memorandum constituted official recognition of the bilateral nature of the problem, as well as the combination of long-range transport with local point-source air pollution into a single international issue. It also constituted a recognition that the 1909 Boundary Waters Treaty and the 1978 Great Lakes Water Quality Agreement had already obligated both nations bilaterally on air quality, as had a number of multilateral agreements to which both were signatory. It recognized the need for compromise with energy needs and, further, it noted (but did not necessarily accept) scientific findings indicating the dangers of acid rain and stated that environmental stress could be (not necessarily would be) increased if action were not taken. In particular, the memorandum found that "the best means to protect the environment from the effects of transboundary air pollution is through the achievement of necessary reductions in pollutant loadings."[50]

It also formally committed the two governments to negotiate a bilateral air quality agreement, with negotiations to begin not later than June 1, 1981, and also "pending conclusion of such an agreement, to take interim actions available under current authority to combat transboundary air pollution."[51] Canada has claimed the Ontario government's control order restricting INCO emissions, the announced reductions by Ontario Hydro of SO_2 emissions over a ten-year period, increased federal and provincial research on acid rain, and the passage in December, 1980, of Bill C-51 as significant contributions to this undertaking. The United States, in turn, claimed increased research and the activation of Section 115 of the U.S. Clean Air Act Amendments of 1977 as its contributions to this mandate. (The passage of Bill C-51 and the activation of Section 115 of the Clean Air Act are related matters that will be described later.)

The memorandum also called for the establishment of the five technical working groups, to be supervised by a coordinating committee, patterned after the model used in laying the groundwork for the Great Lakes Water Quality Agreement of 1972. The technical groups would be composed of government scientists from both nations, while the coordinating committee would be essentially directed by diplomats and senior environmental policymakers. The tasks of the five technical working groups which have been set up are as follows:

> Group 1 — focus on effects, with a mandate to determine what the environment can tolerate, and charged with producing tolerance ranges.
>
> Group 2 — focus on atmospheric chemistry, with a charge to judge ideal atmospheric loadings against actual loadings.
>
> Group 3A — focus on synthesizing information about where loadings are coming from, determine what can be done and what are the appropriate scenarios—the group that pulls it all together.
>
> Group 3B — focus on hardware, and the costs of control, both technical and socioeconomic.
>
> Group 4 — focus on institutions and legal questions.[52]

The memorandum also strengthened the long-standing practice of advance notification and consultation on proposed actions with potential transborder ramifications, specifically including proposed major industrial developments and proposed changes in policy, regulations, or practices with such ramifications.

CONTINUED CANADIAN PRESSURE

Canada, as the more vulnerable and more widely afflicted nation, has throughout been the protagonist in the bilateral acid rain debate, dragging a more cautious and reluctant United States into the fray. And behind Canada's strong leadership in this area, behind its two very concerned environment ministers of recent years—John Fraser (1979–80) and John Roberts (1980–present)—has been Environment Canada's Environmental Protection Service (EPS), the chief air quality regulator at the federal level. Aware of the emotionalism surrounding this issue and the dangers of relying on soft data, EPS has launched a major campaign to improve the reliability of data in all aspects of the subject. Under the umbrella name of *control strategies project,* detailed studies are progressing on emission source assessment to determine the reductions that could be expected from the application of specific abatement technologies and the social and economic consequences of applying various levels of emission reduction both to major emitters and to other sectors of society. A macroscale assessment of the physical and economic benefits that would arise from reduced pollutant loading on the environment is being developed, and analysis is underway of abatement options which will result from the integration and evaluation of the information from the first three components.[53]

This ambitious effort has been coupled with an equally ambitious (and some claim misguided) effort to severely regulate emissions in Canada so that that country's house might be viewed as "in order." EPS has persevered in this stance, which has brought it both praise and criticism. EPS fears not only that acid rain might well do severe and, most importantly, irreversible damage to Canada and its resources but also that the political awareness of the problem might continue to differ between the two countries, leading to a point where ugly political explosions will occur. Former Assistant Deputy Minister Ray Robinson remarked that "ugliness can get in the way of that free movement of goods so important to the business community and to the Canada-U.S. relationship."[54] Robinson also suggested that when you put up serious health problems (debatable in some circles) against cheaper electricity for the U.S. Midwest, the kind of bilateral problem that will ensue will be most difficult for the diplomats to manage or keep under control. He commented:

> If the scientists find that fine particulates emitted by thermal plants and other sources are making life more difficult for people with respiratory illness . . . and possibly shortening their lives as a consequence . . . then I submit to you that pressure to control these emissions will become irresistible.[55]

Former Canadian Environment Minister John Fraser speaks of these same fears, and Environment Minister Roberts often alludes to them in his strong public statements.

Canadian perceptions of this issue are indeed strong, encouraged by extremely heavy media attention (some would say saturation) in all forms, the publication of a number of books in Canada addressing acid rain as a public policy and diplomatic issue,[56] and instigation, to the regret of many serious scientists and, particularly, of those companies which stand to be regulated, of a rather high level of emotion throughout Canadian society. These strong perceptions have appeared to warrant

- rather strong actions by the federal, and by some provincial, governments (such as Quebec), including the distribution by Ottawa of "Stop Acid Rain" leaflets in the United States;
- strong speeches by senior Canadian bureaucrats and politicians to U.S. audiences at every opportunity;
- the rapid passage in one day (all three readings) of Bill C-51;
- an extraordinary Ontario and Quebec intervention in U.S. rule-making processes and EPA decision making;
- an Ontario Hydro reversal on the issue of scrubber installation (since tempered);
- Quebec acid rain research agreements with New York and Vermont;
- the establishment of a joint Quebec–New York acid rain research institute; and
- the placement of acid rain as a top item on the Trudeau-Reagan agenda during President Reagan's state visit to Ottawa in early 1981.

Perhaps most remarkable is the unprecedented establishment in Washington, D.C., of a government-funded lobby, the Canadian Coalition Against Acid Rain (supported by one-third federal, one-third provincial, and one-third nongovernment funding). These are strong initiatives, indeed, all or most of which would likely never have been considered if it had not been for the seriousness or perceived seriousness of this issue and Canada's growing sense of desperation given its vulnerable position.

Of all the nontraditional means Canada has adopted for carrying out the action it feels necessary, perhaps none, however, bears greater significance than its decision in December, 1980, with the acquiescence of the United States and the outgoing Carter administration, to enter into reciprocal legislation, known loosely as "equal access legislation." Such was designed as a new tool in the battle to curb transborder emissions.

RECIPROCAL LEGISLATION

The now infamous Bill C-51, vigorously opposed by some provinces (especially Quebec) as a federal intrusion into their constitutionally protected domain, is an amendment to Canada's federal Clean Air Act which gives Ottawa the power to expand federal authority by regulating air pollution originating in Canada that might pose a health hazard to persons in another country. This represents a direct transfer of power from the provincial to the federal level. Such a transfer is politically very difficult and shows the seriousness of Canada's commitment.

The passage of Canada's statute was designed to trigger a process whereby the United States, following the reciprocity mandate of its Clean Air Act, can control air pollution which might be harmful to persons in other countries which have pollution protection laws. To accomplish this, U.S. federal authorities are enabled to initiate measures requiring state governments to reduce emissions adversely affecting Canada. Thus, an important provision of the U.S. Clean Air Act, which has major ramifications for controlling transborder pollution and which had lain dormant for years because of lack of a needed reciprocal provision in Canada's federal law, could now be activated in Canada's interest.

Passage of Bill C-51 also removed grounds for excuse and delay. It fulfilled requirements of the August 5 Memorandum of Intent to do all possible under existing authority as well as seek new authorities, and also placed the onus on Washington to take strong direct action with U.S. polluters allegedly harming Canada, using the appropriate state governments as the vehicle. Bill C-51, in spite of its controversial nature and implications for sensitive questions of federal-provincial relations, received the unanimous support of all three political parties in the House of Commons, further indication of Canada's seriousness of purpose in this area.

In December, 1980, the Carter administration responded to Ottawa's Bill C-51 (thus validating it) when outgoing EPA administrator Douglas Costle formally acknowledged that the international reciprocity required by Section 115 of the Clean Air Act had been met for administrative action requiring the EPA to make states curb pollution that harms another nation. An official list of states, inevitably to include Ohio, which has been so frequently cited as a major source of acid rain deposition on Canada, was then to be compiled and hearings held. Administrator Costle and the Carter administration then passed from the scene, leaving this international reciprocity legacy to the Reagan administration. (Efforts are under way in the U.S. Congress to repeal Section 115, the basis of the U.S. portion of this reciprocal arrangement.)

Less than two months later, the province of Ontario, which considers itself most afflicted by U.S. emissions, acted under the new arrangement by taking the extraordinary step of initiating legal action (filing an intervention) asking the U.S. EPA to reject proposals from six states (Ohio, Michigan, Indiana, Illinois, West Virginia, and Tennessee) for a relaxation of emission limits governing eighteen power plants (later amended to twenty plants), and also to enforce existing emission limits which are now being exceeded, to review the regulations on emission levels governing all U.S. power plants and to consider permissible emissions from the perspective of total effect on north eastern North America.[57]

The province of Quebec joined in this legal action. This intervention opposes applications to amend state implementation plans (SIPs), under which individual states would be permitted to increase pollutant emissions. Where the EPA disapproves of such an increase, whether because of the intervention or for other reasons, the governor of the state is informed and he must advise the polluter (in this case the appropriate utility) that it cannot increase its pollution. Such intervention is not likely to succeed in reducing emissions but will undoubtedly elevate the issue diplomatically as well as establishing precedent.

> Ontario's position is that the onus should be on an applicant under the State Implementation Plan to demonstrate that there will be no injury from pollution to the territory of another country.[58]

Thus begins a new era in bilateral environmental relations, one involving the direct involvement of provinces, the U.S. federal bureaucracy, and private utilities, all with the full acquiescence of the Canadian diplomatic community, at least if this first example is indicative of things to come. One wonders if U.S. states will soon involve themselves in similar interventions in Canada in this or in a host of other bilateral problem areas.

THE CHALLENGE

Canada's challenge at this point is to get the United States to spend and to regulate at the very time when both seem to be losing support in that country. Or are there other ways to solve the bilateral problem? In addressing the problem on the Great Lakes, Francis et al. wrote of the possibilities of liming acidified waters and breeding acid-resistant stocks of fish, while admitting that these measures exhibit limited utility and that all serious methods of control are likely to be quite expensive.[59] Alternatively, Canada may have the option of linking its own needs for emission control with U.S. desire for Canadian energy and achieve its ends in that manner.

The U.S. EPA has concluded that as much as 20 percent of SO_2 emissions could be eliminated through a utility operational plan that relies on newer, more efficient scrubber-equipped plants for the provision of a baseload of power accompanied by a switch to older, less efficient, uncontrolled polluting plants only at times of high electricity demand when absolutely needed (a technique known as *least emissions dispatching*— LED). Others promote prewashing of coal, better mixes of high- and low-sulfur coal and of coal and oil, and so on among the various techniques put forth. The great national debate in both countries often revolves around the benefit-cost argument, that is, what is (1) the true cost of not controlling the emissions and how does that cost compare with (2) the cost of controlling them? Until a consensus develops on this question, it will be difficult to rally support for the major expenditures that appear to be needed, whether for flue-gas-desulfurization systems (scrubbers) or other techniques to solve the problem. The costs may not be out of reach, however, when those coal-fired power generation costs are averaged over all users. David Hawkins, former assistant administrator of EPA for air quality, estimates that a seven-million-ton SO_2 reduction by 1990 would cost up to $2.7 billion, representing a cost of up to $480 per ton of SO_2 removed, representing an average electricity rate increase of up to 2 percent. If reasonably accurate, this is a much smaller price increase than assumed by most followers of the acid rain debate.[60] (Although a 2 percent average increase may not be high, coal-dependent states such as Ohio and Indiana would undoubtedly experience a greater increase—perhaps 8 or 9 percent—creating proportionately greater economic impacts and thus greater political problems.)

The bilateral as well as the domestic problem of acid precipitation could be alleviated or solved by a number of different SO_2 and NO_x control options, each of which has its own associated advantages and disadvantages.[61] The increased practice of energy conservation and the shifting of energy reliance to non–fossil fuel alternatives would solve the problem, as would expensive flue-gas-desulfurization and denitrifiction methods (scrubbers), while coal prewashing, coal gasification or liquefaction, fluidized bed combustion, coal-limestone injection, and utility pass-through of emission costs to the consumer are among techniques which would either solve or alleviate it. While it is diplomatically unacceptable for one nation to tell another how to achieve control in order to terminate transborder fluxes, it is not out of order for the afflicted nation to conduct research and demonstrate to the offending nation how the problem might be solved. Hence, some attention is being given in the bilateral debate to matters that might normally be regarded as domestic. There is much to be gained by both nations in devoting their joint best efforts to the search for a solution to a problem of such great magnitude and ramification.

By early 1982, the work of these technical working groups established under the Memorandum of Intent had slowed considerably in the face of Reagan administration's lack of interest in moving forward with regulation in this area, and in the midst of the great two-year debate over renewal of the basic federal statute, the U.S. Clean Air Act of 1977, which expired in September, 1981. Canadian pressure is continuing to keep the structure of the working groups and the spirit of the memorandum intact, but Canada is clearly limited in what it can accomplish beyond that. Through its lobbying, its support of the Canadian Coalition Against Acid Rain in Washington, and by other means, Canadian diplomatic strategy has moved toward developing a U.S. constituency composed of U.S. interests in acid-vulnerable states (New England, New York, Wisconsin, Minnesota) to fight the issue in their own self-interest. Additionally, Ottawa has encouraged and supported the provinces (especially Ontario) in their own direct efforts, and even in their legal intervention where this is permitted. It appears at this point, however, that tangible movement on this serious issue must await the discovery of additional evidence, which might in turn persuade new constituencies to join the fray.[62]

The diplomatic issue is now asymmetrical, with a large and vocal Canadian constituency pitted against a still comparatively unaware, uninterested, and seemingly uncaring U.S. population; between a country moving toward environmental regulation (Canada) and one moving toward deregulation (United States). If it expands, this asymmetry could spill over into other areas of the relationship. Though Canada's options in terms of strong response are limited, a residue of bitterness and sourness in Canadian-U.S. relations can result and have long-term implications. A diplomatic remedy, therefore, may be well worth the effort.

In January, 1982, the diplomatic complexity of the acid rain issue took a quantum leap in magnitude, with the growing realization in Canada that Ontario Hydro planned to export substantial electricity through a new sublake cable under Lake Erie to New Jersey and Pennsylvania. The need for this electricity in the U.S. market was occasioned by the loss of the famous Three Mile Island nuclear power plant. The Ontario temptation to provide this export was heightened by the nearly one billion dollars to be earned and, most importantly, the substantial surplus generating capacity extant in Ontario's power plants. The proposal became wedded to the acid rain controversy since this electricity would be generated by Ontario Hydro's huge coal-fired Nanticoke Station, a facility without significant sulfur dioxide controls. Such generation would not only contribute to Canada's own acid deposition (and possibly to that of the northeastern U.S.) but, worst of all, could erode if not demolish Canada's whole negotiating position, to the great embarrassment of Ottawa as well as many Canadian and U.S. environmentalists allied with Ottawa. (Ironically, the coal to be burned at Nanticoke is largely U.S. in origin.) Worse, this particular proposal strengthened the arguments of those U.S. critics raising suspicions as to a Canadian conspiracy to condemn U.S. acid rain emissions as a ploy to send more Canadian energy south of the border. While this conspiracy theory has not been taken seriously in many quarters, the Ontario Hydro proposal created the perception that the theory might have validity, potentially devastating the Canadian position. The major drive in 1982 by both the Canadian government and the Washington-based Canadian Coalition Against Acid Rain to launch a credible large-scale U.S. publicity campaign was also somewhat threatened by this proposed Nanticoke export. The National Energy Board had approved the export, and the federal cabinet in Ottawa must ultimately make the decision between energy sales and diplomatic credibility. It was rescued from that decision when the U.S. buyers withdrew their offer and instead purchased their needed power from Detroit Edison, power largely derived from U.S. coal sources.

The bilateral issue of long-range transport of air pollutants, or its most famous component, acid precipitation, is just beginning and will undoubtedly continue for some years. This chapter is merely a piece out of a much larger, and most complicated story that bears much potential for real, and perhaps permanent, damage to the friendship between the two nations. For that reason alone, and aside from the real or perceived dangers of acid rain itself, this issue deserves maximum public attention. Much in the relationship is at stake and much is in danger of being lost.

It is appropriate to end this survey of transboundary environmental problems with the most serious and current of all major bilateral environmental issues. Part 3 provides a range of possible techniques which may open avenues of resolution to this and other transboundary issues and proposes a formalization of the bilateral environmental relationship.

NOTES

1. PH is a measure of acidity. The lower the pH, the higher the acidity.

2. Local air pollution from identifiable point sources is now regarded as a conventional problem, while recently recognized acid rain may therefore be regarded as unconventional. However, closer familiarity with the transport of upper atmospheric toxics, the least conventional form of air pollution to date, may now render even acid rain conventional. Little is known of these upper atmospheric toxics, however.

3. Gene E. Likens et al., "Acid Rain," *Scientific American* 241, no. 4 (1979):43.

4. Ibid., p. 49.

5. U.S. Environmental Protection Agency, *The Acid Precipitation Problem* (Corvallis, Oregon: Environmental Research Laboratory, 1979), p. 1.

6. Atmospheric Protection Service, Environment Canada, *Blowing in the Wind* (Ottawa, Ontario: Environment Canada, 1979), p. 2.

7. Joseph Dowd, Prepared Statement on behalf of the Edison Electric Institute before the Human Rights and International Organizations and Interamerican Affairs Subcommittees of the U.S. House of Representatives Foreign Affairs Committee, May 20, 1981.

8. United States–Canada Research Consultation Group, *The LRTAP Problem in North America—A Preliminary Overview* (Ottawa and Washington: United States–Canada Research Consultation Group on the Long Range Transport of Air Pollutants, 1979), p. 1. (Hereafter referred to as LRTAP 1979.)

9. United States–Canada Research Consultation Group, *Second Report of*

the United States–Canada Research Consultation Group on the Long Range Transport of Air Pollutants (Ottawa and Washington: United States–Canada Research Consultation Group on the Long Range Transport of Air Pollutants, 1980), pp. 3–4.

10. LRTAP 1979, p. 4.

11. Ibid., p. 6.

12. Environment Canada, *Blowing in the Wind.*

13. American Smelting and Refining Company (ASARCO), *Critique of The LRTAP Problem in North America—A Preliminary Overview* (New York: ASARCO, 1980), p. 2. (Hereafter cited as ASARCO, *Critique.*)

14. LRTAP 1979, p. 11.

15. John M. Wooten, Environmental Manager for Air Quality, Peabody Coal Company, personal correspondence, October 2, 1980.

16. Ibid.

17. F. Frantisak, Noranda Mines, Ltd., Remarks to the Air Pollution Control Association Annual Conference, Montreal, Quebec, June 23, 1980, pp. 5–6.

18. Ibid., p. 6.

19. LRTAP 1979, p. 13. Some Ontario politicians have asserted that the highly industrialized northeastern United States is the source of most acid precipitation affecting Ontario, while the LRTAP study and the U.S. State Department contend that half of the deposition originates in Ontario.

20. Ibid., p. 20.

21. Ibid., p. 25.

22. National Council of the Paper Industry for Air and Stream Improvements, Inc. (NCASI), *Acidic Deposition and Its Effects on Forest Productivity—A Review of the Present State of Knowledge, Research Activities, and Information Needs,* NCASI Technical Bulletin No. 110, January, 1981, pp. 56–57. (Hereafter cited as NCASI, *Acidic Deposition.*)

23. NCASI calls for a research program focusing on the following seven areas:

1. identification of forest soils in important timber growing regions that are sensitive to acidic deposition inputs;
2. effects of acidic deposition on the chemical and biological processes occurring in sensitive forest soils;
3. development of better methods for measuring dry depository rates in forested areas;
4. analysis of throughfall and stemflow chemical composition;
5. monitoring of wet and dry deposition in important timber growing regions;
6. laboratory, greenhouse, and controlled field experiments to elucidate the effects of acidic deposition on sensitive forest soils and commercially important tree species; and
7. assessment of acidic deposition effects in relation to effects of intensive forest management practices [NCASI, *Acidic Deposition,* p. 57].

24. NCASI, *Acidic Deposition,* pp. 40, 42.

25. ASARCO, *Critique,* p. 4.

26. Ibid., p. 6.

27. NCASI, *Acidic Deposition,* pp. 25–26, 45.

28. U.S. energy conservation has actually been more effective than at first seemed possible, but the United States still has a long way to go.

29. Even the sensitive area of Canadian federal-provincial relations has not gone unaffected. Conversations with senior Canadian government officials bear out the fact that former Minister of Environment John Fraser prevailed upon Prime Minister Clark and the rest of the cabinet to use the acid precipitation issue to strengthen federal-provincial relations, namely to demonstrate that there are issues of great concern to the provinces which can best be handled by the federal government working in consort with, and on behalf of, the provinces, rather than leaving the provinces to work singly on their own. See also, John E. Carroll, "Quebec–United States Relations," Paper presented to the Colloquim on Opportunities and Problems in Quebec–U.S. Relations, Harvard University, Cambridge, Massachusetts, September 1–3, 1982.

30. Maxwell Cohen, "The Deadly Rain," *Winnipeg Free Press*, August 13, 1979.

31. International Joint Commission, *An Excerpt from the Annual Report of the Great Lakes Science Advisory Board* (Windsor, Ontario: Great Lakes Regional Office, 1979), p. 20. However, the Canadian government has largely rejected the value of this type of approach, concluding that it may well do more harm than good.

32. Ibid.

33. Ibid., p. 27.

34. Ibid.

35. Harry C. Parrott, Minister of Environment, Province of Ontario, Notes for Remarks to the Action Seminar on Acid Precipitation, Toronto, Ontario, November 3, 1979, p. 1.

36. Ibid., p. 2.

37. Ibid., p. 7.

38. Ibid., p. 8.

39. Ibid., p. 17.

40. This attitude raised concern in the U.S. State Department, especially when coupled with the minister's frequent references in public addresses to there being over two hundred industrial stacks in the United States rising 400 to 1,200 feet, without noting that many of these are nowhere near Canada.

41. There is diplomatic significance to Ontario's decision to either proportion its control effort to movement by the United States or to give it full measure on its own, thus strengthening Ottawa and giving the United States an example.

42. Raymond M. Robinson, "Canadian Perspective on Acid Rain" (Remarks to the Edison Electric Institute Conference on Acid Rain, Boston, Massachusetts, September 26, 1980), p. 5.

43. Ontario Hydro, "Hydro Program to Cut Acid Gas," *Backgrounder,* January 26, 1981, pp. 1–4.

44. John A. Fraser, "Commons Debates," *Hansard,* October 22, 1979, p. 452.

45. John A. Fraser, "Notes for an Address by the Minister of the Environment to the Action Seminar on Acid Precipitation," Toronto, Ontario, November 2, 1979, p. 6.

46. Bill Richards, "U.S. Conversion to Coal Would Add to Acid Rain in Canada," *Washington Post,* February 7, 1980.

47. Canadian Embassy, Note No. 84, Washington, D.C., March 4, 1980, pp. 3–4.

48. John Roberts, Canadian Minister of the Environment, Remarks to the Air Pollution Control Association Annual Meeting, Montreal, Quebec, June 23, 1980, p. 9.

49. Max Baucus, *Congressional Record-Senate*, November 8, 1979, S16202.

50. Governments of Canada and the United States, *Memorandum of Intent Between the Government of Canada and the Government of the United States of America Concerning Transboundary Air Pollution*, Washington, D.C., August 5, 1980, p. 2.

51. Ibid., p. 3.

52. Raymond M. Robinson, Assistant Deputy Minister, Environment Canada, personal interview, Ottawa, September 5, 1980.

53. Raymond M. Robinson, "Acid Rain: The Crossroads for Decision Making," *Journal of the Air Pollution Control Association* 30, no. 5 (1980):108.

54. Robinson interview.

55. Robinson, "Canadian Perspective," p. 11.

56. P. Weller and the Waterloo Public Interest Research Group, *Acid Rain: The Silent Crisis* (Kitchener, Ontario: Between the Lines Press, 1980); Ross Howard and Michael Perley, *Acid Rain: The North American Forecast* (Toronto: Anansi Press, 1980); Gerry Sperlman and Caroline Brown, *Rain of Death* (Edmonton, Alberta: Nu-West Press, 1981).

57. Ontario Ministry of the Environment, "Ontario Takes Legal Action in U.S. to Prevent Acid Rain Emission Increases," press release, Toronto, March 12, 1981, p. 1.

58. Ibid.

59. G. R. Francis et al., *Rehabilitating Great Lakes Ecosystems* (Ann Arbor: Great Lakes Fishery Commission, 1979), pp. 48–49.

60. David Hawkins, "Transboundary Transport of Air Pollutants from Energy Production" (Remarks to the American Bar Association/Canadian Bar Association Conference "Common Boundary, Common Problems: Environmental Consequences of Energy Production," Banff, Alberta, March 20, 1981).

61. For an in-depth treatment of these options as well as further detail on the bilateral aspects of this subject, the reader is referred to John E. Carroll, *Acid Rain: An Issue in Canadian-American Relations* (Washington, D.C.: Canadian-American Committee, National Planning Association; and Montreal: C. D. Howe Institute, 1982). See also *Acid Rain: A Transjurisdictional Problem in Search of Solution*, ed. Peter S. Gold (Buffalo: Canadian-American Center, State University of New York at Buffalo, 1982). See also "Acid Rain Cloud Casts Shadow on Canadian-U.S. Relations" by John E. Carroll, *San Diego Union*, June 28, 1981, p. C-4.

62. For an in-depth analysis of acid rain and Canadian-U.S. relations, see John E. Carroll, *Acid Rain: An Issue in Canadian-American Relations* (Washington, D.C.: Canadian-American Committee, National Planning Association; and Montreal: C. D. Howe Institute, 1982.) See also, John E. Carroll, "Acid Rain Diplomacy: The Need for Bilateral Resolution," *Alternatives* 11, no. 2 (Winter, 1983).

Part 3
The Prescription

CHAPTER 12   Formalizing Environmental
Relations

The chapters of Part 1 described differences between the United States and
Canada which lie at the root of bilateral environmental problems. The case
study chapters of Part 2 have revealed numerous weaknesses in the present
system. Part 1 thus shows a potential for weakness, while Part 2 illustrates
those weaknesses. These weaknesses afflict both governments and harm
business, environmentalist, and border interests. They need to be remedied. This chapter presents a cross-section of remedies which could be
implemented, in whole or in part. Each remedy proposed has its own
associated advantages and disadvantages and must be weighed on its merits
against enduring the costliness of the status quo. The chapter then concludes with the major findings of this study.

But first of all the two societies must be prepared to invest in search-
ing for remedies. Are they? What are the alternatives? What is the status
quo?

The Status Quo

The present state of Canadian-U.S. environmental relations can best be
described as ad hoc with a light patina of order and governance provided by
the Boundary Waters Treaty of 1909 and its vehicle, the International Joint
Commission. As has already been described, however, the treaty and the
commission bind government to very little, indeed only to limited aspects
of boundary water apportionment under certain circumstances.

As noted previously, the formalization of the environmental relationship embodied in the IJC example has not always worked (e.g., Skagit,
Champlain-Richelieu, Poplar water quality). To some extent, this is the
fault of the two governments for writing overly restrictive references or
failing to wholly support the commission (e.g., Poplar water quality). To a
great extent, however, the blame must fall on the commission itself, for
errors in judgment (e.g., Skagit) or for inability to face up to and overcome
national differences (e.g., Champlain-Richelieu). The commission also on
occasion deserves criticism for tardiness in handling references given to it,
although at times this perceived tardiness is unavoidable due to lack of
resources, unclear references, or other problems.

It is true that the IJC mandate is a broad one, touching upon water pollution, air pollution, and, indirectly, on energy generation, navigation, and so on. But its role in these matters is purely advisory, and it may in fact be ignored or even, as in the case of acid rain, kept at the periphery of the issue. It can also be left to die by uninterested governments. Other existing instruments such as the Great Lakes Water Quality Agreements, not being treaties, are not binding under law. Neither are various other agreements and arrangements which are dependent on the good will of both nations, good will which is now being seriously tested by the acid rain dispute. Thus the phrase "ad hoc" appears to be an accurate description. Maintaining the current bilateral environmental relationship on an ad hoc basis has the tangible value of insuring flexibility for both governments in their diplomatic dealings. It gives diplomats an opportunity to exert a greater degree of control, since they can look for elements of a solution which are in the interests of their respective governments on a particular question and at a particular point in time. It is based upon a very limited number of guiding principles in theory, and few if any in fact. From a purely national perspective, there is value in this and it permits maximization of options according to each set of circumstances as they arise. Being flexible, it also allows for easy adjustment not only to changing circumstances but also to changing values and attitudes as they evolve over longer periods of time. Perhaps its greatest disadvantages are uncertainty, unpredictability and the magnifying of issues which might otherwise be handled routinely. Additionally, resources are strained each time a new issue must be tackled, numerous uncertainties about the nature and significance of an issue must be worked out, and many matters are not (and often cannot be) dealt with until they reach a critical advanced stage, by which time it may be too late.

The one value that an ad hoc approach cannot provide—the greatest advantage of a more structured approach—is stability. Aside from the risk of being overwhelmed by events, "adhockery" cannot provide the stability of established rules of the game. This makes a more structured approach of greater real value to those who desire or need an ordered and orderly relationship. Interests which require high levels of predictability and long planning horizons, including corporate planners responsible for long-term investment, value the predictability and orderliness that can be provided by a more structured relationship than exists at present. The same holds true for government agencies making major long-term investment decisions.

Three Alternatives

There are three broad alternatives facing the two peoples. The first might be labeled the "do nothing" approach—maintaining the status quo. This involves making minimal use of the IJC and reacting in a purely ad hoc manner to issues and then only when they reach crisis proportions. It entails letting problems go unattended as long as possible and, to an increasingly dangerous extent, failing to respond, even to loud and clear public outcries and complaints from across the border, thus risking a general spillover into other areas of the bilateral relationship.

The second alternative is the incremental approach used to some extent at various periods in the past, as witness the early Great Lakes Agreement and negotiation; willingness to turn to the IJC on Garrison, Poplar, Skagit, and in a few other areas. However, there was lack of willingness to follow through or to apply such incremental alternatives to issues of major import affecting large geographical areas or significant sectors of the economy. This is a predominantly ad hoc approach with some acceptance of structure and formality.

The third approach looks toward a new order. It is an approach governed by a treaty which would encompass all transboundary environmental issues, consequential and inconsequential, and minimize "adhockery" to the greatest extent. It represents a new direction in the relationship which would submit each new issue to a uniform procedure from the beginning and lay out a formal plan to achieve solution, similar to the direction given by those portions of the Boundary Waters Treaty of 1909 governing allocation of boundary and transboundary waters.[1]

Toward Formalization

How could increased formalization be accomplished? A team of specialists working under explicit principles and procedures, following the example of the IJC, can routinely ferret out problems in their earliest stages, dispose of those which are not significant, recognize those which are, and apply seasoned expertise at all stages (initial fact-finding through negotiation) until the problem is solved, at least to the satisfaction of governments. Ideally, the principles and accepted practices of such an institution should be well known to all parties concerned on both sides of the border to improve both credibility and final acceptability. While some flexibility is lost, the increased objectivity of such a formalized approach might well overcome this disadvantage. It is the purpose of this chapter to treat some of the various ideas and proposals which have been put forth from various quarters to achieve precisely this kind of order and stability.

Based on his long experience with transboundary environmental questions, as legal scholar and as Canadian IJC chairman, Maxwell Cohen has suggested five effective rules for Canada and the United States to follow in any program to avoid conflict.[2] These rules are:

- Do not catch the other side by surprise on a real or perceived interest; and give reasonable notice and provide for meaningful consultation before any serious moves are made.
- Do not prolong by unilateral rhetoric any disagreement over basic

facts; and set up joint fact-finding instruments as early as possible to obtain agreement on disputed facts.

• Try to institutionalize common fact-finding, thus assuring a continuity of tradition and of personnel in the particular dispute settlement process or area.

• Try to anticipate, wisely, through binational perspectives, what soon may be threatening both the national and binational interest of the two countries.

• Where the facts are not in issue and there is a defined legal claim, then consider the resort to binding arbitration or the International Court of Justice.

The Techniques

Many of these rules are incorporated in the ideas and proposals that follow. These proposals are not mutually exclusive and can be viewed as building blocks constructing a total system. They progress generally from less to more institutionalization. Proposals to be considered are common fact-finding, diplomatic early warning and consultation, common control mechanisms, transborder litigation, international environmental mediation, subnational relations, multilateral relations and international environmental standards, role of the nongovernmental sector, new institutions, and an environmental treaty, agreement, or accord.

COMMON FACT-FINDING

Experience suggests that first, joint agreement on facts constitutes two-thirds of the battle to settle transboundary environmental disputes, and solving differences in ideology only one-third; and, second, common data and scientific solutions must precede political solutions. Efforts to agree on common or joint fact-finding will pay rich rewards.

The one institution which stands out par excellence in this area is, of course, the International Joint Commission, already discussed in detail in chapter 3. The IJC has an admirable track record in this, one of its most important tasks, if not its raison d'être. Unquestionably, the commission and its mode of operation have been largely responsible for the credibility of much of what is accepted today by both governments, and the high credibility of its reports has led directly to the negotiated settlement of a large number of disputes since 1909.

The IJC is not the only bilateral institution practicing joint fact-finding, although it is the only one able to stand apart from the two diplomatic communities. In that respect it is indeed unique. The recently constituted Bilateral Consultation Group on the Long-Range Transport of Air Pollutants, assembled in 1979 to fact-find in acid rain and related areas (see chap. 11), is another institutional example. However, it is chaired by representatives from the departments of State and External Affairs and has a less certain future than the IJC, for it does not (as yet) have the protection of a treaty. Nevertheless, it is succeeding in its fact-finding task and is itself making history in the bilateral diplomatic arena.

Other joint fact-finding examples on a smaller and regional scale include the bilateral monitoring group established to alleviate the Poplar River controversy, and the state-provincial group being established with federal blessings to exchange technical data and to monitor effects of the Cabin Creek coal mine on the Flathead River and its tributaries. Minor regional or local examples could also be cited. However, many believe there is potential for expansion of joint fact-finding, especially at the national level.

To the degree that common binational fact-finding is recognized and employed, significant bilateral concerns may be reduced. The greater the degree to which both nations can support and participate in such joint endeavors, whether within the IJC or another vehicle, the more border and international environmental problems can be brought under control and the greater the chance that satisfactory solutions will be found.

Joint fact-finding is insufficient in itself, however, as the IJC experience teaches. It can be applied to a problem too late to be effective. Its findings can be (and sometimes are) ignored. And it does not guarantee bilateral agreement on remedies, even if all do agree on the facts.

DIPLOMATIC EARLY NOTICE AND CONSULTATION

The concept of early notice and consultation, on a nonstructured but jointly agreed basis, is already a favorite among diplomats on both sides of the border.[3] It simply directs that early notice be given to the other nation's diplomats when a development with possible transborder consequences is to be proposed and calls for consultation should either side so request. It is clearly the type of behavior that one would expect between civilized nations. However, it is, surprisingly, not taken for granted in Canadian-U.S. diplomatic relations. On the contrary, it has had to be instituted by informal agreement by the State and External Affairs departments and is viewed as a source of pride by both. This is one mechanism or technique which encounters little or no opposition, at least in theory, and is most often cited as a sign of evolution in the diplomatic relationship. Concerning environmental matters it is clearly viewed as working, and working well. However, there are exceptions. Garrison and Poplar are major examples of where the system failed, while the Cominco lead smelter's leak of mercury into the Columbia River near Trail, British Columbia, with harmful impacts in the U.S. downstream, constitute a more recent example of its failure. There also is some opposition to (or lack of faith in) the concept.[4]

Although seemingly simple, an early notice and consultation procedure is in practice quite difficult to manage. It assumes full cooperation

of private corporations who are concerned about their investments, inflation in land prices, and their proprietary relationship with their competitors
and the public vis-à-vis confidential information. Likewise, it assumes
cooperation of crown corporations and local, state, and provincial governments, all of which must be concerned with the political ramifications of
their decisions as well as their relations with other levels of government.
There are obviously reasons why these various agencies, and others, want
to avoid informing the diplomatic community prematurely, that is to say,
prematurely from their point of view. A further dilemma faces diplomats
who learn of a potential development but have difficulty determining if it
deserves attention—determining whether it is in fact likely to occur, or
whether attention to the subject and formal across-the-border notice might
well raise the issue unnecessarily, causing problems that might not otherwise occur. To investigate every proposal, major or minor, with potential
transborder environmental impacts is costly in scarce dollars and manpower. Thus, the success or failure of an early warning and consultation
system such as now informally exists between the departments of State and
External Affairs revolves around and is dependent, first upon close intergovernmental cooperation to an unprecedented degree, second, upon
corporate-governmental cooperation of a type which is often not possible
for proprietary reasons, and third, upon a high degree of ability to define,
with real foresight, what is or is not likely to become, within a reasonable
amount of time, a true transboundary environmental issue.

Although of lesser value than other bilateral innovations, early notice
and consultation mechanisms are valuable. They signify a high level of
cooperation and sensitivity and can engender a feeling of good faith between the two nations. However, early notice and consultation are not
enough in themselves. They do not lead to agreement on facts, or suggest
solutions to transboundary problems.

COMMON CONTROL MECHANISMS

Mention has already been made of the difference between the U.S. system
of environmental or pollution control standards and the Canadian system of
control objectives. This section reviews the essence of the conflict between
the two and presents a proposal for solution.

By the late 1960s and early 1970s, for a variety of reasons a purist
mentality was beginning to take hold in the United States in reaction to
earlier behavior more tolerant of pollution. This mentality translated into
the enactment between 1960 and 1977 of strong water and air quality
legislation which used as its vehicle stiff emission standards at the source.
These standards were strict and were designed to accommodate expensive
and inevitable litigation (leading in the U.S. system to higher standards).
They had the ultimately impossible goal of terminating all pollution discharges and were predicated on the philosophy that all pollution was harm-

ful and costly and that no pollution could or should be tolerated. The best example of this philosophical approach is the Federal Water Pollution Control Act Amendments of 1972, which demanded an end to all waste discharge, regardless of cost or feasibility, by 1984. It was this set of circumstances which catapulted the U.S. environmentalist movement into the courts and into the limelight, as discussed in chapters 1 and 2.

Canada also followed the fashion and passed extensive pollution control legislation at both the federal and provincial levels during this time. While there was superficial resemblance between the Canadian and U.S. legislation, there were nevertheless important differences. In addition to the much greater provincial authority in Canada relative to the U.S. states, there was also a difference in underlying philosophy. Twenty-three million Canadians living on a vast land mass had not felt pressured as had their U.S. brethren to take extreme action—the Canadian environment was for the most part still relatively clean and pure, and pollution was still easy to escape. Thus Canada applied what it considered to be much more realistic water and air quality objectives while rejecting the more purist (and, in its eyes, unreasonable and unnecessary) emission standards of the United States.

The Canadian approach via objectives considers the overall condition of the body of water or air scheduled to receive pollutants, recognizes it has a pollution-carrying capacity, and then determines how much more pollutant can be accommodated. In contrast to the United States, Canada sees a value to unused carrying capacity and accordingly believes it should be used. Another forseeable Canadian characteristic is a readiness to negotiate, and thus we see provinces negotiating with the proposed polluter to determine what is an appropriate control objective for a given area, and then how much pollutant emission is reasonably allowable in order to reach or preserve this objective. This represents a difference in behavioral preference—negotiation (Canada) versus litigation (United States).

The merits of one approach over the other might well be argued, and many would undoubtedly come down rigidly on one side or the other. However, a case could be made that the standards approach is more suitable for the more heavily populated and polluted United States, while the objectives approach is more suitable for the much less densely populated and polluted land to the north, a land which is also still developing industrially. The problem arises at the border interface, however, for standards and objectives philosophies cannot coexist in the same place, leading to a bilateral dilemma.

On the Great Lakes, the United States has largely given in to the Canadian objectives approach, as it has also on the St. John and St. Croix rivers in Maine and New Brunswick. However, the two nations are at odds over virtually all of the other seventy or so significant transboundary rivers. One Canadian federal view holds that if objectives were in place before differences arose, both upstream and downstream countries could plan river use with assurance that international conflicts over water quality would not arise. The upstream country could then plan and would approve only those projects which would not collectively violate known objectives at the border. Likewise, the country downstream would be assured an agreed water quality as the river crossed the border and could plan accordingly.[5] Overall adoption of one approach or the other, although politically unlikely, would solve the dilemma. However, given the U.S. EPA's opposition to the Canadian system, an appropriate approach outside the Great Lakes might be a compromise whereby the U.S. approach is supported on some transboundary river systems, the Canadian approach on other unconnected systems. As long as there is consistency within systems, problems would be avoided. However, as the United States moves into a new era vis-à-vis government regulation, it may well be that EPA will begin to consider a shift to the more flexible Canadian approach. Such a move would certainly gain the support of U.S. industry and many of the states along the border in addition to alleviating the bilateral problem.[6]

Agreeing to a uniform approach to water pollution control cannot alone solve bilateral environmental problems. Such alone does not address the causes of such problems, nor provide for agreement on the facts. Nor does it provide remedies acceptable to the two governments. It is only a method of reducing the degree of difference by introducing a uniform approach to water quality management.

TRANSBORDER LITIGATION

Demands for equal access to the courts of one nation by the citizens of the other are now reaching crescendo, as are demands for provisions for general private litigation across the border, including established rules of liability and compensation for damages.

In general, U.S. courts have been more tolerant of Canadian-originated litigation than have been the more restrictive Canadian courts. This stems in part from the differing philosophies of the two peoples toward openness in government and rights of private citizens to participate actively in the process rather than leave the full task to their elected representatives. Given the relatively more closed Canadian system, equal reciprocity has not been possible, and thus provision for cross-border private litigation has been slow in coming. The demand is increasing, however, and more research and dialogue are now being devoted to this subject.

The willingness of Canadian courts to exercise jurisdiction in transboundary environmental cases is unclear.[7] It is suspected that a suit by U.S. plaintiffs in Canadian courts to recover damages to U.S. property would fail since Canadian courts would hold they lack power to hear such cases.[8] This stands contrary to the Organization for Economic Cooperation and Development's (OECD) nonbinding recommendation that "member countries remove the obstacles that prevent foreigners injured by transboundary pollution from having access to the members' administrative and judicial systems."[9] Canada is a member of OECD.

U.S. states, acting on behalf of their citizens, have legally binding "long-arm" statutes. Their position was strengthened by the U.S. Supreme Court in the case of *Ohio* v. *Wyandotte Chemical Corporation,* which created a precedent for a state court to assume jurisdiction over a foreign defendant who is causing transboundary pollution injuries.[10] The U.S. Foreign Sovereign Immunities Act of 1976 was in fact enacted to prevent inconsistent application of sovereign immunity in U.S. courts in suits by U.S. citizens against foreign nations.[11] The act defined the conditions under which a foreign nation is subject to U.S. court jurisdiction and granted the judiciary the authority to determine whether these conditions exist.[12] This would be used as a method to establish jurisdiction, with attachment of the defendant's property as the method of enforcement, provided that the plaintiff can prove, by relying on principles of international law that Canada—the foreign country—has breached a duty toward U.S. citizens.[13]

The Bar Associations of Canada and the United States (ABA-CBA) in 1979 jointly published a draft treaty on equal access and remedy in cases of transfrontier pollution.[14] Such a treaty would include a statement of the rights of persons affected, the rights of public and private organizations, provision for notice to persons in the exposed country, and a limitation of rights granted such that

> in no event shall the provisions of this Treaty be construed as granting, *per se,* any greater rights to persons resident or incorporated in the exposed Country than those enjoyed by persons of equivalent condition or status resident or incorporated in the Country of origin.[15]

This ABA-CBA joint effort has thus far fallen on deaf ears in government, but it is an important step toward achieving greater international rights for citizens.

At present, private citizens may encounter at least two obstacles in suing a foreign polluter: jurisdiction, and choice of law.[16] Jurisdictional questions include whether the suit is to be filed where the injury occurred or where the injury-causing act occurred.[17] A choice must be made as to the best law under which to file suit, for some U.S. statutes provide for jurisdiction over an absent defendant and some do not.[18]

There has clearly been a significant trend, in North America and elsewhere, toward settlement of international environmental incidents in terms of civil liability rather than under principles of state responsibility.

This trend will likely continue. There has been a shift of emphasis away from public international dispute settlement and toward adjudication in a local forum. The feasibility of such adjudication in a local forum today is questionable. Thus the consideration of the modification of legal structures in Canadian-U.S. relations to accommodate this shift and resulting new direction is imperative.[19]

INTERNATIONAL ENVIRONMENTAL MEDIATION

Interest in the application of labor mediation techniques to the solution of various environmental problems has been rising in recent years, particularly in the United States, and the bilateral environmental area has not escaped this interest.[20] Both diplomatic communities have been understandably reluctant to embrace this concept in international relations, however. They undoubtedly view it as a way to lose managerial control once a situation has been turned over to a mediator. When one realizes the extremely local nature and very small stakes involved in some of these transborder issues, it is a wonder that international diplomats want to devote their time to these often hard to solve problems instead of welcoming the offer of mediation. However, there is precedent to be considered, for once one issue has been submitted to mediation, demands may increase to submit other more significant issues to mediation as well. While recognizing its potential costs as well as benefits, the mediation approach may be worth a try.[21]

SUBNATIONAL RELATIONS

Constitutional and legislative obstacles exist in both the United States and Canada to the conduct of foreign relations by subnational governmental entities. This issue is an especially sensitive one in Canada, where the power and interests of the provinces present a challenge to Ottawa in this field. The problem is further exacerbated by U.S. recognition that so many of the powers which Washington holds are held north of the border by the provinces. This suggests that there would be a preference for U.S. diplomats and environmental officials to develop direct relations with the provinces if an Ottawa-approved structure could be worked out. Occasionally, U.S. states (for example, Michigan) also strike out on their own, but this is less frequent. Nevertheless, structures are developing, with or without federal blessing, to enable groups of states and provinces to get together periodically.[22]

This is clearly a burgeoning movement and holds promise and pitfalls for the future. In recent years the Canadian and U.S. embassies have each assigned one diplomat to work full-time with subnational governments. Such personnel provide a direct link for provincial and state governments to the diplomatic community. At times, these personnel may also act as devil's advocates in representing subnational interests to their fellow diplomats. Before this effort can be fully effective, however, answers are needed to at least three questions.

What is the nature of the subnational cross-border approach? It might be on any subject and conducted by heads of government, environmental agencies, or, less frequently, through legislative involvement.

What conditions provide an incentive or a disincentive for subnational approaches? Incentive is provided by fear of harm to one government's interests or, less frequently, by perception of an opportunity to achieve a mutually beneficial goal. Disincentive is provided by fear of subnational governments that they are violating federal law or when they lack confidence in the conduct of a transborder relationship.

What are the advantages and disadvantages of subnational relations as a tool for management and problem solving and what is their relationship to U.S. and Canadian foreign policy? Subnational relations offer the advantage of involving people who are much closer to the scene and presumably more knowledgeable as to the problems and opportunities. The disadvantage revolves around failure to consider broader national interests and possibly harming those interests.[23]

> The attempt to resolve difficult problems like . . . environmental protection subnationally may bog down in the same communication failures and bureaucratic infighting that has slowed [the solution of] similar problems in the past. On the other hand, the closeness of the parties and their common understanding of the daily life of the region and the shared needs of its people provide some hope that [Canada and the U.S.] can evolve a unique understanding . . . in a context of equals.[24]

MULTILATERAL RELATIONS AND INTERNATIONAL
ENVIRONMENTAL STANDARDS

The United States and Canada are both signatory to a host of international conventions and agreements, although there are a few which involve one or the other but not both. Of the rare international conventions that involve the environmental relationship, perhaps the principal one is the two countries' joint membership in the Organization for Economic Cooperation and Development (OECD) and the United Nations Economic Commission for Europe (ECE). This latter membership requires that both take all measures necessary to avoid sending air pollutants (and thus acid rain) into the long-range transport air stream. Thus they are both barred from harming each other, and each is barred from harming other signatory nations, notably western Europe. Accusations have been made that President Carter's decision to convert numerous oil-fired power plants to coal without pollution controls was a violation of the U.S. obligation under the ECE international global pollution agreement. Nova Scotia's determination to develop new coal sources and coal-fired power plants without pollution controls but with significant assistance from Ottawa may be a similar violation by Canada.[25]

The OECD appears to be moving in the direction of international environmental standards based upon the levying of emission taxes on pollutants released into the environment and the use of these taxes to finance international interdisciplinary commissions of experts. The task of these commissions would be to improve the environmental performance of existing polluting activities. Further, they would design guidelines for future development, assuring environmental responsibility in balancing growth and environmental protection. Thus they would be both corrective and future-oriented.[26]

Neither the United States nor Canada is close to adopting such a philosophy or concept domestically, so it should not be surprising that they have not moved in this direction bilaterally. Economist Anthony Scott, a former IJC commissioner, supports the institutional aspect of this approach with the establishment of special permanent binational or multinational agencies, but concludes that there is only a limited relationship between economic theory and international management of shared resources.[27] For reasons mainly of national sovereignty, he concludes that the application of economic principles to problems of transnational pollution is not feasible in many instances. It has not been politically feasible because of the difficulty of translating such principles into real world conditions, and also as a result of reluctance to use untried techniques. Such approaches, therefore, may well have to await the arrival of a new era.

NONGOVERNMENTAL INSTRUMENTS

It is possible that no bilateral relationship anywhere in the world has a higher level of involvement by the nongovernmental sector of society than Canadian-U.S. relations. This involvement includes corporate industry, labor unions, citizens organizations, environmentalists, scientific and professional organizations, trade organizations, academic institutions, and, of course, private citizens. In the environmental area, corporate industry and citizen environmental organizations are the principal nongovernmental actors, as described in chapter 1. Media are de facto actors but, since their responsibility is to report the news and not to make it, they would undoubtedly deny a formal acting role.

THE PLACE OF INDUSTRY

With certain exceptions, private industry (and indirectly the people who constitute the markets for the various products produced) finds itself caught in the middle of Canadian-U.S. transboundary environmental problems. Where do its interests fit into the picture? We live in a capitalist private enterprise society (with government intervention and modification) which must be profit-motivated if industry is to succeed, meaningful work is to be provided, and markets are to be served. Hence both countries must show clear interest in, and understanding of, the health and welfare and stability of the corporate sector, just as they are now beginning to demonstrate appropriate interest in the health and welfare of the natural environment, the quality of life, and human health. Both preoccupations are equally deserving of public concern, for both are inextricably intertwined with the future welfare and destiny of society as a whole.

Corporate problems are not limited in impact to the corporate sector of society—they affect all sectors. Thus, ways must be found to integrate the corporate sector into the process of management and decision making in the area of Canadian-U.S. environmental relations. It must be recognized that this sector has a stake in the relationship related both to its own best interests and to those of society as a whole. Further, it has a responsibility to participate, to serve as a key actor, in the general international environmental relationship above and beyond its own immediate specific interests along the border. Business and industry must embrace this responsibility, therefore, and society must recognize business and industry as an active and formal partner with government.

The case studies of Part 2 have demonstrated business and industry's varied involvement in current transboundary environmental problems. This involvement includes offshore oil transport by tanker and outer continental shelf hydrocarbons development in border regions in the Atlantic, Pacific and Arctic oceans. It includes usage of the Great Lakes in the broadest sense (shipping, consumption of the water, use of the water for processing and manufacturing, etc). It includes coal mining in the Rockies and metal mining in the North. And it includes all of those varied industries affected by the bilateral acid rain debate, from raw materials harvesting (timber cutting, coal mining, commercial fishing) to secondary processing (metal smelting, oil refining, pulp and paper making) to the energy producers (coal- and oil-fired power plants) that are the underpinning of many industries. Less directly but no less significantly, industry is affected by the outcome of diplomatic decision making in issues from Garrison (impacting future industrialization options in North Dakota and food processing in both North Dakota and Manitoba); to Poplar (industrial options in Saskatchewan); to Skagit–High Ross Dam (impacts on Seattle electricity rates and thereby industry); and others. Industry's stake is thus clear, but its role is less so. It shares with border and Great Lakes municipalities, crown corporations, and public interest and environmentalist organizations a non-role (in the formal diplomatic sense).

THE PLACE OF LABOR

Labor has played a small role directly and indirectly in the transboundary environmental area, but is affected by diplomatic decision making affecting employment in some areas. For example, decisions occasioned by diplomatic pressures in the acid rain issue could substantially impact jobs in Ohio coal mines, in Ohio and other U.S. midwestern industry dependent on present energy costs, and on employment in locales containing large metal smelters, such as Rouyn-Noranda, Quebec, Sudbury, Ontario, or Thompson and Flin Flon, Manitoba. Likewise, decisions under the Great Lakes Water Quality Agreements could impact employment in the steel industry at Hamilton, Thunder Bay, Cleveland, Gary (Indiana), and elsewhere. Likewise it could impact the automotive industry in Detroit and Windsor. Diplomatic pressures in the Great Lakes toxics area could impact employment in the basin's chemicals industry. Decisions regarding the international impact (or desirability) of large hydroelectric dams in the West and elsewhere can obviously impact the opportunity for employment in a number of industries in varied areas.

In spite of these employment effects, labor has rarely been involved in transboundary issues per se. It has been involved domestically in each country in these issues, but not in their diplomatic or transboundary context. A new role for labor in the transboundary environmental area may be developing with recent events in the Garrison Diversion issue. The North Dakota branch of the AFL-CIO has met with the Manitoba Federation of Labor and is taking unprecedented initiative in getting the two unions to join together to pressure federal, provincial, and state officials to sit down and discuss the problems of the project. They also agreed that their technical and research committees would work jointly on the solution of the problem. North Dakota labor has interests in the enhanced industrialization of North Dakota which would result from the water made available from the project. Manitoba labor interests are specifically concerned about the ability of Manitoba's Souris River Valley to attract food processing plants and associated jobs. They see Garrison as the threat to this achievement.

Should a method be developed for these institutions to participate formally in the conduct of the relationship? If so, how might this participation be instituted?

Given that these actors play a major and ever-increasing role in the bilateral relationship, it might behoove diplomats and environmental officials to seek open, continuous, and formalized linkages with these sets of actors. This would enable them to gain more exposure to a greater variety of interests. Corporate, labor, academic, and environmental interests would provide diplomats with more direct linkage to the thinking of regions and might well achieve better understanding of national needs and differences in both countries.

There are, of course, constitutional obstacles in both countries to direct participation in international negotiation by private citizens. Allowance must also be made for the need for official diplomatic secrecy and confidentiality, sometimes abused but nevertheless necessary, which preclude full involvement of private citizens. However, options are available to guarantee such private involvement and input. The International Joint Commission has demonstrated the utility of nongovernmental scientific involvement through its Great Lakes Science Advisory Board, which includes academics, among others. The interest in public participation during the 1970s, with the establishment of literally thousands of advisory committees to all levels and functions of government, provides numerous examples of this approach.

Thus consideration might be given to the establishment of some type of formal advisory committee in the general area of Canadian-U.S. environmental relations or to the establishment of several such committees serving different aspects of the environmental relationship (international air quality, boundary water allocation, water quality, or subdivision by basin or region, such as prairies, marine and coastal, Great Lakes, and so on). Such advisory committees could be composed of a balanced cross-section of businessmen and industrialists (representing a type of industry or regional collectivity rather than an individual firm); labor; environmentalists (representing a coalition of organizations); academics from a variety of disciplines and types of institutions; and representatives of public affairs groups such as the U.S. League of Women Voters or regional World Affairs Council. Such a group or groups could ideally meet with diplomats and environment officials, who are directly involved in negotiation, on a regular basis, thus assuring input in both directions on a continuing basis. Another technique would be the appointment of nongovernmental members from the foregoing categories to already existing IJC boards. This would insure the same type of cross-fertilization much more quickly through vehicles already functioning.

NEW INSTITUTIONS

. . . [T]here has been an overall failure of United States and Canadian statesmanship in developing the quantity and quality of institutions required for the two countries with so much mutual involvement. I say this . . . after many years of looking at U.S.-Canadian relations and wondering why we have not institutionalized a whole series of relationships which are now handled on an ad hoc, crisis-ridden basis, yielding to the headlines and to the politics and pressures of the moment.

One can understand the reluctance of the professional diplomat to institutionalize. Diplomacy has the virtue of privacy. . . . One can make a pretty good thesis that institutionalizing a situation, by increasing the level of accountability, reduces the role of privacy, and the role of the diplomat, at the end of the road. . . . I suggest that we are on the edge of the need to think through a network of permanent advisory and fact-finding bodies, in areas ranging from energy to investment, that will do for those sectors of our relationship what . . . the Boundary Waters Treaty has done for the very sensitive issue of fresh water.[28]

There are many who are not as optimistic about an institutional solution to current transborder environmental problems. Scott writes that the likelihood of establishing further transborder resource management agencies is remote. He cites examples of Canada's turning away from international involvement and toward national assertiveness and suggests that the IJC survives because it is a product of an earlier era before water resource issues became crucial to either country. (One may infer from this, therefore, that we could not establish the IJC today.) Scott further believes that the new global air pollution problem requires an international agency with greater powers and scope than anything now in existence. He suggests that

to be effective, such an agency would, with its mixed membership, have to be empowered to override and overrule the decisions of local, state, and provincial elected bodies with respect to their air pollution regulations. Furthermore, its effectiveness in the long run would depend on the extent to which it had affected industrial and residential location, with attendant effects on employment, transport, roads, and land values. Can one really conceive of either the United States or Canada accepting decision making by a non-elected body with a foreign component in its membership on such politically sensitive subjects?[29]

In contrast to this pessimism is the optimism of John Fraser and Maxwell Cohen, optimism born out of what they perceive to be the inevitability of events. Fraser, Canada's former environment minister, speaks of continental environmental policies (even when allusion to any kind of continental policy is very often anathema to Canadian politicians) and of the need for new "high tribunals" to adjudicate environmental disputes which cannot otherwise be settled.
Fraser has stated,

No doubt the long debate over a continental policy of resource use will continue, but there should no longer be any debate as to the essential necessity of a continental environmental policy. . . . If we are to achieve a continental environmental policy, each nation must make a domestic decision, not so much to surrender sovereignty, but to recognize that certain acts that may be totally legal within the traditional concept of sovereignty can cause such damage to another state that they amount to acts against the sovereignty of that state. . . . . Rather than limiting itself to border problems, this concept recognizes the common law principle that you must conduct yourself so that you do not harm your neighbor and recognizes that an act in one place can have harmful affect on a neighbor thousands of miles away.[30]

Realistically, he recognizes that

it is doubtful whether either country has accepted the proposition that its otherwise sovereign activities, apart from treaty or agreement, can be challenged in advance as potentially destructive of the environment of its neighbor. Until such a concept is translated into a joint international policy administered by an effective international mechanism, nations will continue to act unilaterally.[31]

Fraser considers international environmental degradation to be a breach of sovereignty[32] and suggests that to remedy it calls for an environmental accord between the two nations. He would mandate that any government project (including one aided by government) which can have an across-the-border effect must be approved by a "high tribunal." This mechanism would be totally public and would grant standing to citizens of both countries. Rules of procedure would need to be developed and an adversary process established to keep it open and accessible. An appeal system and mechanism for review would also be required, as would the power of amendment to enable it to adapt to changing times and conditions. A basic question to be answered is whether such a tribunal should issue binding orders or be advisory, recognizing that this involves the bottom line of sovereignty. Fraser suggests that a formula could be worked out whereby if a majority of one national section favored an order, then that country at least should be bound by it. He recommends that the finest people available must be appointed to the tribunal and the two nations must be prepared to go along with their findings. It should have the authority of a court and, importantly, be looked up to as such. He recommends that it have initiatory authority (which the IJC does not have), with a simple majority being sufficient to initiate, and that the departments of State and External Affairs should not play any role—it thus should be independent of the diplomatic community.[33]

The Fraser concept of a high tribunal, as well as the belief that environmental degradation is a violation of sovereignty, are departures from tradition and perhaps rather ambitious. It is well to remember, however, that John Fraser is not an ivory-tower academic or a pie-in-the-sky idealist but rather a practicing politician and a very successful one at that.

Maxwell Cohen has also been a consistent champion of an institutional solution to current disputes. Not only has he encouraged the role of the IJC, but he has been a strong advocate of applying the IJC concept more broadly. For example, he has suggested a Joint Seaward Boundaries Commission "as indispensable for the continuing management of common environmental-resource situations offshore on the Atlantic, the Pacific and the Arctic."[34] Opting for institutionalization, he remarks,

> The national interest now demands systematic and hard-headed exploration of procedures to stabilize Canadian-U.S. relations in the many areas where the present knee jerk response is clearly not enough.[35]

Cohen, as an international legal scholar, is particularly concerned about ways of overcoming the U.S. constitutional problem of required Senate ratification of executive branch treaties and Canadian constitutional problems inherent in federal-provincial relations. Viewing both sides, he points out,

> It is a nice question whether the U.S. Senate "partnerships" with the Executive in treaty-making presents greater difficulties than the limitations on Canadian federal power that requires provincial legislative cooperation to make Canada's international obligations effective in the resource areas.[36]

Further institutionalization, with established rules of procedure for the institutions created, is one way around these obstacles.[37]

One further institutional initiative to be mentioned is the joint proposal of the Canadian (CBA) and American Bar associations (ABA) to establish an arbitration commission, or tribunal, to make judgment on what the lawyers call intractable international problems. Its decisions would be binding on all governments, which would presumably require a treaty. What this proposal ignores, however, is that the IJC is itself empowered under Article X of the Boundary Waters Treaty of 1909 to sit as a panel of arbitration and issue binding decisions in that capacity. However, in all the IJC's history, neither government has yet seen fit to use it in this way. Therefore, is there any reason to believe that the governments would use the CBA-ABA–proposed arbitral body to solve any dispute? It is rather unlikely. The only difference in the bar association proposal is that a new body separate from the commission would be formed, but any prediction as to its chance of success given the unwillingness to use even the relatively innocuous IJC for this purpose must be pessimistic indeed.[38]

What impact would Fraser's high tribunal, Cohen's Joint Seaward Boundaries Commission, or the CBA-ABA's arbitration commission have on the resolution of the transboundary environmental problems discussed in Part 2? Assuming such an institution had the full confidence of both governments, one might safely assume that none of these issues would have experienced the frequently significant delay, vacillation, and non-definitive conclusions that have characterized so many of them to date. Diplomats would probably not have lost their ultimate authority to review the outcome of these issues but would have had a much lesser role to play in their resolution. Interests affected by the disposition of the individual cases, whether subnational governments, private corporations, or individual citizens, would have had clearer knowledge of what to expect at each stage of the process (i.e., reduced uncertainty). Finally, assuming that the tribunal or commission had the respect and confidence of society in general as well as the full support of the two governments, its findings and decisions would be not only palatable but acceptable, in contrast to the status quo where clear decisions, when they are made, rarely receive broad acceptance.

It is not possible, of course, to prove that such institutions could solve all or even a majority of the transboundary environmental issues described in the case study chapters. Much would depend on the structure of the body, the makeup of its personnel, the support both governments were willing to give to it, and, perhaps most importantly, the willingness of society to invest in such an endeavor to enable it to be deserving of broad faith and confidence. The mere establishment of more institutions is insufficient in and of itself.

AN ENVIRONMENTAL TREATY, AGREEMENT, OR ACCORD

All of the preceding problems of process and technique are elements which would have to be dealt with if an environmental treaty, agreement, or accord were to be struck between the United States and Canada. Likewise, the concepts and mechanisms described in the preceding institutional section merit consideration as vehicles to implement the provisions of any such treaty, agreement, or accord. Any of these alternatives would be difficult to achieve, and the chances of establishing such a mechanism would be dependent on the public perception of the costs now being incurred or likely to be incurred in its absence.

We have seen that a boundary waters treaty strongest on apportionment of water at the border, but with clear water quality and some air quality jurisdiction, already exists for the interior and freshwater environments. We have seen that some progress is now being made toward an air quality agreement. We have seen that the demand is increasing for a solution of numerous marine disputes. The way of achieving a strong general treaty may ultimately be found in a long-term pooling of narrower agreements, both those that exist in the inland environments and those that may be developed in the marine environment. This assumes that a sufficient number of these narrower agreements are in fact implemented and that Canadian provincial objections can be overcome and U.S. congressional approval obtained.

Some of the foregoing are ambitious proposals, but the findings of this study suggest that the costs of maintaining the status quo are prohibitive.

Summary and Review of Findings

The major findings of this study may be grouped under six headings. These categories relate to:

- the nature of the bilateral environmental relationship and the nature of the issues;
- the Boundary Waters Treaty and the International Joint Commission;
- the response of the Canadian and U.S. federal governments to these issues and to each other;
- the impact of provinces and states on bilateral environmental relations;
- the impact of nongovernmental actors and institutions;
- and the future role of bilateral environmental disputes in the overall Canadian-U.S. relationship.

THE NATURE OF THE BILATERAL ENVIRONMENTAL
RELATIONSHIP AND THE NATURE OF THE ISSUES

Canadian-U.S. bilateral environmental relations are a microcosm of the bilateral relationship as a whole. Both nations are dependent upon and vulnerable to each other, although the United States has a greater capacity to affect (and inflict harm upon) Canada than vice versa. Canada is much more aware of this interdependence and assigns higher priority to it than does the United States. Finally, what are to most U.S. residents rather unimportant and regional problems are often to Canadians matters of extremely high import and of national significance. Hence, there is a bilateral imbalance even greater than what geographic or demographic differences might suggest. Changing historical trends and the evolution of attitudes at different rates in the two countries further exacerbate these differences.

Transboundary environmental disputes between the two countries are dominated by water apportionment, water quality, and air quality issues. At risk are not only environmental and ecological values but also energy self-sufficiency, economic development, jobs, and inflation. Both nations impact each other, although the magnitude of the U.S. impact on Canada is greater than vice versa. Issues vary from local border differences with direct impacts limited to the local border region (though often with wider implications) to those with broad continental impacts, or at least perceived as having such impacts. All of the issues have economic development components, and nearly all have energy components. Disputes have occurred in all parts of the continental and maritime transborder regions, and will continue to develop in all regions. From a predominance of water apportionment disputes in the early history of the environmental relationship, the focus has now shifted to environmental quality types of disputes. Air quality problems are likely to play an increasingly important role.

The trend is clearly toward increased numbers of disputes. It is toward greater complexity and magnitude of impact on greater numbers of people and on broader areas (including regions removed from the border). The stakes are now much higher. Energy will continue to be a central focus, and the apportionment of good quality water will likely reemerge as a significant cause of problems in coming years. New issues will emerge in yet to be developed regions, particularly along the Alaska–British Columbia–Yukon frontier. Differing political priorities and differing attitudes toward government regulation will ensure further exacerbation of all of these issues and of the environmental relationship in general.

THE BOUNDARY WATERS TREATY AND THE
INTERNATIONAL JOINT COMMISSION

The Boundary Waters Treaty of 1909 was a remarkable document for its day, in that it showed foresight in identifying pollution as a future bilateral problem. It was further significant in that it represented a rare surrender of sovereignty by a major power, the United States. The treaty was written in a very broad manner, and can be used to accomplish many tasks. Likewise its vehicle, the IJC, has a broad mandate, as long as the governments are willing to entrust it with responsibility appropriate to its mandate.

The International Joint Commission has experienced over seventy years of extraordinary success in operating as a single unitary agency supported by two national governments, and has achieved and maintained a collegial behavior honoring the spirit of the treaty. With few exceptions, its commissioners have impartially served the broad interests of the bilateral relationship as embodied in the Boundary Waters Treaty of 1909, rather than the narrow partisan interests of the government of the day. For the most part the two governments have respected this separate identity of the commission. Nevertheless, there have been problems, ranging from government's failure to use the commission to its full potential (including failure to use the arbitral function); to the writing of unnecessarily vague or sometimes overly restrictive references to the commission, and failure to support the commission with sufficient resources; to feelings of distrust between commissioners and diplomats; to tardiness of the commission in doing its work, occasional errors in judgment, and having such pride in its

record that it is occasionally unwilling to admit to differences along national lines. In addition to the remedying of these larger problems, international reforms of a specific nature are called for to insure a better product
from the IJC process, most notably a method of broader input into the
decision-making processes of the boards of inquiry upon which the commission is dependent.

Neither the treaty nor the commission has been used to its fullest
potential. It is quite conceivable that the model established can be used
both to provide a foundation for a new expanded environmental treaty or
agreement and likewise to provide a foundation for formalizing the relationship in other areas (e.g., fisheries, sea and shore boundaries, energy,
and even trade). Such usage, however, requires greater faith by both
governments than has heretofore been demonstrated in the utility of such
an instrument.

THE RESPONSE OF THE TWO FEDERAL GOVERNMENTS
TO THE ISSUES AND TO EACH OTHER

In earlier years the United States was most interested in adherence to the
Harmon Doctrine relative to water apportionment and in protecting its
upstream rights under this doctrine. Both nations were interested in max-
imizing joint collaboration on large-scale resource development projects.
Neither was particularly concerned about environmental quality or transborder pollution.

With rapidly rising expectations over environmental protection in the
United States and formalization of those expectations in legislation and
institutions, U.S. residents became concerned over increasing Canadian
development of border region resources. For a time in the 1970s, the
United States championed environmental protection initiatives against a
host of actual and planned Canadian near-border developments that impacted, or were thought to impact, U.S. environments. The United States
also took note of Canada's lack of binding environmental protection statutes, refusal to mandate sulfur scrubbers and other expensive controls to
curb pollution, and its general interest in frontier development, seemingly
at all costs.

By the dawn of the 1980s, however, positions had changed and the
Canadian defensive reaction to the U.S. environmentalist thinking of the
1970s changed to offensive efforts to curb much greater perceived U.S.
threats against Canada's environment. The United States then entered a
period of reaction and response which continues to this day. Likewise, the
increased U.S. emphasis on regulation in the 1960s and 1970s has been
reversed at the same time as belief in regulation (especially at the federal
level) appears to be increasing in Canada.

The U.S. government's environmental response to Canada now reflects the general U.S. attitude of unawareness and lack of interest. The

Canadian government's environmental response to the United States reflects the much higher priority Canada assigns to U.S. relations. There is even awareness in Canada that these bilateral environmental issues cannot be contained and will spill over into other vital areas of diplomatic relations.

THE IMPACT OF PROVINCES AND STATES ON
BILATERAL ENVIRONMENTAL RELATIONS

Both Canadian provinces and U.S. states represent their own interests by playing a role internally in influencing their respective federal governments on transboundary issues and externally in influencing events across the border. However, due to differences in the structure of the U.S. and Canadian systems of government, their role and impact is not equal— Canadian provinces are much more significant. At times the provincial and Ottawa governments work in consort forming a strong united Canadian front. At other times, the two are at odds, with the province either strongly pressuring Ottawa or resisting its pressure and thus presenting a divided front to the United States. Under certain circumstances strong Canadian federal-provincial differences can complicate and even further exacerbate international differences. Since the provinces in many respects have greater fundamental authority over air and water resources, there is even a temptation for the United States to seek to deal more directly with the provinces than with Ottawa. Occasionally, provinces also prefer to deal directly with the United States state or federal interests. They have the expertise and the resources to do so, if not the legal mandate.

The impact of U.S. federal-state differences on the transboundary environmental relationship is more subtle and often superseded by differences between the executive and legislative branches at the federal level. The latter, particularly as it involves the leadership of congressional delegations of affected states, can be an expression of federal-state differences.

As long as there are serious regional differences in either country and as long as the political power of the Canadian provinces remains strong, federal-subfederal relations problems will continue to complicate Canadian-U.S. transboundary environmental problems. The formal incorporation of provincial and state government into the process of negotiation and consultation from an early stage may be the only way of avoiding this at times serious problem.

THE IMPACT OF NONGOVERNMENTAL ACTORS
AND INSTITUTIONS

The most important nongovernmental actors are corporate industry and citizen environmentalist groups. Labor and academia play a lesser role.

The interests of the corporate sector are particularly at stake in the diplomatic disputes over acid rain, Great Lakes levels and water quality, toxic disposal, energy development at near-border locations, hydrocarbons development and transport in the maritime border regions and High Arctic, and water resource development in the dry interior. The corporate sector with its long-term planning and investment horizons has the most to gain from certainty and predictability. It has much to lose from the present system of "adhockery" and resultant unpredictability with its inherent risks. Corporate sector involvement in bilateral environmental issues has thus far been limited and narrow but will most certainly increase as more corporate interests are affected.

The citizen environmentalist sector is much stronger and better organized in the United States than in Canada. It is only very recently, however, with the advent of transborder threats to U.S. national parks and wilderness areas and with the advent of acid rain as a bilateral issue, that such groups became actively involved with the international (in contrast to the domestic) aspects of transboundary environmental issues. Environmentalists on both sides of the border are increasingly aware of each other and are increasingly involved in cross-border strategy and planning and the formation of alliances to achieve mutual goals. While there is every reason to expect that such cross-border relationships will continue to mature, the real basis of such relationships, of both national and regional groups, will be dominated by single issue interests, as such single issues develop along the border. When time and the right circumstances coalesce, the existence of an international border is soon forgotten and the groups work jointly. With Canadian federal and provincial government support of Washington and Toronto environmentalists on the acid rain issue, bilateral environmentalism has entered a new era as environmentalists now become a direct vehicle of government in the latter's effort to achieve a diplomatic goal.

From time to time various groups representing academia, labor, and other sectors of society become active bilaterally. This activity has yet to reach a level of significance, with the possible exception of academia's involvement on the Great Lakes water quality issues. As the stakes increase, the role of all nongovernmental interest groups will increase, and greater numbers of people in both countries will become much more aware of the Canadian-U.S. international environment.

THE ROLE OF BILATERAL ENVIRONMENTAL
DISPUTES IN FUTURE CANADIAN-U.S. RELATIONS

The ultimate question concerns the role of these transboundary environmental disputes in the overall Canadian-U.S. bilateral relationship. The disputes are increasing quantitatively. They are geographically widespread. They are increasing in complexity. The stakes involved in their settlement are rising dramatically. For the most part, these disputes are not being solved or even, for that matter, contained. They are leaving a residue of bad feelings in both countries, and particularly in Canada. They are leading to talk of retaliation. They are likely to spill over into other important areas of the bilateral relationship. They will thereby poison the historic feeling of friendship, admiration, and respect maintained by the peoples of both nations for many generations. Such a loss of goodwill inevitably will be costly to both nations, nations and peoples who clearly have a need for each other, who must share a continent and its commons, its air and its water. Maintaining the status quo, therefore, is unacceptable—a new approach must be tried. Given the furor over acid rain and continental air quality, continuing serious problems in the Great Lakes, increasing problems in the maritime border, and inevitably intensified competition for water in the continental interior, the time for new approaches is now. The irony is that there has long been a willingness to make a commitment. What is now needed is a willingness to implement that commitment.

THE CHOICE: FLEXIBILITY VERSUS STABILITY

Recognizing the advantages of flexibility inherent in an ad hoc approach to bilateral environmental relations, such as at present, and the advantages of stability in a more structured or institutionalized approach which could be put into place, one must question if we can afford the luxury of maintaining the present system. Has the present ad hoc approach already been overwhelmed? Can we continue to play without rules? Or, do we need stability, predictability, established rules of the road, clearer principles, and mechanisms to carry them out? It would appear, given the emergence on the current agenda of many and varied issues, together with the enormous potential for more issues lying ahead, that the present approach will not suffice, as desirable as it may have been in the past. Too many important issues are not being solved, will not go away, are creating frustration, bitterness, and even cynicism in some quarters, because of society's failure to achieve solution. The potential for serious and unnecessary damage looms large. *The only answer increasingly appears to be clear rules, known to a great number of people in both nations, by which the two governments and all other actors will play. Such rules would clearly set out the governance of all types of transboundary environmental issues and how all parties are to discharge their duties. Institutional mechanisms will perforce need to be established to implement such a goal. It is not the intent of this book to endorse a treaty, agreement, or specific mechanism, but it is intended to suggest that a more formalized solution appears inevitably to be needed. Would that this could be avoided, but can it be? The answer appears to be no.*

Institutions can be designed to achieve greater equality in the relationship. Such is certainly in Canada's best interest and gives that country an added incentive to overcome federal-provincial problems in the interests of both levels of government. It may well be that such equalization is in the long-range best interests of the United States too, given U.S. interest in a healthy economy and stable nation north of the border.

WHERE WE HAVE BEEN AND WHERE
WE ARE GOING

The preceding chapters have demonstrated that bilateral environmental issues or disputes arise either in the conventional near-border scenarios or in the increasingly common megaregional scenarios, such as in the case of acid rain. Wherever one nation is seen to enjoy all the benefits of a given development while the other nation is seen to bear only, although not necessarily all, the costs, the basis exists for a transboundary environmental problem. It has also been shown that issues are as likely to arise regarding water quantity or quality or air quality depending on the circumstances at hand. They may arise in as many different ways as there are issues:

- from diplomatic initiation (rare except when there are conventional sovereignty questions);
- from political initiation (when politicians are either under so much pressure from constituents that they cannot ignore a problem or when they see an opportunity to benefit politically by exploiting an issue);
- from private (and/or organized) citizen initiation when environmental, property, or health problems or fears are present or, increasingly, when jobs are at stake; or
- from media initiation (which technically does not occur, but the role of the media in focusing on an issue for an extended period of time can be so significant as to make it appear media-initiated).

Once initiated, issues have been seen to grow rapidly or slowly, reach a peak, deescalate for a short or long period, sometimes be resurrected and sometimes be put on the back burner indefinitely. Rarely do they appear to be resolved in the true sense of the word.

Diplomats, who have ultimate authority and responsibility in this area, often believe that the best they can hope to achieve, especially without jointly agreed upon rules, is management of the issue. In the words of one diplomat, "we do not resolve problems, we only manage them, and can hope to do no more."

The reason diplomats have such modest objectives is because this type of bilateral issue—the environmental—is more intractable than many other types. Intractability arises from the intangible nature of many of the values at stake and the failure of both societies to define these intangible values in adequate and manageable terms. For example, neither the United States nor Canada has yet come to grips with an acceptable definition of clean air or clean water or, for that matter, quality of life. This lack of ability to define is not only characteristic of the international level but is also present at the domestic level. Until an answer to the question "Clean for what?" can be given, the definition will elude us. Numbers associated with quality often cannot be readily assigned, thus true benefit-cost ratios cannot be developed and, under such circumstances, appeasement of the afflicted population or interests is all that can be expected.

The likelihood that these environmental disputes will increase in both number and intensity, not to mention complexity, in coming years is certain. Portions of the border which have not as yet experienced such problems will undoubtedly be drawn into the fray. Certain provinces and more states which presently have little or no experience in transborder environmental problems will soon be new actors, reluctantly or otherwise, to be joined by a host of previously uninvolved businesses and environmental groups. Diplomats trained for other, and perhaps more exotic, work in other parts of the globe will be called in and will attempt reluctantly to find solutions. Must we tolerate such problems? Must we accept the status quo? Can we afford to devote our already limited ecological, political, business, or diplomatic expertise to these questions when they could be more routinely disposed of, releasing this expertise to tackle matters of greater consequence in other areas of private and public affairs?

The preceding chapters have demonstrated constant behavior patterns on each side of the border (for example, a similarity of views toward water on the prairies, or universal interest in keeping future options open). They may have also, however, demonstrated that both countries are capable of changing direction. Canada has evolved from a highly prodevelopment-oriented nation to a development orientation tempered with environmental concern. Relatively speaking the United States evolved from prodevelopmental to proenvironmentalist but may be moving back again toward more prodevelopmental attitudes. Indeed, the two nations could well experience a reversal of roles in their national evolution. For these reasons it is impossible at this time to determine definitively if the actual differences in objectives are narrowing or widening. The answer depends on the type of issue under consideration, the regional location, and, most importantly, the time frame. Many have written or implied that Canada exists as a reaction to the United States. If true, the existence of a state of difference is mandated by the necessities of national pride and separate identity, but the degree of that difference is controlled by the realities of geography, history, and economics—in the environmental sphere as it is in all others.

A free and open border such as that separating the two great North American powers allows for the movement of both ideas and anxieties. This free movement encourages the development of approaches and solutions which can benefit either or both peoples. Canada's circumstances and needs may not only require differing approaches from those of the United

States. They may provide the latter with a distinctly different (though North American) window on the world.[39] In turn, the values the United States provides to Canada are varied, the most valuable of which may be the encouragement to Canada to develop, maintain, and appreciate its own separate identity.

The numbers and variations of environmental issues that can arise along the 5,525-mile Canadian-U.S. border are virtually unlimited, and will certainly tax the best efforts of those who attempt to resolve or even to manage them. More attention will clearly be needed if these disputes are to be prevented from eroding the good health of the general relationship. Specific techniques, such as the extraterritorial application of the U.S. National Environmental Policy Act (NEPA) process (with Canadian reciprocity), bilateral environmental impact statements, and even foreign policy impact statements may well merit serious consideration. In the foreign policy area it must be recognized that there are often foreign relations costs associated with these disputes, separate and apart from environmental and economic costs.

The experience of Canadian-U.S. environmental relations tells us that, as the issues become more intractable, we are able to agree upon fewer principles.[40] The issues simply do not allow it. Establishment of principles thus appears to be in indirect proportion to the intractability of issues, for the greater the intractability, the less chance for principle. There has been no evolution of principles in this area, leaving little likelihood that these problems will be solved in the future under current scenarios.

Can the United States afford the luxury of continuing its "take it for granted" pattern of behavior with respect to Canada indefinitely? Can it continue to afford its ignorance of affairs north of the border and the aspirations of the people who live there? Can Canada continue to afford the burden of responding as it must to these increasing and ever more serious environmental issues, of permitting them thereby to compete with and displace other bilateral issues critical to its well-being? Can industry continue to afford the costs of diplomatic involvement in their plans and endeavors, to risk being caught in the middle between the political positions of Washington, Ottawa, and various state and provincial governments? Can the border environment and population or, indeed, the North American ecosystem, continue to endure the ecological and health problems which are compounded by failure to achieve international solutions to these continuing disputes? These questions are yet to be answered.

The time has come for Washington to consider seriously a comprehensive, all-encompassing "Canada Policy." Such is a recognition of serious present and developing strains in the relationship, a recognition that the United States has too long ignored the health of the relationship. It is also a recognition that such a comprehensive policy is of great importance to the long-range interests of the United States and that the United States can continue to ignore Canada only at its own peril. Bilateral environmental relations would, of course, be part of such a "Canada Policy," and a change in the approach to containing these issues in the direction heretofore described might well be an important part of such a new vision in Washington. To fail to give serious consideration to such a policy may well be costly to the United States.

The words of economist Anthony Scott are especially apt at this juncture:

We must grow up to a new phase of our boundary relations, and be prepared to improve the environment while at the same time compensating the losers from the necessary steps and policies. The issue is whether Canada and the U.S. are going to depend on general goodwill or generosity to bring about improvements or deal with new threats, or are going to be more hard-nosed, demanding quality and paying for it.[41]

Transboundary environmental relations are going to be one of, if not the, major bilateral concern(s) in Canadian-U.S. relations during the remaining decades of this century. It behooves citizens of both nations to see that they do not get out of control.

NOTES

1. In order to understand and prepare for the consequences of alternative no. 3, a major conference of invited expertise from within and outside government might be convened. Such a conference would have as its purpose the identification and analysis of the impacts of this alternative. From this conference, a formal working group of scholars, diplomats, bureaucrats, businessmen, and citizen environmentalists might be assembled to plan and draft the outlines of the treaty document itself for submission to both societies. It would then be up to the people of Canada and the United States to decide the issue.

2. Maxwell Cohen, "The Patterns of Settlement—Canada, the United States and the International Joint Commission" (Remarks to the Conference Board in Canada, November 9, 1976), p. 13.

3. One diplomat has noted that foreign ministries are always outward-looking and don't pay attention to domestic matters, and domestic agencies are not concerned or aware of matters across the border. This gap must be filled by embassies and consulates. The central question in early warning and consultation, however, is timing.

4. One provincial official rejects early warning and consultation, believing that if you put it on an "as it happens" basis, something will slip through, while if you have a review meeting every six months to go over the border, you waste a lot of time. Canada has applied its greater diplomatic staff depth to the study of some potential problems which never materialize, a cost in dollars and manpower.

5. James Bruce, "Water Quality Issues in Boundary and Transboundary Waters" (Remarks to the American Society of Civil Engineers Conference, Hollywood, Florida, October 27–31, 1980), p. 4. Bruce further advocates instituting such water quality objectives before problems arise, for it is then much easier. He describes the method as follows: You look at the downstream uses and determine what quality you need to protect them. Then the upstream area agrees to provide water at the border of that agreed upon quality. Both upstream and downstream jurisdictions would get 50 percent of the assimilative capacity, and Bruce contends that this is the only way to deal with interjurisdictionally divided water resources. (James Bruce, Assistant Deputy Minister, Environment Canada, interview, Ottawa, October 20, 1980.)

6. Environment Canada's Ralph Pentland has cited a need for boundary water quality objectives at all streams on a bilateral basis, with IJC involvement where matters become so difficult that the diplomats cannot handle it (Ralph Pentland, Environment Canada, interview, Ottawa, September 18, 1980).

7. Donald C. Arbitlit, "The Plight of American Citizens Injured by Transboundary River Pollution," *Ecology Law Quarterly* 8, no. 2 (1979):339–70.

8. Ibid., p. 342.

9. Ibid., p. 343.

10. Ibid., p. 345.

11. The major goal of FSIA was to modify the so-called restrictive principle of sovereign immunity under which "the immunity of a foreign state is 'restricted' to suits involving a foreign state's public acts (jure imperii) and does not extend to suits based on its commercial or private acts (jure gestionis)." Ibid., p. 354.

12. Specifically, the act provides that a foreign state is subject to jurisdiction in an American court for actions based on the commercial activities of foreign states and for certain noncommercial torts.

13. Arbitlit, "The Plight of American Citizens," p. 359.

14. American Bar Association and Canadian Bar Association, *Settlement of International Disputes Between Canada and the USA* (Dallas, Texas and Calgary, Alberta: American Bar Association/Canadian Bar Association, 1979).

15. Ibid., p. xv.

16. Stephen C. McCaffrey, "Pollution of Shared Natural Resources: Legal and Trade Implications," (Proceedings of the 71st Annual Meeting of the American Society of International Law, San Francisco, California, April 21–23, 1977), p. 56.

17. Ibid.

18. 495 F. 2d 213 (6th Cir. 1974); Stephen C. McCaffrey, "Legal Remedies for Existing or Threatened Pollution Damage in Canada and the United States" (Study prepared for the Environment Directorate, Organization for Economic Cooperation and Development, June, 1979). The same author recently completed a major study on this subject for the Environment Directorate of the OECD, and the International Union for the Conservation of Nature and Natural Resources (IUCN) also completed its comprehensive document, *Survey of Current Developments in International Environmental Law* (Alexander Charles Kiss, IUCN Environmental Policy and Law Paper No. 10 [Morges, Switzerland: International Union for the Conservation of Nature and Natural Resources, 1976]).

19. On the subject of state-to-state compensation, there is less feeling today

that compensation and payoffs can take care of environmental damages. The Canadian and Manitoba governments could not likely survive politically if they accepted a payoff solution for Garrison; however, the chance of a structural compensation solution is likely greater than the chance of a monetary compensation solution.

20. Initiative in this direction has been taken on the U.S. side, one of the few instances of initiatives in Canadian-U.S. environmental relations originating south of the border. A Washington-based group, Environmental Mediation International (EMI) has been responsible, thus far unsuccessfully.

21. Robert Stein of Environmental Mediation International has stated that mediation may be appropriate and even preferred by the parties in circumstances where:

a. Each party affected by a claim considers a mutually satisfactory settlement to be in its self-interest;

b. Each party considers the need for such a settlement to be a matter of urgent priority;

c. Each party recognizes the need for, and is prepared to make, reasonable compromises to achieve a settlement because it believes that a mutually acceptable settlement serves its long-term self-interest. [Robert Stein, "Canada and the United States, Common Boundary/Common Problems, Environmental Consequences of Energy Production" (Paper presented to the American Bar Association/Canadian Bar Association Conference, Banff, Alberta, 1981), pp. 12–13.]

But, there are practical problems with mediation, including the protectiveness of government toward its own prerogatives, the conviction on the part of the aggrieved party that its rights are being violated, making it less compromising, and the desire of corporations to use their own lawyers in court. Mediation, to be utilized, would have to overcome all three.

22. The oldest and best known is the New England Governors–Eastern Canadian Premiers annual meetings, facilitated through the New England Regional Commission (since 1982 through the New England Governors Conference) and the Council of Maritime Premiers. Additionally, Washington and Alaska now meet annually with British Columbia and the Yukon Territory, and Ontario and the Great Lakes states and Alberta and the northern Rocky Mountain states (spearheaded by Montana) are moving in this direction. See also Kenneth M. Curtis, "Shaping the Canadian-United States Dialogue—A Truly Democratic Process," Paper presented to the Harvard Center for International Affairs, Harvard University, Cambridge, Massachusetts, April 27, 1982. These concepts are being further developed in a new book by Kenneth M. Curtis and John E. Carroll, tentatively titled *The Promise and the Challenge: Canadian-American Relations in the 80s*.

23. Questions modified after Patrick Heffernan, "Subnational Relations in Transborder Management Under the FMCA of 1976," Center for International Studies, Massachusetts Institute of Technology, 1980, p. 4.

24. Ibid., p. 37.

25. Based on interviews with U.S. State Department and Nova Scotia officials.

26. John D. Cumberland, "Establishment of International Environmental

Standards—Some Economic and Related Aspects,'' in *Problems in Transfrontier Pollution* (Paris: Organization for Economic Cooperation and Development, 1972), p. 225.

27. Anthony Scott, "Economic Aspects of Transnational Pollution," in *Problems in Transfrontier Pollution*, p. 14.

28. Maxwell Cohen, "The International Joint Commission—United States and Canada," in *Entente Cordiale? Part II: Bilateral Commissions and International Legal Methods of Adjustment*, Proceedings of the 68th Annual Meeting of the American Society of International Law, April, 1974, p. 237.

29. Anthony Scott, "Fisheries, Pollution, and Canadian-American Transnational Relations," *International Organization* 28, no. 4 (1974):847–48.

30. John A. Fraser, Remarks on a Continental Environmental Policy, n.d., pp. 11–12.

31. Ibid., pp. 12–13.

32. British Columbia law professor Charles Bourne disagrees, saying it depends on the extent of damage and the meaning of damage (interview, Vancouver, British Columbia, November 27, 1980).

33. John A. Fraser, M.P., interview, House of Commons, Ottawa, September 1, 1980 (and on numerous other occasions).

34. Maxwell Cohen, "Canada-U.S. Irritabilities are Rising," *Ottawa Journal*, n.d.

35. Ibid.

36. Maxwell Cohen, "Transboundary Environmental Attitudes and Policy—Some Canadian Perspectives" (Paper presented to the Harvard Center for International Affairs, Harvard University, Cambridge, Massachusetts, October 21, 1980), p. 55.

37. Cohen has noted that we now have three categories of techniques in bilateral diplomacy: regular diplomacy; increasing bilateral institutions; and managing the multiplicity of relations by nongovernmental entities, essentially a coordinative role. Much of the latter could conceivably be institutionalized into the second category (Maxwell Cohen, interview, Ottawa, September 10, 1980).

38. On the subject of process, attorney Roger Leed believes we must develop a mechanism to take into account extraterritorial environmental impacts, develop a process to formalize these considerations, and perhaps incorporate such in a protocol or treaty (interview, Seattle, Washington, December 3, 1980).

See also Great Lakes Fishery Commission, *Proposal: A Joint Strategic Plan for Management of Great Lakes Fisheries* (Ann Arbor: Great Lakes Fishery Commission, 1980), pp, 9–10. The Great Lakes Fishery Commission presents a model or process for joint management which can be adopted by existing institutions. It consists of an effort to identify obstacles to past efforts, suggest broad strategies of resolution, and propose procedures to initiate implementation. It recognizes that international, intergovernmental, and interagency consensus is necessary; that agencies must be considered; and that means of measuring and predicting effects of decisions must be developed. These four strategies, carried out by thirteen different strategic procedures, underlie proposed extension of the joint international managerial work of the commission and present to those considering new institutions a model for a technique which may optimize the chance for success in a difficult area of endeavor. See also John E. Carroll and Newell B. Mack, "On Living Together in North America: Canada, the United States, and International Environmental Relations," *Denver Journal of International Law and Policy* II, no. 1 (Winter, 1982).

39. Based on a concept developed by Dr. John S. Dickey, former president of Dartmouth College, in his book, *Canada and the American Presence* (New York: New York University Press, 1975).

40. The word *principles* as used here refers to principles or tenets of international behavior, which can be used to guide such behavior. Examples include equitable sharing of water along geographical or demographic criteria, provision of superior sovereignty rights to upstream states (e.g., Harmon Doctrine), reference of unsolvable disputes to arbitration, etc. Such principles have their basis in science, philosophy, law, and political and economic theory.

41. Anthony Scott, University of British Columbia, personal correspondence, Vancouver, British Columbia, December 29, 1980.

Afterword

Space limitations in this book did not permit treatment of all transboundary environmental issues. Perhaps most notable of those omitted must be two in the potential category: the once proposed Dickey-Lincoln Dam in northern Maine, which would inundate Quebec acreage and deprive Quebec loggers of some employment in the Maine woods while providing downstream electricity benefits (and minor water pollution) to New Brunswick; and the large 4,000 Mw Ontario Hydro Nanticoke coal-fired power plant, which will impact New York and Pennsylvania air quality but about which U.S. concerns are yet to be raised. In the historical category, two Maine–New Brunswick issues must be mentioned: the St. Croix River water pollution issue, now largely rectified under IJC auspices; and the St. John River flooding and water quality issues, which are now also largely under control with the help of the bilateral St. John River Water Quality Committee. The Lake Osoyoos pollution and flooding issue in Washington–British Columbia also deserves mention as one which is now essentially solved. Finally, the successful IJC management of the levels of Lake of the Woods (in Minnesota, Ontario, and Manitoba) to satisfy a diversity of interests, including hydro power generation, over a long period of time (since 1912—the commission's oldest effort) and amelioration of water quality problems on the nearby Rainy River and air pollution crossing the boundary in both directions at International Falls, Minnesota, and Fort Frances, Ontario, are deserving of note.

Some of these issues, such as the St. Croix River, St. John River, and Lake of the Woods issues, do constitute success stories in the resolution of Canadian-U.S. transboundary environmental problems. They are not earth-shaking events, but most would agree they have been satisfactorily resolved. It is possible, therefore, that they may offer lessons or provide models useful in the solution of larger, more complex issues. For these reasons, they deserve further study.

Appendixes

Boundary Waters Treaty of 1909

Chronology of Events Leading to the Boundary Waters Treaty of 1909

1783 *Treaty of Paris* conceded territorial jurisdiction of boundary waters on each side. The middle of the St. Lawrence from 45 degrees latitude, West, middle of Lakes Ontario, Erie, Huron and connecting channels up to Lake Superior became boundary lines which ran into Lake Superior, Lake of the Woods, etc.

1794 *Jay Treaty* provided for freedom of passage by land or water in respective territories and freedom of navigation and trade (1796 explanatory annex).

1814 *Treaty of Ghent* assigned Commissioners to clarify points about boundary lines. Their designations and decisions to be considered as "final and conclusive."

1817 *Rush-Bagot Agreement* limited armament on the Great Lakes, Lake Ontario and Lake Champlain.

1842 *Webster-Ashburton Treaty* provided ". . . all the water communications, and all the usual portages along the line from Lake Superior to the Lake of the Woods, and also Grand Portage from the shore of Lake Superior to the Pigeon River, as now actually used, shall be free and open to the use of the subjects and citizens of both countries."

"In order to promote the interests and encourage the industry of all the inhabitants of both countries watered by the River St. John and its tributaries . . . the navigation of the said river shall be free and open to both parties. . . ."

". . . the channels in the River St. Lawrence on both sides of Long Sault Islands and of Barnhart Island, the Channels in the Detroit River on both sides of the Island Bois Blanc, and between that Island and both the Canadian and American shores, and all the several channels and passages between the various islands lying near the junction of the River St.

From International Joint Commission, Vade Mecum (International Joint Commission, Ottawa, Ontario, Canada, 1969).

Clair with the lake of that name shall be equally free and open to the ships vessels and boats of both parties''.

1846 *Northwest, Boundary Treaty (Oregon Treaty)* gave British subjects right to navigate Columbia River to the Ocean.

1854 *Treaty* provided a reciprocal basis for U.S. navigation of Canadian part of the St. Lawrence and for British subjects of Lake Michigan (terminated 1866).

1871 *Treaty of Washington* provided navigation of the St. Lawrence "should forever remain free and open for the purpose of commerce . . ." to U.S., Britain and Canada; also for freedom of Welland, St. Lawrence and other Canadian canals to U.S. citizens, and U.S. granting of similar rights in St. Clair Flats Canal and influence with State governments for other canals to British subjects; also for free navigation of Rivers Yukon, Porcupine and Stikine.

1895 *Harmon Doctrine* (United States dispute with Mexico over Rio Grand waters) enunciated viz. no duty on U.S. to deny to its inhabitants use of water lying wholly within U.S. despite resultant reduction in volume below boundary—such duty being inconsistent with sovereign jurisdiction over national domain.

1905 *International Waterways Commission* appointed (for investigation only).

1906 *International Waterways Commission* (on Minnesota Canal and Power Company application) recommended principles governing use of boundary waters and creation of permanent commission with wider powers.

1906 First proposal (by George Gibbons) of one treaty to settle differences re all international rivers.

1907–8 Washington negotiations—Bryce-Root; Gibbons-Clinton and Anderson.

January 11, 1909, signature of *Boundary Waters Treaty* in Washington.

May 5, 1910, exchange of ratifications between U.S. and Britain.

1911 Boundary Waters Act; U.S. legislation 1910 and 1911.

January 10, 1912, first meeting of International Joint Commission.

TREATY
BETWEEN THE UNITED STATES AND GREAT BRITAIN RELATING TO BOUNDARY WATERS, AND QUESTIONS ARISING BETWEEN THE UNITED STATES AND CANADA.

The United States of America and His Majesty the King of the United Kingdom of Great Britain and Ireland and of the British Dominions beyond the Seas, Emperor of India, being equally desirous to prevent disputes regarding the use of boundary waters and to settle all questions which are now pending between the United States and the Dominion of Canada involving the rights, obligations, or interests of either in relation to the other or to the inhabitants of the other, along their common frontier, and to make provision for the adjustment and settlement of all such questions as may hereafter arise, have resolved to conclude a treaty in furtherance of these ends, and for that purpose have appointed as their respective plenipotentiaries:

The President of the United States of America, Elihu Root, Secretary of State of the United States; and

His Britannic Majesty, the Right Honourable James Bryce, O.M., his Ambassador Extraordinary and Plenipotentiary at Washington;

Who, after having communicated to one another their full powers, found in good and due form, have agreed upon the following articles:

PRELIMINARY ARTICLE

For the purposes of this treaty boundary waters are defined as the waters from main shore to main shore of the lakes and rivers and connecting waterways, or the portions thereof, along which the international boundary between the United States and the Dominion of Canada passes, including all bays, arms, and inlets thereof, but not including tributary waters which in their natural channels would flow into such lakes, rivers, and waterways, or waters flowing from such lakes, rivers, and waterways, or the waters of rivers flowing across the boundary.

ARTICLE I

The High Contracting Parties agree that the navigation of all navigable boundary waters shall forever continue free and open for the purposes of commerce to the inhabitants and to the ships, vessels, and boats of both countries equally, subject, however, to any laws and regulations of either country, within its own territory, not inconsistent with such privilege of free navigation and applying equally and without discrimination to the inhabitants, ships, vessels, and boats of both countries.

It is further agreed that so long as this treaty shall remain in force, this same right of navigation shall extend to the waters of Lake Michigan and to all canals connecting boundary waters, and now existing or which may hereafter be constructed on either side of the line. Either of the High Contracting Parties may adopt rules and regulations governing the use of such canals within its own territory and may charge tolls for the use thereof, but all such rules and regulations and all tolls charged shall apply alike to the subjects or citizens of the High Contracting Parties and the ships, vessels, and boats of both of the High Contracting Parties, and they shall be placed on terms of equality in the use thereof.

Reprinted courtesy of the International Joint Commission.

Each of the High Contracting Parties reserves to itself or to the several State Governments on the one side and the Dominion or Provincial Governments on the other as the case may be, subject to any treaty provisions now existing with respect thereto, the exclusive jurisdiction and control over the use and diversion, whether temporary or permanent, of all waters on its own side of the line which in their natural channels would flow across the boundary or into boundary waters; but it is agreed that any interference with or diversion from their natural channel of such waters on either side of the boundary, resulting in any injury on the other side of the boundary, shall give rise to the same rights and entitle the injured parties to the same legal remedies as if such injury took place in the country where such diversion or interference occurs; but this provision shall not apply to cases already existing or to cases expressly covered by special agreement between the parties hereto.

It is understood, however, that neither of the High Contracting Parties intends by the foregoing provision to surrender any right, which it may have, to object to any interference with or diversions of waters on the other side of the boundary the effect of which would be productive of material injury to the navigation interests on its own side of the boundary.

It is agreed that, in addition to the uses, obstructions, and diversions heretofore permitted or hereafter provided for by special agreement between the Parties hereto, no further or other uses or obstructions or diversions, whether temporary or permanent, of boundary waters on either side of the line, affecting the natural level or flow of boundary waters on the other side of the line shall be made except by authority of the United States or the Dominion of Canada within their respective jurisdictions and with the approval, as hereinafter provided, of a joint commission, to be known as the International Joint Commission.

The foregoing provisions are not intended to limit or interfere with the existing rights of the Government of the United States on the one side and the Government of the Dominion of Canada on the other, to undertake and carry on governmental works in boundary waters for the deepening of channels, the construction of breakwaters, the improvement of harbours, and other governmental works for the benefit of commerce and navigation, provided that such works are wholly on its own side of the line and do not materially affect the level or flow of the boundary waters on the other, nor are such provisions intended to interfere with the ordinary use of such waters for domestic and sanitary purposes.

The High Contracting Parties agree that, except in cases provided for by special agreement between them, they will not permit the construction or maintenance on their respective sides of the boundary of any remedial or protective works or any dams or other obstructions in waters flowing from boundary waters or in waters at a lower level than the boundary in rivers flowing across the boundary, the effect of which is to raise the natural level of waters on the other side of the boundary unless the construction or maintenance thereof is approved by the aforesaid International Joint Commission.

It is further agreed that the waters herein defined as boundary waters and waters flowing across the boundary shall not be polluted on either side to the injury of health or property on the other.

ARTICLE V

The High Contracting Parties agree that it is expedient to limit the diversion of waters from the Niagara River so that the level of Lake Erie and the flow of the stream shall not be appreciably affected. It is the desire of both Parties to accomplish this object with the least possible injury to investments which have already been made in the construction of power plants on the United States side of the river under grants of authority from the State of New York, and on the Canadian side of the river under licences authorized by the Dominion of Canada and the Province of Ontario.

So long as this treaty shall remain in force, no diversion of the waters of the Niagara River above the Falls from the natural course and stream thereof shall be permitted except for the purposes and to the extent hereinafter provided.

The United States may authorize and permit the diversion within the State of New York of the waters of said river above the Falls of Niagara, for power purposes, not exceeding in the aggregate a daily diversion at the rate of twenty thousand cubic feet of water per second.

The United Kingdom, by the Dominion of Canada, or the Province of Ontario, may authorize and permit the diversion within the Province of Ontario of the waters of said river above the Falls of Niagara, for power purposes, not exceeding in the aggregate a daily diversion at the rate of thirty-six thousand cubic feet of water per second.

The prohibitions of this article shall not apply to the diversion of water for sanitary or domestic purposes, or for the service of canals for the purposes of navigation.

NOTE: The third, fourth and fifth paragraphs of Article V were terminated by the Canada-United States Treaty of February 27, 1950 concerning the diversion of the Niagara River.

ARTICLE VI

The High Contracting Parties agree that the St. Mary and Milk Rivers and their tributaries (in the State of Montana and the Provinces of Alberta and Saskatchewan) are to be treated as one stream for the purposes of irrigation and power, and the waters thereof shall be apportioned equally between the two countries, but in making such equal apportionment more than half may be taken from one river and less than half from the other by either country so as to afford a more beneficial use to each. It is further agreed that in the division of such waters during the irrigation season, between the 1st of April and 31st of October, inclusive, annually, the United States is entitled to a prior appropriation of 500 cubic feet per second of the waters of the Milk River, or so much of such amount as constitutes three-fourths of its natural flow, and that Canada is entitled to a prior appropriation of 500 cubic feet per second of the flow of St. Mary River, or so much of such amount as constitutes three-fourths of its natural flow.

The channel of the Milk River in Canada may be used at the convenience of the United States for the conveyance, while passing through Canadian territory, of waters diverted from the St. Mary River. The provisions of Article II of this treaty shall apply to any injury resulting to property in Canada from the conveyance of such waters through the Milk River.

The measurement and apportionment of the water to be used by each country shall from time to time be made jointly by the properly constituted reclamation officers of the United States and the properly constituted irrigation officers of His Majesty under the direction of the International Joint Commission.

ARTICLE VII

The High Contracting Parties agree to establish and maintain an International Joint Commission of the United States and Canada composed of six commissioners, three on the part of the United States appointed by the President thereof, and three on the part of the United Kingdom appointed by His Majesty on the recommendation of the Governor in Council of the Dominion of Canada.

ARTICLE VIII

This International Joint Commission shall have jurisdiction over and shall pass upon all cases involving the use or obstruction or diversion of the waters with respect to which under Articles III and IV of this Treaty the approval of this Commission is required, and in passing upon such cases the Commission shall be governed by the following rules or principles which are adopted by the High Contracting Parties for this purpose:

The High Contracting Parties shall have, each on its own side of the boundary, equal and similar rights in the use of the waters hereinbefore defined as boundary waters.

The following order of precedence shall be observed among the various uses enumerated hereinafter for these waters, and no use shall be permitted which tends materially to conflict with or restrain any other use which is given preference over it in this order of precedence:

(1) Uses for domestic and sanitary purposes;

(2) Uses for navigation, including the service of canals for the purposes of navigation;

(3) Uses for power and for irrigation purposes.

The foregoing provisions shall not apply to or disturb any existing uses of boundary waters on either side of the boundary.

The requirement for an equal division may in the discretion of the Commission be suspended in cases of temporary diversions along boundary waters at points where such equal division can not be made advantageously on account of local conditions, and where such diversion does not diminish elsewhere the amount available for use on the other side.

The Commission in its discretion may make its approval in any case conditional upon the construction of remedial or protective works to compensate so far as possible for the particular use or diversion proposed, and in such cases may require that suitable and adequate provision, approved by the Commission, be made for the protection and indemnity against injury of any interests on either side of the boundary.

In cases involving the elevation of the natural level of waters on either side of the line as a result of the construction or maintenance on the other side of remedial or protective works or dams or other obstructions in boundary waters or in waters flowing therefrom or in waters below the boundary in rivers flowing across the boundary, the Commission shall require, as a condition of its approval thereof, that suitable and adequate provision, approved by it, be made for the protection and indemnity of all interests on the other side of the line which may be injured thereby.

The majority of the Commissioners shall have power to render a decision. In case the Commission is evenly divided upon any question or matter presented to it for decision, separate reports shall be made by the Commissioners on each side to their own Government. The High Contracting Parties shall thereupon endeavour to agree upon an adjustment of the question or matter of difference, and if an agreement is reached between them, it shall be reduced to writing in the form of a protocol, and shall be communicated to the Commissioners, who shall take such further proceedings as may be necessary to carry out such agreement.

ARTICLE IX

The High Contracting Parties further agree that any other questions or matters of difference arising between them involving the rights, obligations, or interests of either in relation to the other or to the inhabitants of the other, along the common frontier between the United States and the Dominion of Canada, shall be referred from time to time to the International Joint Commission for examination and report, whenever either the Government of the United States or the Government of the Dominion of Canada shall request that such questions or matters of difference be so referred.

The International Joint Commission is authorized in each case so referred to examine into and report upon the facts and circumstances of the particular questions and matters referred, together with such conclusions and recommendations as may be appropriate, subject, however, to any restrictions or exceptions which may be imposed with respect thereto by the terms of the reference.

Such reports of the Commission shall not be regarded as decisions of the questions or matters so submitted either on the facts or the law, and shall in no way have the character of an arbitral award.

The Commission shall make a joint report to both Governments in all cases in which all or a majority of the Commissioners agree, and in case of disagreement the minority may make a joint report to both Governments, or separate reports to their respective Governments.

In case the Commission is evenly divided upon any question or matter referred to it for report, separate reports shall be made by the Commissioners on each side to their own Government.

ARTICLE X

Any questions or matters of difference arising between the High Contracting Parties involving the rights, obligations, or interests of the United States or of the Dominion of Canada either in relation to each other or to their respective inhabitants, may be referred for decision to the International Joint Commission by the consent of the two Parties, it being understood that on the part of the United States any such action will be by and with the advice and consent of the Senate, and on the part of His Majesty's Government with the consent of the Governor General in Council. In each case so referred, the said Commission is authorized to examine into and report upon the facts and circumstances of the particular questions any matters referred, together with such conclusions and recommendations as may be appropriate, subject, however, to any restrictions or exceptions which may be imposed with respect thereto by the terms of the reference.

A majority of the said Commission shall have power to render a decision or finding upon any of the questions or matters so referred.

If the said Commission is equally divided or otherwise unable to render a decision or finding as to any questions or matters so referred, it shall be the duty of the Commissioners to make a joint report to both Governments, or separate reports to their respective Governments, showing the different conclusions arrived at with regard to the matters or questions so referred, which questions or matters shall thereupon be referred for decision by the High Contracting Parties to an umpire chosen in accordance with the procedure prescribed in the fourth, fifth and sixth paragraphs of Article XLV of the Hague Convention for the pacific settlement of international disputes, dated

October 18, 1907. Such umpire shall have power to render a final decision with respect to those matters and questions so referred on which the Commission failed to agree.

ARTICLE XI

A duplicate original of all decisions rendered and joint reports made by the Commission shall be transmitted to and filed with the Secretary of State of the United States and the Governor General of the Dominion of Canada, and to them shall be addressed all communications of the Commission.

ARTICLE XII

The International Joint Commission shall meet and organize at Washington promptly after the members thereof are appointed, and when organized the Commission may fix such times and places for its meetings as may be necessary, subject at all times to special call or direction by the two Governments. Each Commissioner upon the first joint meeting of the Commission after his appointment, shall, before proceeding with the work of the Commission, make and subscribe a solemn declaration in writing that he will faithfully and impartially perform the duties imposed upon him under this treaty, and such declaration shall be entered on the records of the proceedings of the Commission.

The United States and Canadian sections of the Commission may each appoint a secretary, and these shall act as joint secretaries of the Commission at its joint sessions, and the Commission may employ engineers and clerical assistants from time to time as it may deem advisable. The salaries and personal expenses of the Commission and of the secretaries shall be paid by their respective Governments, and all reasonable and necessary joint expenses of the Commission, incurred by it, shall be paid in equal moieties by the High Contracting Parties.

The Commission shall have power to administer oaths to witnesses, and to take evidence on oath whenever deemed necessary in any proceeding, or inquiry, or matter within its jurisdiction under this treaty, and all parties interested therein shall be given convenient oportunity to be heard, and the High Contracting Parties agree to adopt such legislation as may be appropriate and necessary to give the Commission the powers above mentioned on each side of the boundary, and to provide for the issue of subpœnas and for compelling the attendance of witnesses in procedings before the Commission. The Commission may adopt such rules of procedure as shall be in accordance with justice and equity, and may make such examination in person and through agents or employees as may be deemed advisable.

ARTICLE XIII

In all cases where special agreements between the High Contracting Parties hereto are referred to in the foregoing articles, such agreements are understood and intended to include not only direct agreements between the High Contracting Parties, but also any mutual arrangement between the United States and the Dominion of Canada expressed by concurrent or reciprocal legislation on the part of Congress and the Parliament of the Dominion.

ARTICLE XIV

The present treaty shall be ratified by the President of the United States of America, by and with the advice and consent of the Senate thereof, and by His Britannic Majesty. The ratifications shall be exchanged at Washington as soon as possible and the treaty shall take effect on the date of the exchange of its ratifications. It shall remain in force for five years, dating from the day of exchange of ratifications, and thereafter until terminated by twelve months' written notice given by either High Contracting Party to the other.

In faith whereof the respective plenipotentiaries have signed this treaty in duplicate and have hereunto affixed their seals.

Done at Washington the 11th day of January, in the year of our Lord one thousand nine hundred and nine.

(Signed) ELIHU ROOT   [SEAL]

(Signed) JAMES BRYCE   [SEAL]

AND WHEREAS the Senate of the United States by their resolution of March 3, 1909, (two-thirds of the Senators present concurring therein) did advise and consent to the ratification of the said Treaty with the following understanding, to wit:

"Resolved further, as a part of this ratification, That the United States approves this treaty with the understanding that nothing in this treaty shall be construed as affecting, or changing, any existing territorial or riparian rights in the water, or rights of the owners of lands under water, on either side of the international boundary at the rapids of the St. Mary's river at Sault Ste. Marie, in the use of the waters flowing over such lands, subject to the requirements of navigation in boundary waters and of navigation canals, and without prejudice to the existing right of the United States and Canada, each to use the waters of the St. Mary's river, within its own territory, and further, that nothing in this treaty shall be construed to interfere with the drainage of wet swamp and overflowed lands into streams flowing into boundary waters, and that this interpretation will be mentioned in the ratification of this treaty as conveying the true meaning of the treaty, and will, in effect, form part of the treaty;"

AND WHEREAS the said understanding has been accepted by the Government of Great Britain, and the ratifications of the two Governments of the said treaty were exchanged in the City of Washington, on the 5th day of May, one thousand nine hundred and ten;

Now, THEREFORE, be it known that I, William Howard Taft, President of the United States of America, have caused the said treaty and the said understanding, as forming a part thereof, to be made public, to the end that the same and every article and clause thereof may be observed and fulfilled with good faith by the United States and the citizens thereof.

In testimony whereof, I have hereunto set my hand and caused the seal of the United States to be affixed.

Done at the City of Washington this thirteenth day of May in the year of our Lord one thousand nine hundred and ten,

[SEAL]   and of the Independence of the United States of America the one hundred and thirty-fourth.

Wm H Taft

By the President:

    P C Knox
    Secretary of State.

On proceeding to the exchange of the ratifications of the treaty signed at Washington on January 11, 1909, between the United States and Great Britain, relating to boundary waters and questions arising along the boundary between the United States and the Dominion of Canada, the undersigned plenipotentiaries, duly authorized thereto by their respective Governments, hereby declare that nothing in this treaty shall be construed as affecting, or changing, any existing territorial, or riparian rights in the water, or rights of the owners of lands under water, on either side of the international boundary at the rapids of the St. Mary's River at Sault Ste. Marie, in the use of the waters flowing over such lands, subject to the requirements of navigation in boundary waters and of navigation canals, and without prejudice to the existing right of the United States and Canada, each to use the waters of the St. Mary's River, within its own territory; and further, that nothing in this treaty shall be construed to interfere with the drainage of wet, swamp, and overflowed lands into streams flowing into boundary waters, and also that this declaration shall be deemed to have equal force and effect as the treaty itself and to form an integral part thereto.

The exchange of ratifications then took place in the usual form.

In witness whereof, they have signed the present Protocol of Exchange and have affixed their seals thereto.

DONE at Washington this 5th day of May, one thousand nine hundred and ten.

PHILANDER C KNOX   [SEAL]

JAMES BRYCE   [SEAL]

Great Lakes Water Quality
Agreements of 1978 and 1972

AGREEMENT BETWEEN CANADA AND THE UNITED STATES OF AMERICA
ON GREAT LAKES WATER QUALITY, 1978

The Government of Canada and the Government of the
United States of America,

Having in 1972 entered into an Agreement on Great Lakes
Water Quality;

Reaffirming their determination to restore and enhance
water quality in the Great Lakes System;

Continuing to be concerned about the impairment of
water quality on each side of the boundary to an extent that is
causing injury to health and property on the other side, as
described by the International Joint Commission;

Reaffirming their intent to prevent further pollution
of the Great Lakes Basin Ecosystem owing to continuing population
growth, resource development and increasing use of water;

Reaffirming in a spirit of friendship and cooperation
the rights and obligations of both countries under the Boundary
Waters Treaty, signed on January 11, 1909, and in particular
their obligation not to pollute boundary waters;

Continuing to recognize the rights of each country in
the use of its Great Lakes waters;

Having decided that the Great Lakes Water Quality
Agreement of April 15, 1972 and subsequent reports of the
International Joint Commission provide a sound basis for new and
more effective cooperative actions to restore and enhance water
quality in the Great Lakes Basin Ecosystem;

Recognizing that restoration and enhancement of the
boundary waters can not be achieved independently of other parts
of the Great Lakes Basin Ecosystem with which these waters
interact;

Concluding that the best means to preserve the aquatic
ecosystem and achieve improved water quality throughout the Great
Lakes System is by adopting common objectives, developing and
implementing cooperative programs and other measures, and

assigning special responsibilities and functions to the
International Joint Commission;

Have agreed as follows:

ARTICLE I

DEFINITIONS

As used in this Agreement:

(a) "Agreement" means the present Agreement as distinguished from the Great Lakes Water Quality Agreement of April 15, 1972;

(b) "Annex" means any of the Annexes to this Agreement, each of which is attached to and forms an integral part of this Agreement;

(c) "Boundary waters of the Great Lakes System" or "boundary waters" means boundary waters, as defined in the Boundary Waters Treaty, that are within the Great Lakes System;

(d) "Boundary Waters Treaty" means the Treaty between the United States and Great Britain Relating to Boundary Waters, and Questions Arising Between the United States and Canada, signed at Washington on January 11, 1909;

(e) "Compatible regulations" means regulations no less restrictive than the agreed principles set out in this Agreement;

(f) "General Objectives" are broad descriptions of water quality conditions consistent with the protection of the beneficial uses and the level of environmental quality which the Parties desire to secure and which will provide overall water management guidance;

(g) "Great Lakes Basin Ecosystem" means the interacting components of air, land, water and living organisms, including man, within the drainage basin of the St. Lawrence River at or upstream from the point at which this river becomes the international boundary between Canada and the United States;

(h) "Great Lakes System" means all of the streams, rivers, lakes and other bodies of water that are within the drainage basin on the St. Lawrence River at or upstream from the point at which this river becomes the international boundary between Canada and the United States;

(i) "Harmful quantity" means any quantity of a substance that if discharged into receiving water would be inconsistent with the achievement of the General and Specific Objectives;

(j) "Hazardous polluting substance" means any element or compound identified by the Parties which, if discharged in any quantity into or upon receiving waters or adjoining shorelines, would present an imminent and substantial danger to public health or welfare; for this purpose, "public health or welfare" encompasses all factors affecting the health and welfare of man including but not limited to human health, and the conservation and protection of flora and fauna, public and private property, shorelines and beaches;

(k) "International Joint Commmission" or "Commission" means the International Joint Commission established by the Boundary Waters Treaty;

(l) "Monitoring" means a scientifically designed system of continuing standardized measurements and observations and the evaluation thereof;

(m) "Objectives" means the General Objectives adopted pursuant to Article III and the Specific Objectives adopted pursuant to Article IV of this Agreement;

(n) "Parties" means the Government of Canada and the Government of the United States of America;

(o) "Phosphorus" means the element phosphorus present as a constituent of various organic and inorganic complexes and compounds;

(p) "Research" means development, demonstration and other research activities but does not include monitoring and surveillance of water or air quality;

(q) "Science Advisory Board" means the Great Lakes Science Advisory Board of the International Joint Commission established pursuant to Article VIII of this Agreement;

(r) "Specific Objectives" means the concentration or quantity of a substance or level of effect that the Parties agree, after investigation, to recognize as a maximum or minimum desired limit for a defined body of water or portion thereof, taking into account the beneficial uses or level of environmental quality which the Parties desire to secure and protect;

(s) "State and Provincial Governments" means the Governments of the States of Illinois, Indiana, Michigan, Minnesota, New York, Ohio, Wisconsin and the Commonwealth of Pennsylvania, and the Government of the Province of Ontario;

(t) "Surveillance" means specific observations and measurements relative to control or management;

(u) "Terms of Reference" means the Terms of Reference for the Joint Institutions and the Great Lakes Regional Office established pursuant to this Agreement, which are attached to and form an integral part of this Agreement;

(v) "Toxic substance" means a substance which can cause death, disease, behavioural abnormalties, cancer, genetic mutations, physiological or reproductive malfunctions or physical deformities in any organism or its offspring, or which can become poisonous after concentration in the food chain or in combination with other substances;

(w) "Tributary waters of the Great Lakes System" or "tributary waters" means all the waters within the Great Lakes System that are not boundary waters;

(x) "Water Quality Board" means the Great Lakes Water Quality Board of the International Joint Commission established pursuant to Article VIII of this Agreement.

ARTICLE II

PURPOSE

The purpose of the Parties is to restore and maintain the chemical, physical, and biological integrity of the waters of the Great Lakes Basin Ecosystem. In order to achieve this purpose, the Parties agree to make a maximum effort to develop programs, practices and technology necessary for a better understanding of the Great Lakes Basin Ecosystem and to eliminate or reduce to the maximum extent practicable the discharge of pollutants into the Great Lakes System.

Consistent with the provisions of this Agreement, it is the policy of the Parties that:

(a) The discharge of toxic substances in toxic amounts be prohibited and the discharge of any or all persistent toxic substances be virtually eliminated;

(b) Financial assistance to construct publicly owned waste treatment works be provided by a combination of local, state, provincial, and federal participation; and

(c) Coordinated planning processes and best management practices be developed and implemented by the respective jurisdictions to ensure adequate control of all sources of pollutants.

ARTICLE III

GENERAL OBJECTIVES

The Parties adopt the following General Objectives for the Great Lakes System. These waters should be:

(a) Free from substances that directly or indirectly enter the waters as a result of human activity and that will settle to form putrescent or otherwise objectionable sludge deposits, or that will adversely affect aquatic life or waterfowl;

(b) Free from floating materials such as debris, oil, scum, and other immiscible substances resulting from human activities in amounts that are unsightly or deleterious;

(c) Free from materials and heat directly or indirectly entering the water as a result of human activity that alone, or in combination with other materials, will produce colour, odour, taste, or other conditions in such a degree as to interfere with beneficial uses;

(d) Free from materials and heat directly or indirectly entering the water as a result of human activity that alone, or in combination with other materials, will produce conditions that are toxic or harmful to human, animal, or aquatic life; and

(e) Free from nutrients directly or indirectly entering the waters as a result of human activity in amounts that create growths of aquatic life that interfere with beneficial uses.

ARTICLE IV

SPECIFIC OBJECTIVES

1.   The Parties adopt the Specific Objectives for the boundary waters of the Great Lakes System as set forth in Annex 1, subject to the following:

(a) The Specific Objectives adopted pursuant to this Article represent the minimum levels of water quality desired in the boundary waters of the Great Lakes System and are not intended to preclude the establishment of more stringent requirements.

(b) The determination of the achievement of Specific Objectives shall be based on statistically valid sampling data.

(c) Notwithstanding the adoption of Specific Objectives, all reasonable and practicable measures shall be taken to maintain or improve the existing water quality in those areas of the boundary waters of the Great Lakes System where such water quality is better than that prescribed by the Specific Objectives, and in those areas having outstanding natural resource value.

(d) The responsible regulatory agencies shall not consider flow augmentation as a substitute for adequate treatment to meet the Specific Objectives.

(e) The Parties recognize that in certain areas of inshore waters natural phenomena exist which, despite the best efforts of the Parties, will prevent the achievement of some of the Specific Objectives. As early as possible, these areas should be identified explicitly by the appropriate jurisdictions and reported to the International Joint Commission.

(f) Limited use zones in the vicinity of present and future municipal, industrial and tributary point source discharges shall be designated by the responsible regulatory agencies within which some of the Specific Objectives may not apply. Establishment of these zones shall not be considered a substitute for adequate treatment or control of discharges at their source. The size shall be minimized to the greatest possible degree, being no larger than that attainable by all reasonable and practicable effluent treatment measures. The boundary of a limited use zone shall not transect the international boundary. Principles for the designation of limited use zones are set out in Annex 2.

2.   The Specific Objectives for the boundary waters of the Great Lakes System or for particular portions thereof shall be kept under review by the Parties and by the International Joint Commission, which shall make appropriate recommendations.

3.   The Parties shall consult on:

(a) The establishment of Specific Objectives to protect beneficial uses from the combined effects of pollutants; and

(b) The control of pollutant loading rates for each lake basin to protect the integrity of the ecosystem over the long term.

ARTICLE V

STANDARDS, OTHER REGULATORY REQUIREMENTS, AND RESEARCH

1.   Water quality standards and other regulatory
requirements of the Parties shall be consistent with the
achievement of the General and Specific Objectives. The Parties
shall use their best efforts to ensure that water quality
standards and other regulatory requirements of the State and
Provincial Governments shall similarly be consistent with the
achievement of these Objectives. Flow augmentation shall not be
considered as a substitute for adequate treatment to meet water
quality standards or other regulatory requirements.

2.   The Parties shall use their best efforts to ensure
that:

    (a) The principal research funding agencies in both
         countries orient the research programs of their
         organizations in response to research priorities
         identified by the Science Advisory Board and
         recommended by the Commission; and

    (b) Mechanisms be developed for appropriate cost-effective
         international cooperation.

ARTICLE VI

PROGRAMS AND OTHER MEASURES

1.   The Parties shall continue to develop and implement
programs and other measures to fulfil the purpose of this
Agreement and to meet the General and Specific Objectives. Where
present treatment is inadequate to meet the General and Specific
Objectives, additional treatment shall be required. The programs
and measures shall include the following:

    (a) Pollution from Municipal Sources. Programs for the
         abatement, control and prevention of municipal
         discharges and urban drainage into the Great Lakes
         System. These programs shall be completed and in
         operation as soon as practicable, and in the case of
         municipal sewage treatment facilities no later than
         December 31, 1982. These programs shall include:

         (i)   Construction and operation of waste treatment
               facilities in all municipalities having sewer
               systems to provide levels of treatment consistent
               with the achievement of phosphorus requirements
               and the General and Specific Objectives, taking
               into account the effects of waste from other
               sources;

         (ii) Provision of financial resources to ensure prompt
               construction of needed facilities;

         (iii) Establishment of requirements for construction and
               operating standards for facilities;

         (iv) Establishment of pre-treatment requirements for
               all industrial plants discharging waste into
               publicly owned treatment works where such
               industrial wastes are not amenable to adequate
               treatment or removal using conventional municipal
               treatment processes;

         (v)   Development and implementation of practical
               programs for reducing pollution from storm,
               sanitary, and combined sewer discharges; and

(vi)　Establishment of effective enforcement programs to
　　　　　　 ensure that the above pollution abatement
　　　　　　 requirements are fully met.

(b)　Pollution from Industrial Sources. Programs for the
　　　abatement, control and prevention of pollution from
　　　industrial sources entering the Great Lakes System.
　　　These programs shall be completed and in operation as
　　　soon as practicable and in any case no later than
　　　December 31, 1983, and shall include:

　　　(i)　Establishment of waste treatment or control
　　　　　　 requirements expressed as effluent limitations
　　　　　　 (concentrations and/or loading limits for specific
　　　　　　 pollutants where possible) for all industrial
　　　　　　 plants, including power generating facilities, to
　　　　　　 provide levels of treatment or reduction or
　　　　　　 elimination of inputs of substances and effects
　　　　　　 consistent with the achievement of the General and
　　　　　　 Specific Objectives and other control
　　　　　　 requirements, taking into account the effects of
　　　　　　 waste from other sources;

　　　(ii)　Requirements for the substantial elimination of
　　　　　　 discharges into the Great Lakes System of
　　　　　　 persistent toxic substances;

　　　(iii)　Requirements for the control of thermal
　　　　　　 discharges;

　　　(iv)　Measures to control the discharge of radioactive
　　　　　　 materials into the Great Lakes System;

　　　(v)　Requirements to minimize adverse environmental
　　　　　　 impacts of water intakes;

　　　(vi)　Development and implementation of programs to meet
　　　　　　 industrial pre-treatment requirements as specified
　　　　　　 under sub-paragraph (a) (iv) above; and

　　　(vii)　Establishment of effective enforcement programs to
　　　　　　 ensure the above pollution abatement requirements
　　　　　　 are fully met.

(c)　Inventory of Pollution Abatement Requirements.
　　　Preparation of an inventory of pollution abatement
　　　requirements for all municipal and industrial
　　　facilities discharging into the Great Lakes System in
　　　order to gauge progress toward the earliest practicable
　　　completion and operation of the programs listed in
　　　sub-paragraphs (a) and (b) above. This inventory,
　　　prepared and revised annually, shall include compliance
　　　schedules and status of compliance with monitoring and
　　　effluent restrictions, and shall be made available to
　　　the International Joint Commission and to the public.
　　　In the initial preparation of this inventory, priority
　　　shall be given to the problem areas previously
　　　identified by the Water Quality Board.

(d)　Eutrophication. Programs and measures for the
　　　reduction and control of inputs of phosphorus and other
　　　nutrients, in accordance with the provisions of Annex
　　　3.

(e)　Pollution from Agricultural, Forestry and Other Land
　　　Use Activities. Measures for the abatement and control
　　　of pollution from agricultural, forestry and other land
　　　use activities including:

(i)   Measures for the control of pest control products used in the Great Lakes Basin to ensure that pest control products likely to have long-term deleterious effects on the quality of water or its biota be used only as authorized by the responsible regulatory agencies; that inventories of pest control products used in the Great Lakes Basin be established and maintained by appropriate agencies; and that research and educational programs be strengthened to facilitate integration of cultural, biological and chemical pest control techniques;

(ii)   Measures for the abatement and control of pollution from animal husbandry operations, including encouragement to appropriate agencies to adopt policies and regulations regarding utilization of animal wastes, and site selection and disposal of liquid and solid wastes, and to strengthen educational and technical assistance programs to enable farmers to establish waste utilization, handling and disposal systems;

(iii)   Measures governing the hauling and disposal of liquid and solid wastes, including encouragement to appropriate regulatory agencies to ensure proper location, design, and regulation governing land disposal, and to ensure sufficient, adequately trained technical and administrative capability to review plans and to supervise and monitor systems for application of wastes on land;

(iv)   Measures to review and supervise road salting practices and salt storage to ensure optimum use of salt and all-weather protection of salt stores in consideration of long-term environmental impact;

(v)   Measures to control soil losses from urban and suburban as well as rural areas;

(vi)   Measures to encourage and facilitate improvements in land use planning and management programs to take account of impacts on Great Lakes water quality;

(vii)   Other advisory programs and measures to abate and control inputs of nutrients, toxic substances and sediments from agricultural, forestry and other land use activities; and

(viii)   Consideration of future recommendations from the International Joint Commission based on the Pollution from Land Use Activities Reference.

(f)   Pollution from Shipping Activities. Measures for the abatement and control of pollution from shipping sources, including:

(i)   Programs and compatible regulations to prevent discharges of harmful quantities of oil and hazardous polluting substances, in accordance with Annex 4;

    (ii)   Compatible regulations for the control of discharges of vessel wastes, in accordance with Annex 5;

  (iii)   Such compatible regulations to abate and control pollution from shipping sources as may be deemed desirable in the light of continuing reviews and studies to be undertaken in accordance with Annex 6;

   (iv)   Programs and any necessary compatible regulations in accordance with Annexes 4 and 5, for the safe and efficient handling of shipboard generated wastes, including oil, hazardous polluting substances, garbage, waste water and sewage, and for their subsequent disposal, including the type and quantity of reception facilities and, if applicable, treatment standards; and

    (v)   Establishment by the Canadian Coast Guard and the United States Coast Guard of a coordinated system for aerial and surface surveillance for the purpose of enforcement of regulations and the early identification, abatement and clean-up of spills of oil, hazardous polluting substances, or other pollution.

(g)   Pollution from Dredging Activities. Measures for the abatement and control of pollution from all dredging activities, including the development of criteria for the identification of polluted sediments and compatible programs for disposal of polluted dredged material, in accordance with Annex 7. Pending the development of compatible criteria and programs, dredging operations shall be conducted in a manner that will minimize adverse effects on the environment.

(h)   Pollution from Onshore and Offshore Facilities. Measures for the abatement and control of pollution from onshore and offshore facilities, including programs and compatible regulations for the prevention of discharges of harmful quantities of oil and hazardous polluting substances, in accordance with Annex 8.

(i)   Contingency Plan. Maintenance of a joint contingency plan for use in the event of a discharge or the imminent threat of a discharge of oil or hazardous polluting substances, in accordance with Annex 9.

(j)   Hazardous Polluting Substances. Implementation of Annex 10 concerning hazardous polluting substances. The Parties shall further consult from time to time for the purpose of revising the list of hazardous polluting substances and of identifying harmful quantities of these substances.

(k)   Persistent Toxic Substances. Measures for the control of inputs of persistent toxic substances including control programs for their production, use, distribution and disposal, in accordance with Annex 12.

(l)   Airborne Pollutants. Programs to identify pollutant sources and relative source contributions, including the more accurate definition of wet and dry deposition rates, for those substances which may have significant adverse effects on environmental quality including the indirect effects of impairment of tributary water quality through atmospheric deposition in drainage basins. In cases where significant contributions to Great Lakes pollution from atmospheric sources are identified, the Parties agree to consult on appropriate remedial programs.

(m) <u>Surveillance and Monitoring.</u> Implementation of a coordinated surveillance and monitoring program in the Great Lakes System, in accordance with Annex 11, to assess compliance with pollution control requirements and achievement of the Objectives, to provide information for measuring local and whole lake response to control measures, and to identify emerging problems.

2.   The Parties shall develop and implement such additional programs as they jointly decide are necessary and desirable to fulfil the purpose of this Agreement and to meet the General and Specific Objectives.

ARTICLE VII

POWERS, RESPONSIBILITIES AND FUNCTIONS OF THE INTERNATIONAL JOINT COMMISSION

1.   The International Joint Commission shall assist in the implementation of this Agreement. Accordingly, the Commission is hereby given, by a Reference pursuant to Article IX of the Boundary Waters Treaty, the following responsibilities:

(a) Collation, analysis and dissemination of data and information supplied by the Parties and State and Provincial Governments relating to the quality of the boundary waters of the Great Lakes System and to pollution that enters the boundary waters from tributary waters and other sources;

(b) Collection, analysis and dissemination of data and information concerning the General and Specific Objectives and the operation and effectiveness of the programs and other measures established pursuant to this Agreement;

(c) Tendering of advice and recommendations to the Parties and to the State and Provincial Governments on problems of and matters related to the quality of the boundary waters of the Great Lakes System including specific recommendations concerning the General and Specific Objectives, legislation, standards and other regulatory requirements, programs and other measures, and intergovernmental agreements relating to the quality of these waters;

(d) Tendering of advice and recommendations to the Parties in connection with matters covered under the Annexes to this Agreement;

(e) Provision of assistance in the coordination of the joint activities envisaged by this Agreement;

(f) Provision of assistance in and advice on matters related to research in the Great Lakes Basin Ecosystem, including identification of objectives for research activities, tendering of advice and recommendations concerning research to the Parties and to the State and Provincial Governments, and dissemination of information concerning research to interested persons and agencies;

(g) Investigations of such subjects related to the Great
Lakes Basin Ecosystem as the Parties may from time to
time refer to it.

2.   In the discharge of its responsibilities under this
Reference, the Commission may exercise all of the powers
conferred upon it by the Boundary Waters Treaty and by any
legislation passed pursuant thereto including the power to
conduct public hearings and to compel the testimony of witnesses
and the production of documents.

3.   The Commission shall make a full report to the Parties
and to the State and Provincial Governments no less frequently
than biennially concerning progress toward the achievement of the
General and Specific Objectives including, as appropriate,
matters related to Annexes to this Agreement. This report shall
include an assessment of the effectiveness of the programs and
other measures undertaken pursuant to this Agreement, and advice
and recommendations. In alternate years the Commission may
submit a summary report. The Commission may at any time make
special reports to the Parties, to the State and Provincial
Governments and to the public concerning any problem of water
quality in the Great Lakes System.

4.   The Commission may in its discretion publish any
report, statement or other document prepared by it in the
discharge of its functions under this Reference.

5.   The Commission shall have authority to verify
independently the data and other information submitted by the
Parties and by the State and Provincial Governments through such
tests or other means as appear appropriate to it, consistent with
the Boundary Waters Treaty and with applicable legislation.

6.   The Commission shall carry out its responsibilities
under this Reference utilizing principally the services of the
Water Quality Board and the Science Advisory Board established
under Article VIII of this Agreement. The Commission shall also
ensure liaison and coordination between the institutions
established under this Agreement and other institutions which may
address concerns relevant to the Great Lakes Basin Ecosystem,
including both those within its purview, such as those Boards
related to Great Lakes levels and air pollution matters, and
other international bodies, as appropriate.

ARTICLE VIII

JOINT INSTITUTIONS AND REGIONAL OFFICE

1.   To assist the International Joint Commission in the
exercise of the powers and responsibilities assigned to it under
this Agreement, there shall be two Boards:

(a) A Great Lakes Water Quality Board which shall be the
principal advisor to the Commission. The Board shall
be composed of an equal number of members from Canada
and the United States, including representatives from
the Parties and each of the State and Provincial
Governments; and

(b) A Great Lakes Science Advisory Board which shall
provide advice on research to the Commission and to the
Water Quality Board. The Board shall further provide
advice on scientific matters referred to it by the
Commission, or by the Water Quality Board in
consultation with the Commission. The Science Advisory

Board shall consist of managers of Great Lakes research programs and recognized experts on Great Lakes water quality problems and related fields.

2.   The members of the Water Quality Board and the Science Advisory Board shall be appointed by the Commission after consultation with the appropriate government or governments concerned. The functions of the Boards shall be as specified in the Terms of Reference appended to this Agreement.

3.   To provide administrative support and technical assistance to the two Boards, and to provide a public information service for the programs, including public hearings, undertaken by the International Joint Commission and by the Boards, there shall be a Great Lakes Regional Office of the International Joint Commission. Specific duties and organization of the Office shall be as specified in the Terms of Reference appended to this Agreement.

4.   The Commission shall submit an annual budget of anticipated expenses to be incurred in carrying out its responsibilities under this Agreement to the Parties for approval. Each Party shall seek funds to pay one-half of the annual budget so approved, but neither Party shall be under an obligation to pay a larger amount than the other toward this budget.

ARTICLE IX

SUBMISSION AND EXCHANGE OF INFORMATION

1.   The International Joint Commission shall be given at its request any data or other information relating to water quality in the Great Lakes System in accordance with procedures established by the Commission.

2.   The Commission shall make available to the Parties and to the State and Provincial Governments upon request all data or other information furnished to it in accordance with this Article.

3.   Each Party shall make available to the other at its request any data or other information in its control relating to water quality in the Great Lakes System.

4.   Notwithstanding any other provision of this Agreement, the Commission shall not release without the consent of the owner any information identified as proprietary information under the law of the place where such information has been acquired.

ARTICLE X

CONSULTATION AND REVIEW

1.   Following the receipt of each report submitted to the Parties by the International Joint Commission in accordance with paragraph 3 of Article VII of this Agreement, the Parties shall consult on the recommendations contained in such report and shall consider such action as may be appropriate, including:

    (a) The modification of existing Objectives and the adoption of new Objectives;

    (b) The modification or improvement of programs and joint measures; and

(c) The amendment of this Agreement or any Annex thereto.

Additional consultations may be held at the request of either Party on any matter arising out of the implementation of this Agreement.

2. When a Party becomes aware of a special pollution problem that is of joint concern and requires an immediate response, it shall notify and consult the other Party forthwith about appropriate remedial action.

3. The Parties shall conduct a comprehensive review of the operation and effectiveness of this Agreement following the third biennial report of the Commission required under Article VII of this Agreement.

ARTICLE XI

IMPLEMENTATION

1. The obligations undertaken in this Agreement shall be subject to the appropriation of funds in accordance with the constitutional procedures of the Parties.

2. The Parties commit themselves to seek:

(a) The appropriation of the funds required to implement this Agreement, including the funds needed to develop and implement the programs and other measures provided for in Article VI of this Agreement, and the funds required by the International Joint Commission to carry out its responsibilities effectively;

(b) The enactment of any additional legislation that may be necessary in order to implement the programs and other measures provided for in Article VI of this Agreement; and

(c) The cooperation of the State and Provincial Governments in all matters relating to this Agreement.

ARTICLE XII

EXISTING RIGHTS AND OBLIGATIONS

Nothing in this Agreement shall be deemed to diminish the rights and obligations of the Parties as set forth in the Boundary Waters Treaty.

ARTICLE XIII

AMENDMENT

1. This Agreement, the Annexes, and the Terms of Reference may be amended by agreement of the Parties. The Annexes may also be amended as provided therein, subject to the requirement that such amendments shall be within the scope of this Agreement. All such amendments to the Annexes shall be confirmed by an exchange of notes or letters between the Parties through diplomatic channels which shall specify the effective date or dates of such amendments.

2.   All amendments to this Agreement, the Annexes, and the Terms of Reference shall be communicated promptly to the International Joint Commission.

ARTICLE XIV

ENTRY INTO FORCE AND TERMINATION

        This Agreement shall enter into force upon signature by the duly authorized representatives of the Parties, and shall remain in force for a period of five years and thereafter until terminated upon twelve months' notice given in writing by one of the Parties to the other.

ARTICLE XV

SUPERSESSION

        This Agreement supersedes the Great Lakes Water Quality Agreement of April 15, 1972, and shall be referred to as the "Great Lakes Water Quality Agreement of 1978".

IN WITNESS WHEREOF the undersigned representatives, duly authorized by their respective Governments, have signed this Agreement.

DONE in duplicate at Ottawa in the English and French languages, both versions being equally authentic, this 22nd day of *November* 1978.

EN FOI DE QUOI, les représentants soussignées, dûment authorisés par leur Gouvernement respectif, ont signé le présent Accord.

FAIT en double exemplaire à Ottawa en français et en anglais, chaque version faisant également foi, ce 22ème jour de *novembre* 1978.

For the Government of Canada
Pour le Gouvernement du Canada

For the Government of the
United States of America
Pour le Gouvernement des
Etats-Unis d'Amérique

UNITED STATES AND CANADA

Great Lakes Water Quality

*Agreement, with annexes and texts and terms of reference,
signed at Ottawa April 15, 1972;
Entered into force April 15, 1972.*

International Joint Commission
United States and Canada
1974

AGREEMENT BETWEEN
THE UNITED STATES OF AMERICA AND CANADA
ON GREAT LAKES WATER QUALITY[1]

The Government of the United States of America and
the Government of Canada,

Determined to restore and enhance water quality in
the Great Lakes System;

Seriously concerned about the grave deterioration of
water quality on each side of the boundary to an extent that is
causing injury to health and property on the other side, as
described in the 1970 report of the International Joint
Commission on Pollution of Lake Erie, Lake Ontario and the
International Section of the St. Lawrence River;

Intent upon preventing further pollution of the Great
Lakes System owing to continuing population growth, resource
development and increasing use of water;

Reaffirming in a spirit of friendship and cooperation
the rights and obligations of both countries under the Boundary
Waters Treaty signed on January 11, 1909,[2] and in particular their
obligation not to pollute boundary waters;

Recognizing the rights of each country in the use of
its Great Lakes waters;

Satisfied that the 1970 report of the International
Joint Commission provides a sound basis for new and more effective
cooperative actions to restore and enhance water quality in the
Great Lakes System;

Convinced that the best means to achieve improved water
quality in the Great Lakes System is through the adoption of
common objectives, the development and implementation of
cooperative programs and other measures, and the assignment of
special responsibilities and functions to the International
Joint Commission;

Have agreed as follows:

------

1 As found at TIAS 7312
2 TS 548;36 Stat. 2448.

DEFINITIONS

As used in this Agreement:

(a) "Boundary waters of the Great Lakes System" or
     "boundary waters" means boundary waters, as defined
     in the Boundary Waters Treaty, that are within the
     Great Lakes System;

(b) "Boundary Waters Treaty" means the Treaty between
     the United States and Great Britain Relating to
     Boundary Waters, and Questions Arising Between the
     United States and Canada, signed at Washington on
     January 11, 1909;

(c) "Compatible regulations" means regulations no less
     restrictive than agreed principles;

(d) "Great Lakes System" means all of the streams, rivers,
     lakes and other bodies of water that are within the
     drainage basin of the St. Lawrence River at or up-
     stream from the point at which this river becomes the
     international boundary between Canada and the United
     States;

(e) "Harmful quantity" means any quantity of a substance
     that if discharged into receiving waters would be
     inconsistent with the achievement of the water quality
     objectives;

(f) "Hazardous polluting substance" means any element or
     compound identified by the Parties which, when dis-
     charged in any quantity into or upon receiving waters
     or adjoining shorelines, presents an imminent and
     substantial danger to public health or welfare; for
     this purpose, "public health or welfare" encompasses
     all factors affecting the health and welfare of man
     including but not limited to human health, and the
     conservation and protection of fish, shellfish,
     wildlife, public and private property, shorelines and
     beaches;

(g) "International Joint Commission" or "Commission" means
     the International Joint Commission established by the
     Boundary Waters Treaty;

(h) "Phosphorus" means the element phosphorus present as a
     constituent of various organic and inorganic complexes
     and compounds;

(i) "Specific water quality objective" means the level of
     a substance or physical effect that the Parties agree,
     after investigation, to recognize as a maximum or
     minimum desired limit for a defined body of water or
     portion thereof, taking into account the beneficial
     uses of the water that the Parties desire to secure
     and protect;

(j) "State and Provincial Governments" means the Governments
     of the States of Illinois, Indiana, Michigan, Minnesota,
     New York, Ohio, Pennsylvania, and Wisconsin, and the
     Government of the Province of Ontario;

(k)   "Tributary waters of the Great Lakes System" or
       "tributary waters" means all the waters of the Great
       Lakes System that are not boundary waters;

(1)   "Water quality objectives" means the general water
       quality objectives adopted pursuant to Article II of
       this Agreement and the specific water quality
       objectives adopted pursuant to Article III of this
       Agreement.

ARTICLE II

GENERAL WATER QUALITY OBJECTIVES

The following general water quality objectives for the
boundary waters of the Great Lakes System are adopted. These
waters should be:

(a)   Free from substances that enter the waters as a result
       of human activity and that will settle to form putrescent
       or otherwise objectionable sludge deposits, or that will
       adversely affect aquatic life or waterfowl;

(b)   Free from floating debris, oil, scum and other floating
       materials entering the waters as a result of human
       activity in amounts sufficient to be unsightly or
       deleterious;

(c)   Free from materials entering the waters as a result of
       human activity producing colour, odour or other con-
       ditions in such a degree as to create a nuisance;

(d)   Free from substances entering the waters as a result of
       human activity in concentrations that are toxic or
       harmful to human, animal or aquatic life;

(e)   Free from nutrients entering the waters as a result of
       human activity in concentrations that create nuisance
       growths of aquatic weeds and algae.

ARTICLE III

SPECIFIC WATER QUALITY OBJECTIVES

1.   The specific water quality objectives for the boundary
waters of the Great Lakes System set forth in Annex 1 are
adopted.

2.   The specific water quality objectives may be modified
and additional specific water quality objectives for the boundary
waters of the Great Lakes System or for particular sections
thereof may be adopted by the Parties in accordance with the
provisions of Articles IX and XII of this Agreement.

3.   The specific water quality objectives adopted pursuant
to this Article represent the minimum desired levels of water
quality in the boundary waters of the Great Lakes System and are
not intended to preclude the establishment of more stringent
requirements.

Notwithstanding the adoption of specific water quality objectives, all reasonable and practicable measures shall be taken to maintain the levels of water quality existing at the date of entry into force of this Agreement in those areas of the boundary waters of the Great Lakes System where such levels exceed the specific water quality objectives.

ARTICLE IV

STANDARDS AND OTHER REGULATORY REQUIREMENTS

Water quality standards and other regulatory requirements of the Parties shall be consistent with the achievement of the water quality objectives. The Parties shall use their best efforts to ensure that water quality standards and other regulatory requirements of the State and Provincial Governments shall similarly be consistent with the achievement of the water quality objectives.

ARTICLE V

PROGRAMS AND OTHER MEASURES

1.   Programs and other measures directed toward the achievement of the water quality objectives shall be developed and implemented as soon as practicable in accordance with legislation in the two countries. Unless otherwise agreed, such programs and other measures shall be either completed or in process of implementation by December 31, 1975. They shall include the following:

(a) Pollution from Municipal Sources. Programs for the abatement and control of discharges of municipal sewage into the Great Lakes System including:

(i)   construction and operation in all municipalities having sewer systems of waste treatment facilities providing levels of treatment consistent with the achievement of the water quality objectives, taking into account the effects of waste from other sources;

(ii) provision of financial resources to assist prompt construction of needed facilities;

(iii) establishment of requirements for construction and operating standards for facilities;

(iv) measures to find practical solutions for reducing pollution from overflows of combined storm and sanitary sewers;

(v)   monitoring, surveillance and enforcement activities necessary to ensure compliance with the foregoing programs and measures.

(b) Pollution from Industrial Sources. Programs for the
abatement and control of pollution from industrial
sources, including:

   (i)   establishment of waste treatment or control
         requirements for all industrial plants discharging
         waste into the Great Lakes System, to provide
         levels of treatment or reduction of inputs of
         substances and effects consistent with the achieve-
         ment of the water quality objectives, taking into
         account the effects of waste from other sources;

   (ii) requirements for the substantial elimination of
         discharges into the Great Lakes System of mercury
         and other toxic heavy metals;

   (iii) requirements for the substantial elimination of
         discharges into the Great Lakes System of toxic
         persistent organic contaminants;

   (iv) requirements for the control of thermal discharges;

   (v)   measures to control the discharge of radioactive
         materials into the Great Lakes System;

   (vi) monitoring, surveillance and enforcement activities
         necessary to ensure compliance with the foregoing
         requirements and measures.

(c) Eutrophication. Measures for the control of inputs of
phosphorus and other nutrients including programs to
reduce phosphorus inputs, in accordance with the provisions of Annex 2.

(d) Pollution from Agricultural, Forestry and Other Land
Use Activities. Measures for the abatement and control
of pollution from agricultural, forestry and other land
use activities, including:

   (i)   measures for the control of pest control products
         with a view to limiting inputs into the Great Lakes
         System, including regulations to ensure that pest
         control products judged to have long term
         deleterious effects on the quality of water or its
         biotic components shall be used only as authorized
         by the responsible regulatory agencies, and that
         pest control products shall not be applied directly
         to water except in accordance with the requirements
         of the responsible regulatory agencies;

   (ii) measures for the abatement and control of pollution
         from animal husbandry operations, including en-
         couragement to appropriate regulatory agencies to
         adopt regulations governing site selection and
         disposal of liquid and solid wastes in order to
         minimize the loss of pollutants to receiving waters;

(iii)   measures governing the disposal of solid wastes and contributing to the achievement of the water quality objectives, including encouragement to appropriate regulatory agencies to ensure proper location of land fill and land dumping sites and regulations governing the disposal on land of hazardous polluting substances;

(iv)   advisory programs and measures that serve to abate and control inputs of nutrients and sediments into receiving waters from agricultural, forestry and other land use activities.

(e)   <u>Pollution from Shipping Activities</u>. Measures for the abatement and control of pollution from shipping sources, including:

(i)   programs and compatible regulations for vessel design, construction and operation, to prevent discharges of harmful quantities of oil and hazardous polluting substances, in accordance with the principles set forth in Annex 3;

(ii)   compatible regulations for the control of vessel waste discharges in accordance with the principles set forth in Annex 4;

(iii)   such compatible regulations to abate and control pollution from shipping sources as may be deemed desirable in the light of studies to be undertaken in accordance with the terms of references set forth in Annex 5;

(iv)   programs for the safe and efficient handling of shipboard generated wastes, including oil, hazardous polluting substances, garbage, waste water and sewage, and their subsequent disposal, including any necessary compatible regulations relating to the type, quantity and capacity of shore reception facilities;

(v)   establishment of a coordinated system for the surveillance and enforcement of regulations dealing with the abatement and control of pollution from shipping activities.

(f)   <u>Pollution from Dredging Activities</u>. Measures for the abatement and control of pollution from dredging activities, including the development of criteria for the identification of polluted dredged spoil and compatible programs for disposal of polluted dredged spoil, which shall be considered in the light of the review provided for in Annex 6; pending the development of compatible criteria and programs, dredging operations shall be conducted in a manner that will minimize adverse effects on the environment.

(g)   <u>Pollution from Onshore and Offshore Facilities</u>. Measures for the abatement and control of pollution from onshore and offshore facilities, including programs and compatible regulations for the prevention of discharges of harmful quantities of oil and hazardous polluting substances, in accordance with the principles set forth in Annex 7.

(h) Contingency Plan. Maintenance of a joint contingency plan for use in the event of a discharge or the imminent threat of a discharge of oil or hazardous polluting substances, in accordance with the provisions of Annex 8.

(i) Hazardous Polluting Substances. Consultation within one year from the date of entry into force of this Agreement for the purpose of developing an Annex identifying hazardous polluting substances; the Parties shall further consult from time to time for the purpose of identifying harmful quantities of these substances and of reviewing the definition of "harmful quantity of oil" set forth in Annexes 3 and 7.

2.   The Parties shall develop and implement such additional programs as they jointly decide are necessary and desirable for the achievement of the water quality objectives.

3.   The programs and other measures provided for in this Article shall be designed to abate and control pollution of tributary waters where necessary or desirable for the achievement of the water quality objectives for the boundary waters of the Great Lakes System.

ARTICLE VI

POWERS, RESPONSIBILITIES AND FUNCTIONS OF
THE INTERNATIONAL JOINT COMMISSION

1.   The International Joint Commission shall assist in the implementation of this Agreement. Accordingly, the Commission is hereby given, pursuant to Article IX of the Boundary Waters Treaty, the following responsibilities:

(a) Collation, analysis and dissemination of data and information supplied by the Parties and State and Provincial Governments relating to the quality of the boundary waters of the Great Lakes System and to pollution that enters the boundary waters from tributary waters;

(b) Collection, analysis and dissemination of data and information concerning the water quality objectives and the operation and effectiveness of the programs and other measures established pursuant to this Agreement;

(c) Tendering of advice and recommendations to the Parties and to the State and Provincial Governments on problems of the quality of the boundary waters of the Great Lakes System, including specific recommendations concerning the water quality objectives, legislation, standards and other regulatory requirements, programs and other measures, and intergovernmental agreements relating to the quality of these waters;

(d) Provision of assistance in the coordination of the joint activities envisaged by this Agreement, including such matters as contingency planning and consultation on special situations;

(e) Provision of assistance in the coordination of Great Lakes water quality research, including identification of objectives for research activities, tendering of advice and recommendations concerning research to the Parties and to the State and Provincial Governments and dissemination of information concerning research to interested persons and agencies;

(f) Investigations of such subjects related to Great Lakes water quality as the Parties may from time to time refer to it. At the time of signature of this Agreement, the Parties are requesting the Commission to enquire into and report to them upon:

    (i) pollution of the boundary waters of the Great Lakes System from agricultural, forestry and other land use activities, in accordance with the terms of reference attached to this Agreement;

    (ii) actions needed to preserve and enhance the quality of the waters of Lake Huron and Lake Superior in accordance with the terms of reference attached to this Agreement.

2.   In the discharge of its responsibilities under this Agreement, the Commission may exercise all of the powers conferred upon it by the Boundary Waters Treaty and by any legislation passed pursuant thereto, including the power to conduct public hearings and to compel the testimony of witnesses and the production of documents.

3.   The Commission shall make a report to the Parties and to the State and Provincial Governments no less frequently than annually concerning progress toward the achievement of the water quality objectives. This report shall include an assessment of the effectiveness of the programs and other measures undertaken pursuant to this Agreement, and advice and recommendations. The Commission may at any time make special reports to the Parties, to the State and Provincial Governments and to the public concerning any problem of water quality in the Great Lakes System.

4.   The Commission may in its discretion publish any report, statement or other document prepared by it in the discharge of its functions under this Agreement.

5.   The Commission shall have authority to verify independently the data and other information submitted by the Parties and by the State and Provincial Governments through such tests or other means as appear appropriate to it, consistent with the Boundary Waters Treaty and with applicable legislation.

ARTICLE VII

JOINT INSTITUTIONS

1.   The International Joint Commission shall establish a Great Lakes Water Quality Board to assist it in the exercise of the powers and responsibilities assigned to it under this Agreement. Such Board shall be composed of an equal number of

members from Canada and the United States, including representation from the Parties and from each of the State and Provincial
Governments. The Commission shall also establish a Research
Advisory Board in accordance with the terms of reference attached
to this Agreement. The members of the Great Lakes Water
Quality Board and the Research Advisory Board shall be appointed
by the Commission after consultation with the appropriate government or governments concerned. In addition, the Commission shall
have the authority to establish as it may deem appropriate such
subordinate bodies as may be required to undertake specific tasks,
as well as a regional office, which may be located in the basin
of the Great Lakes System, to assist it in the discharge of its
functions under this Agreement. The Commission shall also
consult the Parties about the site and staffing of any regional
office that might be established.

2.   The Commission shall submit an annual budget of anticipated expenses to be incurred in carrying out its responsibilities
under this Agreement to the Parties for approval. Each Party
shall seek funds to pay one-half of the annual budget so approved,
but neither Party shall be under an obligation to pay a larger
amount than the other toward this budget.

ARTICLE VIII

SUBMISSION AND EXCHANGE OF INFORMATION

1.   The International Joint Commission shall be given at
its request any data or other information relating to the quality
of the boundary waters of the Great Lakes System in accordance
with procedures to be established, within three months of the
entry into force of this Agreement or as soon thereafter as
possible, by the Commission in consultation with the Parties and
with the State and Provincial Governments.

2.   The Commission shall make available to the Parties and
to the State and Provincial Governments upon request all data or
other information furnished to it in accordance with this
Article.

3.   Each Party shall make available to the other at its
request any data or other information in its control relating
to the quality of the waters of the Great Lakes System.

4.   Notwithstanding any other provision of this Agreement,
the Commission shall not release without the consent of the
owner any information identified as proprietary information under
the law of the place where such information has been acquired.

ARTICLE IX

CONSULTATION AND REVIEW

1.   Following the receipt of each report submitted to the
Parties by the International Joint Commission in accordance
with paragraph 3 of Article VI of this Agreement, the Parties
shall consult on the recommendations contained in such report
and shall consider such action as may be appropriate, including:

(a) The modification of existing water quality objectives and the adoption of new objectives;

(b) The modification or improvement of programs and joint measures;

(c) The amendment of this Agreement or any annex thereto.

Additional consultations may be held at the request of either Party on any matter arising out of the implementation of this Agreement.

2.   When a Party becomes aware of a special pollution ‧ problem that is of joint concern and requires an immediate response, it shall notify and consult the other Party forthwith about appropriate remedial action.

3.   The Parties shall conduct a comprehensive review of the operation and effectiveness of this Agreement during the fifth year after its coming into force.   Thereafter, further comprehensive reviews shall be conducted upon the request of either Party.

ARTICLE X

IMPLEMENTATION

1.   The obligations undertaken in this Agreement shall be subject to the appropriation of funds in accordance with the constitutional procedures of the Parties.

2.   The Parties commit themselves to seek:

(a) The appropriation of the funds required to implement this Agreement, including the funds needed to develop and implement the programs and other measures provided for in Article V, and the funds required by the International Joint Commission to carry out its responsibilities effectively;

(b) The enactment of any additional legislation that may be necessary in order to implement the programs and other measures provided for in Article V;

(c) The cooperation of the State and Provincial Governments in all matters relating to this Agreement.

ARTICLE XI

EXISTING RIGHTS AND OBLIGATIONS

Nothing in this Agreement shall be deemed to diminish the rights and obligations of the Parties as set forth in the Boundary Waters Treaty.

ARTICLE XII

AMENDMENT

This Agreement and the Annexes thereto may be amended by agreement of the Parties. The Annexes may also be amended as provided therein, subject to the requirement that such amendments shall be within the scope of this Agreement.

ARTICLE XIII

ENTRY INTO FORCE AND TERMINATION

This Agreement shall enter into force upon signature by the duly authorized representatives of the Parties, and shall remain in force for a period of five years and thereafter until terminated upon twelve months' notice given in writing by one of the Parties to the other.

IN WITNESS WHEREOF the Representatives of the two Governments have signed this Agreement.

DONE in two copies at Ottawa this fifteenth day of April 1972 in English and French, each version being equally authentic.

[1]

[2]

For the Government of the United States of America
Pour le Gouvernement des Etats-Unis d'Amérique

[3]

[4]

For the Government of Canada

---

[1] Richard Nixon
[2] William P. Rogers
[3] P. E. Trudeau
[4] Mitchell Sharp

ECE Draft Convention on Long-Range Transboundary Air Pollution (1979)

APPENDIX A: CONVENTION ON LONG-RANGE TRANSBOUNDARY AIR POLLUTION

The Parties to the present Convention,

Determined to promote relations and co-operation in the field of environmental protection,

Aware of the significance of the activities of the United Nations Economic Commission for Europe in strengthening such relations and co-operation, particularly in the field of air pollution including long-range transport of air pollutants,

Recognizing the contribution of the Economic Commission for Europe to the multilateral implementation of the pertinent provisions of the Final Act of the Conference on Security and Co-operation in Europe,

Cognizant of the references in the chapter on environment of the Final Act of the Conference on Security and Co-operation in Europe calling for co-operation to control air pollution and its effects, including long-range transport of air pollutants, and to the development through international co-operation of an extensive programme for the monitoring and evaluation of long-range transport of air pollutants, starting with sulphur dioxide and with possible extension to other pollutants,

Considering the pertinent provisions of the Declaration of the United Nations Conference on the Human Environment, and in particular principle 21, which expresses the common conviction that States have, in accordance with the Charter of the United Nations and the principles of international law, the sovereign right to exploit their own resources pursuant to their own environmental policies, and the responsibility to ensure that activities within their jurisdiction or control do not cause damage to the environment of other States or of areas beyond the limits of national jurisdiction,

Recognizing the existence of possible adverse effects, in the short and long term, of air pollution including transboundary air pollution,

Reprinted courtesy of the United Nations Economic Commission for Europe.

Concerned that a rise in the level of emissions of air pollutants within the region as forecast may increase such adverse effects,

Recognizing the need to study the implications of the long-range transport of air pollutants and the need to seek solutions for the problems identified,

Affirming their willingness to reinforce active international co-operation to develop appropriate national policies and by means of exchange of information, consultation, research and monitoring, to co-ordinate national action for combating air pollution including long-range transboundary air pollution,

Have agreed as follows:

DEFINITIONS

Article 1

For the purposes of the present Convention:

(a) *"air pollution"* means the introduction by man, directly or indirectly, of substances or energy into the air resulting in deleterious effects of such a nature as to endanger human health, harm living resources and ecosystems and material property and impair or interfere with amenities and other legitimate uses of the environment, and "air pollutants" shall be construed accordingly;

(b) *"long-range transboundary air pollution"* means air pollution whose physical origin is situated wholly or in part within the area under the national jurisdiction of one state and which has adverse effects in the area under the jurisdiction of another state at such a distance that it is not generally possible to distinguish the contribution of individual emission sources or groups of sources.

FUNDAMENTAL PRINCIPLES

Article 2

The Contracting Parties, taking due account of the facts and problems involved, are determined to protect man and his environment against air pollution and shall endeavour to limit and, as far as possible, gradually reduce and prevent air pollution including long-range transboundary air pollution.

Article 3

The Contracting Parties, within the framework of the present Convention, shall by means of exchanges of information, consultation, research and monitoring, develop without undue delay policies and strategies which shall serve as a means of combating the discharge of air pollutants, taking into account efforts already made at national and international levels.

Article 4

The Contracting Parties shall exchange information on and review their policies, scientific activities and technical measures aimed at combating, as far as possible, the discharge of air pollutants which may have adverse effects, thereby contributing to the reduction of air pollution including long-range transboundary air pollution.

Article 5

Consultations shall be held, upon request, at an early stage between, on the one hand, Contracting Parties which are actually affected by or exposed to a significant risk of long-range transboundary air pollution and, on the other hand, Contracting Parties within which and subject to whose jurisdiction a significant contribution to long-range transboundary air pollution originates or could originate, in connexion with activities carried on or contemplated therein.

AIR QUALITY MANAGEMENT

Article 6

Taking into account Articles 2 to 5, the ongoing research, exchange of information and monitoring and the results thereof, the cost and effectiveness of local and other remedies and, in order to combat air pollution, in particular that originating from new or rebuilt installations, each Contracting Party undertakes to develop the best policies and strategies including air quality management systems and, as part of them, control measures compatible with balanced development, in particular by using the best available technology which is economically feasible and low- and non-waste technology.

RESEARCH AND DEVELOPMENT

Article 7

The Contracting Parties, as appropriate to their needs, shall initiate and co-operate in the conduct of research into and/or development of:

(a) existing and proposed technologies for reducing emissions of sulphur compounds and other major air pollutants, including technical and economic feasibility, and environmental consequences;

(b) instrumentation and other techniques for monitoring and measuring emission rates and ambient concentrations of air pollutants;

(c) improved models for a better understanding of the transmission of long-range transboundary air pollutants;

(d) the effects of sulphur compounds and other major air pollutants on human health and the environment, including agriculture, forestry, materials, aquatic and other natural ecosystems and visibility, with a view to establishing a scientific basis for dose/effect relationships designed to protect the environment;

(e) the economic, social and environmental assessment of alternative measures for attaining environmental objectives including the reduction of long-range transboundary air pollution;

(f) education and training programmes related to the environmental aspects of pollution by sulphur compounds and other major air pollutants.

EXCHANGE OF INFORMATION

Article 8

The Contracting Parties, within the framework of the Executive Body referred to in article 10 and bilaterally, shall, in their common interests, exchange available information on:

(a) data on emissions at periods of time to be agreed upon, of agreed air pollutants, starting with sulphur dioxide, coming from grid-units of agreed size; or on the fluxes of agreed air pollutants, starting with sulphur dioxide, across national borders, at distances and at periods of time to be agreed upon;

(b) major changes in national policies and in general industrial development, and their potential impact, which would be likely to cause significant changes in long-range transboundary air pollution;

(c) control technologies for reducing air pollution relevant to long-range transboundary air pollution;

(d) the projected cost of the emission control of sulphur compounds and other major air pollutants on a national scale;

(e) meteorological and physico-chemical data relating to the processes during transmission;

(f) physico-chemical and biological data relating to the effects of long-range transboundary air pollution and the extent of the damage[§] which these data indicate can be attributed to long-range transboundary air pollution;

(g) national, subregional and regional policies and strategies for the control of sulphur compounds and other major air pollutants.

---

§The present Convention does not contain a rule on State liability as to damage.

IMPLEMENTATION AND FURTHER DEVELOPMENT OF THE
CO-OPERATIVE PROGRAMME FOR THE MONITORING AND
EVALUATION OF THE LONG-RANGE TRANSMISSION OF
AIR POLLUTANTS IN EUROPE

Article 9

The Contracting Parties stress the need for the implementation of the existing "Co-operative programme for the monitoring and evaluation of the long-range transmission of air pollutants in Europe" (hereinafter referred to as EMEP) and, with regard to the further development of this programme, agree to emphasize;

(a) the desirability of Contracting Parties joining in and fully implementing EMEP which, as a first step, is based on the monitoring of sulphur dioxide and related substances;

(b) the need to use comparable or standardized procedures for monitoring whenever possible;

(c) the desirability of basing the monitoring programme on the framework of both national and international programmes. The establishment of monitoring stations and the collection of data shall be carried out under the national jurisdiction of the country in which the monitoring stations are located;

(d) the desirability of establishing a framework for a co-operative environmental monitoring programme, based on and taking into account present and future national, sub-regional, regional and other international programmes;

(e) the need to exchange data on emissions at periods of time to be agreed upon, of agreed air pollutants, starting with sulphur dioxide, coming from grid-units of agreed size; or on the fluxes of agreed air pollutants, starting with sulphur dioxide, across national borders, at distances and at periods of time to be agreed upon. The method, including the model, used to determine the fluxes, as well as the method, including the model, used to determine the transmission of air pollutants based on the emissions per grid-unit, shall be made available and periodically reviewed, in order to improve the methods and the models;

(f) their willingness to continue the exchange and periodic updating of national data on total emissions of agreed air pollutants, starting with sulphur dioxide;

(g) the need to provide meteorological and physico-chemical data relating to processes during transmission;

(h) the need to monitor chemical components in other media such as water, soil and vegetation, as well as a similar monitoring programme to record effects on health and environment;

(i) the desirability of extending the national EMEP networks to make them operational for control and surveillance purposes.

EXECUTIVE BODY

Article 10

1. The representatives of the Contracting Parties shall, within the framework of the Senior Advisers to ECE Governments on Environmental Problems, constitute the Executive Body of the present Convention, and shall meet at least annually in that capacity.

2. The Executive Body shall:

- (a) review the implementation of the present Convention;
- (b) establish, as appropriate, workings groups to consider matters related to the implementation and development of the present Convention and to this end to prepare appropriate studies and other documentation and to submit recommendations to be considered by the Executive Body;
- (c) fulfill such other functions as may be appropriate under the provisions of the present Convention.

3. The Executive Body shall utilize the Steering Body for the EMEP to play an integral part in the operation of the present Convention, in particular with regard to data collection and scientific co-operation.

4. The Executive Body, in discharging its functions, shall, when it deems appropriate, also make use of information from other relevant international organizations.

SECRETARIAT

Article 11

The Executive Secretary of the Economic Commission for Europe shall carry out, for the Executive Body, the following secretariat functions:

- (a) to convene and prepare the meetings of the Executive Body;
- (b) to transmit to the Contracting Parties reports and other information received in accordance with the provisions of the present Convention;
- (c) to discharge the functions assigned by the Executive Body.

AMENDMENTS TO THE CONVENTION

Article 12

1. Any Contracting Party may propose amendments to the present Convention.

2. The text of proposed amendments shall be submitted in writing to the Executive Secretary of the Economic Commission for Europe, who shall communicate them to all Contracting Parties. The Executive Body shall discuss proposed amendments at its next annual meeting provided that such proposals have been circulated by the Executive Secretary of the Economic Commission for Europe to the Contracting Parties at least ninety days in advance.

3. An amendment to the present Convention shall be adopted by consensus of the representatives of the Contracting Parties, and shall enter into force for the Contracting Parties which have accepted it on the nine-tieth day after the date on which two-thirds of the Contracting Parties have deposited their instruments of acceptance with the depositary. Thereafter, the amendment shall enter into force for any other Contracting Party on the ninetieth day after the date on which that Contracting Party deposits its instrument of acceptance of the amendment.

SETTLEMENT OF DISPUTES

Article 13

If a dispute arises between two or more Contracting Parties to the present Convention as to the interpretation or application of the Conven-tion, they shall seek a solution by negotiation or by any other method of dispute settlement acceptable to the parties to the dispute.

SIGNATURE

Article 14

1. The present Convention shall be open for signature at the United Nations Office at Geneva from 13 to 16 November 1979 on the occasion of the High-level Meeting within the framework of the Economic Commis-sion for Europe on the Protection of the Environment, by the member States of the Economic Commission for Europe as well as States having consultative status with the Economic Commission for Europe, pursuant to paragraph 8 of Economic and Social Council resolution 36 (IV) of 28 March 1947, and by regional economic integration organizations, con-stituted by sovereign States members of the Economic Commission for Europe, which have competence in respect of the negotiation, conclusion and application of international agreements in matters covered by the pres-ent Convention.

2. In matters within their competence, such regional economic inte-gration organizations shall, on their own behalf, exercise the rights and fulfil the responsibilities which the present Convention attributes to their member States. In such cases, the member States of these organizations shall not be entitled to exercise such rights individually.

RATIFICATION, ACCEPTANCE, APPROVAL AND
ACCESSION

Article 15

1. The present Convention shall be subject to ratification, acceptance or approval.

2. The present Convention shall be open for accession as from 17 November 1979 by the States and organizations referred to in Article 14, paragraph 1.

3. The instruments of ratification acceptance, approval or accession shall be deposited with the Secretary-General of the United Nations, who will perform the functions of the depositary.

ENTRY INTO FORCE

Article 16

1. The present Convention shall enter into force on the ninetieth day after the date of deposit of the twenty-fourth instrument of ratification, acceptance, approval or accession.

2. For each Contracting Party which ratifies, accepts or approves the present Convention or accedes thereto after the deposit of the twenty-fourth instrument of ratification, acceptance, approval or accession, the Convention shall enter into force on the ninetieth day after the date of deposit by such Contracting Party of its instrument of ratification, acceptance, approval or accession.

WITHDRAWAL

Article 17

At any time after five years from the date on which the present Convention has come into force with respect to a Contracting Party, that Contracting Party may withdraw from the Convention by giving written notification to the depositary. Any such withdrawal shall take effect on the ninetieth day after the date of its receipt by the depositary.

AUTHENTIC TEXTS

Article 18

The original of the present Convention, of which the English, French and Russian texts are equally authentic, shall be deposited with the Secretary-General of the United Nations.

United States–Canada
Memorandum of Intent (on
Transboundary Air Pollution)
of August 5, 1980

**Memorandum of Intent between the
Government of Canada and the Government
of the United States of America concerning
Transboundary Air Pollution**

The Government of Canada and the Government of the United States
of America.

Share a concern about actual and potential damage resulting from
transboundary air pollution, (which is the short and long range transport of
air pollutants between their countries), including the already serious problem of acid rain;

Recognize this is an important and urgent bilateral problem as it
involves the flow of air pollutants in both directions across the international
boundary, especially the long range transport of air pollutants;

Share also a common determination to combat transboundary air
pollution in keeping with their existing international rights, obligations,
commitments and cooperative practices, including those set forth in the
1909 Boundary Waters Treaty, the 1972 Stockholm Declaration on the
Human Environment, the 1978 Great Lakes Water Quality Agreement, and
the 1979 ECE Convention on Long Range Transboundary Air Pollution;

Undertook in July 1979 to develop a bilateral cooperative agreement
on air quality which would deal effectively with transboundary air
pollution;

Are resolved as a matter of priority both to improve scientific understanding of the long range transport of air pollutants and its effects and to
develop and implement policies, practices and technologies to combat its
impact;

Reprinted courtesy of the Canadian Department of External Affairs.

Are resolved to protect the environment in harmony with measures to meet energy needs and other national objectives;

Note scientific findings which indicate that continued pollutant loadings will result in extensive acidification in geologically sensitive areas during the coming years, and that increased pollutant loadings will accelerate this process;

Are concerned that environmental stress could be increased if action is not taken to reduce transboundary air pollution;

Are convinced that the best means to protect the environment from the effects of transboundary air pollution is through the achievement of necessary reductions in pollutant loadings;

Are convinced also that this common problem requires cooperative action by both countries;

Intend to increase bilateral cooperative action to deal effectively with transboundary air pollution, including acid rain.

In particular, the Government of Canada and the Government of the United States of America intend:

1. to develop a bilateral agreement which will reflect and further the development of effective domestic control programs and other measures to combat transboundary air pollution;
2. to facilitate the conclusion of such an agreement as soon as possible; and,
3. pending conclusion of such an agreement, to take interim actions available under current authority to combat transboundary air pollution.

The specific undertakings of both Governments at this time are outlined below.

Interim Actions

1. *Transboundary Air Pollution Agreement*

Further to their Joint Statement of July 26, 1979, and subsequent bilateral discussions, both Governments shall take all necessary steps forthwith:

(a) to establish a Canada/United States Coordinating Committee which will undertake preparatory discussions immediately and commence formal negotiations no later than June 1, 1981, of a cooperative agreement on transboundary air pollution; and
(b) to provide the necessary resources for the Committee to carry out its work, including the working group structure as set forth in the Annex.

Members will be appointed to the work groups by each Government as soon as possible.

2. Control Measures

To combat transboundary air pollution both Governments shall:

(a) develop domestic air pollution control policies and strategies, and as necessary and appropriate, seek legislative or other support to give effect to them;
(b) promote vigorous enforcement of existing laws and regulations as they require limitation of emissions from new, substantially modified and existing facilities in a way which is responsive to the problems of transboundary air pollution; and
(c) share information and consult on actions being taken pursuant to (a) and (b) above.

3. Notification and Consultation

Both Governments shall continue and expand their long-standing practice of advance notification and consultation on proposed actions involving a significant risk or potential risk of causing or increasing transboundary air pollution, including:

(a) proposed major industrial development or other actions which may cause significant increases in transboundary air pollution; and
(b) proposed changes of policy, regulations or practices which may significantly affect transboundary air pollution.

4. Scientific Information, Research and Development

In order to improve understanding of their common problem and to increase their capability for controlling transboundary air pollution both Governments shall:

(a) exchange information generated in research programs being undertaken in both countries on the atmospheric aspects of the transport of air pollutants and on their effects on aquatic and terrestrial ecosystems and on human health and property;
(b) maintain and further develop a coordinated program for monitoring and evaluation of the impacts of transboundary air pollution, including the maintenance of a Canada/United States sampling network and

exchange of data on current and projected emissions of major air pollutants; and

(c) continue to exchange information on research to develop improved technologies for reducing emissions of major air pollutants of concern.

The Memorandum of Intent will become effective on signature and will remain in effect until revised by mutual agreement.

DONE in duplicate at Washington, this fifth day of August, 1980, in the English and French languages, both texts being equally authoritative.

For the Government   For the Government of the
of Canada: United States of America:

Annex: Work Group Structure for
Negotiation of a Transboundary Air Pollution
Agreement

I. *Purpose*

To establish technical and scientific work groups to assist in preparations for and the conduct of negotiations on a bilateral transboundary air pollution agreement. These groups shall include:

1. Impact Assessment Work Group
2. Atmospheric Modelling Work Group
3A. Strategies Development and Implementation Work Group
3B. Emissions, Costs and Engineering Assessment Subgroup
4. Legal, Institutional Arrangements and Drafting Work Group

II. *Terms of Reference*

A. GENERAL

1. The Work Groups shall function under the general direction and policy guidance of a Canada/United States Coordinating Committee co-chaired by the Department of External Affairs and the Department of State.

2. The Work Groups shall provide reports assembling and analyzing information and identifying measures as outlined in Part B below, which will provide the basis of proposals for inclusion in a transboundary air pollution agreement. These reports shall be provided by January 1982 and shall be based on available information.

3. Within one month of the establishment of the Work Groups, they shall submit to the Canada/United States Coordinating Committee a work plan to accomplish the specific tasks outlined in Part 8, below. Additionally, each Work Group shall submit an interim report by January 15, 1981.

4. During the course of negotiations and under the general direction and policy guidance of the Coordinating Committee, the Work Groups shall assist the Coordinating Committee as required.

5. Nothing in the foregoing shall preclude subsequent alteration of the tasks of the Work Groups or the establishment of additional Work Groups as may be agreed upon by the Governments.

The specific tasks of the Work Groups are set forth below.

1. *Impact Assessment Work Group*

The Group will provide information on the current and projected impact of air pollutants on sensitive receptor areas, and prepare proposals for the "Research, Modelling and Monitoring" element of an agreement.

In carrying out this work, the Group will:

—identify and assess physical and biological consequences possibly related to transboundary air pollution;
—determine the present status of physical and biological indicators which characterize the ecological stability of each sensitive area identified;
—review available data bases to establish more accurately historic adverse environmental impacts;
—determine the current adverse environmental impact within identified sensitive areas—annual, seasonal and episodic;
—determine the release of residues potentially related to transboundary air pollution, including possible episodic release from snowpack melt in sensitive areas;
—assess the years remaining before significant ecological changes are sustained within identified sensitive areas;
—propose reductions in the air pollutant deposition rates—annual, seasonal and episodic—which would be necessary to protect identified sensitive areas; and
—prepare proposals for the "Research, Modelling and Monitoring" element of an agreement.

2. *Atmospheric Modelling Work Group*

The Group will provide information based on cooperative atmospheric modelling activities leading to an understanding of the transport of air pollutants between source regions and sensitive areas, and prepare proposals for the "Research, Modelling and Monitoring" element of an agreement. As a first priority the Group will by October 1, 1980 provide initial guidance on suitable atmospheric transport models to be used in preliminary assessment activities.

In carrying out its work, the Group will:

—identify source regions and applicable emission data bases;
—evaluate and select atmospheric transport models and data bases to be used;
—relate emissions from the source regions to loadings in each identified sensitive area;
—calculate emission reductions required from source regions to achieve proposed reductions in air pollutant concentration and deposition rates which would be necessary in order to protect sensitive areas;
—assess historic trends of emissions, ambient concentrations and atmospheric deposition trends to gain further insights into source receptor relationships for air quality, including deposition; and
—prepare proposals for the "Research, Modelling and Monitoring" element of an agreement.

3A. *Strategies Development and Implementation Work Group*

The Group will identify, assess and propose options for the "Control" element of an agreement. Subject to the overall direction of the Coordinating Committee, it will be responsible also for coordination of the activities of Work Groups I and II. It will have one subgroup.

In carrying out its work, the Group will:

—prepare various strategy packages for the Coordinating Committee designed to achieve proposed emission reductions;
—coordinate with other Work Groups to increase the effectiveness of these packages;
—identify monitoring requirements for the implementation of any tentatively agreed-upon emission-reduction strategy for each country;
—propose additional means to further coordinate the air quality programs of the two countries; and
—prepare proposals relating to the actions each Government would need to take to implement the various strategy options.

3B. *Emissions, Costs and Engineering Assessment Subgroup*

This Subgroup will provide support to the development of the "Control" element of an agreement. It will also prepare proposals for the Applied Research and Development" element of an agreement.

In carrying out its work, the Subgroup will:

—identify control technologies, which are available presently or in the near future, and their associated costs;

—review available data bases in order to establish improved historical emission trends for defined source regions;

—determine current emission rates from defined source regions;

—project future emission rates from defined source regions for most probable economic growth and pollution control conditions;

—project future emission rates resulting from the implementation of proposed strategy packages, and associated costs of implementing the proposed strategy packages; and

—prepare proposals for the "Applied Research and Development" element of an agreement.

4. *Legal, Institutional and Drafting Work Group*

The Group will:

—develop the legal elements of an agreement such as notification and consultation, equal access, non-discrimination, liability and compensation;

—propose institutional arrangements needed to give effect to an agreement and monitor its implementation; and

—review proposals of the Work Groups and refine language of draft provisions of an agreement.

Montana Senate Resolution on Transboundary Air Pollution (1977)

A JOINT RESOLUTION OF THE SENATE AND THE HOUSE OF REPRESENTATIVES OF THE STATE OF MONTANA REQUESTING THE UNITED STATES CONGRESS AND THE PRESIDENT OF THE UNITED STATES TO INITIATE NEGOTIATIONS WITH CANADA FOR A BOUNDARY TREATY FOR AIR QUALITY PROTECTION.

WHEREAS, it is essential to quality environment that clean air be protected; and

WHEREAS, potential air quality degradation could occur as a result of industrial projects in Canada and the United States close to their respective borders, over which a joint control does not exist; and

WHEREAS, there exists a successful mechanism in the Boundary Waters Treaty of 1909 to alleviate similar problems relating to water quality; and

WHEREAS, the International Joint Commission created by that treaty has authority over international boundary water pollution situations; and

WHEREAS, that Commission or one similar to it should have jurisdiction over international air quality problems but presently only has limited authority under the ''air reference'' provision.

NOW, THEREFORE, BE IT RESOLVED BY THE SENATE AND THE HOUSE OF REPRESENTATIVES OF THE STATE OF MONTANA:

(1) That the United States Congress and the President of the United States are urged to instruct the United States Department of State to begin negotiations with the Canadian government to eventually adopt a treaty

---

Reprinted courtesy of the Montana Senate.

between the two countries to mutually protect with best available technologies the airsheds of both countries.

(2) That the governments of the United States and Canada are urged to establish and send delegations to an international commission, similar in scope to the International Joint Commission now in existence to protect boundary waters, to ensure the health and welfare of their citizens through nondegradation of air quality by projects on either side of the border.

(3) That the Secretary of State send a copy of this resolution to the President of the United States, the Montana Congressional Delegation, and the United States Department of State.

APPENDIX 6   # Skagit-High Ross Order

INTERNATIONAL JOINT COMMISSION

IN THE MATTER OF THE APPLICATION OF THE CITY OF SEATTLE
FOR AUTHORITY TO RAISE THE WATER LEVEL OF THE SKAGIT
RIVER APPROXIMATELY 130 FEET AT THE INTERNATIONAL
BOUNDARY BETWEEN THE UNITED STATES AND CANADA

SUPPLEMENTARY ORDER
OTTAWA, ONTARIO
APRIL 28, 1982

WHEREAS the Commission is committed to the provision of the
Boundary Waters Treaty calling for the prevention of disputes along the
common boundary.

WHEREAS the parties have not, since the Commission's Minute of
October 9, 1981, engaged in direct negotiation on the question of an
alternative to the High Ross Dam;

WHEREAS a negotiated solution to this matter requires an immediate
total commitment, by both parties, to the process of negotiation;

WHEREAS the report of the Special Advisors to the Commission dated
April 2, 1982 demonstrates that reasonable alternatives to the raising of
High Ross Dam are available;

WHEREAS the Commission cannot pursue further action unless the
Governments of the United States and Canada are willing to formally
support and be full participants in the process of settling the matter;

WHEREAS the formal participation of Governments is imperative if
there is to be any degree of certainty that a negotiated solution will be
effected;

Reprinted courtesy of the International Joint Commission.

WHEREAS the participation of the Governments of the United States and Canada is required in order to facilitate both the planning and completion of domestic regulatory and legislative actions and bilateral arrangements that will be required to implement any negotiated settlement. These actions might include but would not necessarily be limited to: National Energy Board of Canada licencing; Washington State revenue bond legislation; transmission arrangements with U.S. and Canadian utilities, and adjustments related to the Columbia River Treaty;

WHEREAS the Commission determines that the Boundary Waters Treaty of 1909 confers on it continuing jurisdiction in respect of Orders made by it, but that this continuing jurisdiction does not necessarily carry with it the obligation to exercise such jurisdiction;

WHEREAS the Commission has reviewed the Request in the application of the Province of British Columbia dated August 14, 1980;

WHEREAS the Commission has reviewed and considered all arguments and materials filed pursuant to the British Columbia Request in the application;

THEREFORE the Commission is of the view that the British Columbia Request in the Application and all arguments and materials presented pursuant to that Request do not constitute sufficient grounds to persuade it to exercise its jurisdiction as requested therein, and accordingly declines to grant the relief sought.

Notwithstanding the Commission's decision above on the Province's Request, the Commission also decides that in light of the views of the Governments of Canada and British Columbia and the Commission's responsibility under the Treaty to prevent disputes, and under present circumstances, the Canadian Skagit Valley should not be flooded beyond its current level provided that appropriate compensation in the form of money, energy or any other means is made to the City for the loss of a valuable and reliable source of electric power which would result if the Ross Dam project is not completed.

THEREFORE the Commission, after careful consideration and in the exercise of its continuing jurisdiction over the matter, decides to take the following extraordinary action:

    (a) Seattle is hereby ordered to maintain the level of the Skagit River at the International Boundary at or below elevation 1602.5' for a period of one year from the date of this Order.

(b) The Commission will appoint a Special Board composed of two members of the Commission, who shall serve as Co-Chairmen, and two non-governmental experts. The Commission will invite the Government of the United States, the Government of Canada, the City of Seattle, and the Province of British Columbia to each nominate a representative to be a member of the Board. This Board will co-ordinate, facilitate and review on a continuing basis, activities directed to achieving and implementing a negotiated, mutually acceptable agreement between the City and the Province and to provide status reports regarding such progress to the Commission every four months.

The Commission retains jurisdiction over the subject matter of the 1942 Order of Approval, and may make such further Order or Orders relating thereto as may be necessary in the judgment of the Commission.

Dated at the City of Ottawa this 28th Day of April, 1982.

E. Richmond Olson   Robert C. McEwen

Charles M. Bédard   L. Keith Bulen

Index